W0227651

# Facts in Mesmerism

*With Reasons for a*
*Dispassionate Inquiry into it*

CHAUNCY HARE TOWNSHEND

CAMBRIDGE
UNIVERSITY PRESS

CAMBRIDGE UNIVERSITY PRESS

Cambridge, New York, Melbourne, Madrid, Cape Town, Singapore,
São Paolo, Delhi, Dubai, Tokyo, Mexico City

Published in the United States of America by Cambridge University Press, New York

www.cambridge.org
Information on this title: www.cambridge.org/9781108025898

© in this compilation Cambridge University Press 2011

This edition first published 1840
This digitally printed version 2011

ISBN 978-1-108-02589-8 Paperback

# CAMBRIDGE LIBRARY COLLECTION

*Books of enduring scholarly value*

## Spiritualism and Esoteric Knowledge

Magic, superstition, the occult sciences and esoteric knowledge appear regularly in the history of ideas alongside more established academic disciplines such as philosophy, natural history and theology. Particularly fascinating are periods of rapid scientific advances such as the Renaissance or the nineteenth century which also see a burgeoning of interest in the paranormal among the educated elite. This series provides primary texts and secondary sources for social historians and cultural anthropologists working in these areas, and all who wish for a wider understanding of the diverse intellectual and spiritual movements that formed a backdrop to the academic and political achievements of their day. It ranges from works on Babylonian and Jewish magic in the ancient world, through studies of sixteenth-century topics such as Cornelius Agrippa and the rapid spread of Rosicrucianism, to nineteenth-century publications by Sir Walter Scott and Sir Arthur Conan Doyle. Subjects include astrology, mesmerism, spiritualism, theosophy, clairvoyance, and ghost-seeing, as described both by their adherents and by sceptics.

## Facts in Mesmerism

Chauncy Hare Townshend (1798–1868), poet and collector, was a well-connected friend of Robert Southey and Charles Dickens. He became fascinated with Mesmerism while in Germany and went on to popularise it in England. This book, first published in 1840, was his passionate defence of Mesmerism. Developed in the late eighteenth century by Franz Mesmer, Mesmerism was a kind of hypnosis based on the theory of animal magnetism. With its spiritual associations and uncanny effects, it was an extremely controversial topic in the nineteenth century and its practitioners were widely considered fraudsters. Townshend describes in detail the mental states Mesmerism induces, which he identifies as similar to a state of sleepwalking. Perhaps most fascinating are the eye-witness accounts describing experiments carried out by Townshend on the continent, in which he hypnotized his subjects into feeling his own sensations and knowing things they could not know.

Cambridge University Press has long been a pioneer in the reissuing of out-of-print titles from its own backlist, producing digital reprints of books that are still sought after by scholars and students but could not be reprinted economically using traditional technology. The Cambridge Library Collection extends this activity to a wider range of books which are still of importance to researchers and professionals, either for the source material they contain, or as landmarks in the history of their academic discipline.

Drawing from the world-renowned collections in the Cambridge University Library, and guided by the advice of experts in each subject area, Cambridge University Press is using state-of-the-art scanning machines in its own Printing House to capture the content of each book selected for inclusion. The files are processed to give a consistently clear, crisp image, and the books finished to the high quality standard for which the Press is recognised around the world. The latest print-on-demand technology ensures that the books will remain available indefinitely, and that orders for single or multiple copies can quickly be supplied.

The Cambridge Library Collection will bring back to life books of enduring scholarly value (including out-of-copyright works originally issued by other publishers) across a wide range of disciplines in the humanities and social sciences and in science and technology.

# FACTS

IN

# MESMERISM,

WITH

## REASONS

FOR

A DISPASSIONATE INQUIRY INTO IT.

BY THE

## REV. CHAUNCY HARE TOWNSHEND, A.M.

LATE OF TRINITY HALL, CAMBRIDGE.

---

" A great perturbation in nature! — to receive at once the benefit of
sleep, and do the effects of watching. In this slumbery agitation, besides
her walking and. other actual performances, what, at any time, have you
heard her say ? " — SHAKSPEARE.

" I'll charm your blood with pleasing heaviness,
Making such difference 'twixt sleep and 'wake,
As is the difference 'twixt night and day, —
The hour before the heavenly-harness'd team
Begins its golden progress in the east."      SHAKSPEARE.

---

## LONDON:

LONGMAN, ORME, BROWN, GREEN, & LONGMANS,

PATERNOSTER-ROW.

1840.

# INTRODUCTORY EPISTLE.

TO JOHN ELLIOTSON, M. D. CANTAB. F.R.S.

My DEAR SIR,

On seeing that I dedicate the following pages
to *you*, the world will, perhaps, be kind enough to
say — " Here is a Coalition ! " I not the less fearlessly place the work under your auspices ; trusting
that some persons at least may conceive that two may
be of one mind on a subject, and yet guiltless of a
conspiracy against Church or State — Truth or
Science. And what if we are ranged under the same
banner ? Union is not Treason ; — and I trust that
there is no harm in our being equally impressed with
a conviction of the reality of Mesmerism, and equally
animated by a resolution to disclose honestly that
which we know certainly. Here, then, is our coalition — if any one so chooses to term it, — a coalition
to defend truth — and not to spread imposture.
Perhaps, however, it may be as well to state (lest we
should leave too much to the sagacity of those who
smell a plot in every thing), that I have pursued my

mesmerical researches entirely apart from yourself;
nay, on the other side of the Channel; and that, yet
more, we were but recently made acquainted, by
letter only — through the medium of a mutual friend,
who is not a mesmeriser, but a public functionary and
man of letters. Again : let it be known that I never
had the pleasure of seeing you till within the last two
months; though, indeed, I must add, that I think it a
loss not to have made your acquaintance earlier. It
was said of some great man — Burke, I think, — that
you could not stand up with him under a shed during
a shower of rain, without finding out that you were
in company with a fine genius. The remark may be
applied, with some diversity, in most cases. Our first
impressions of persons are often an instinctive judgment, of which our after feelings towards them are
only a developement; and, with regard to yourself in
particular, I am not singular in remarking, that to
converse with you, but for a quarter of an hour, is to
carry away a pledge of your honour as a gentleman,
and of your sincerity as a man of principle; and this,
independently of the knowledge that, to the cause of
truth, you have made every sacrifice except that of
integrity. I must speak my opinion, — though at
the risk of being suspected of flattery. *You*, at least,
shall not have the credit of countenancing the pane-
gyric, as this letter will only be read by you on the
day it is given to the public.

These things ought to be known : — for they are testimony to mesmerism. Here are two persons, in different countries, wholly unconnected, setting out on an inquiry by different paths, and yet meeting, at length, in one common conclusion and point of union. As regards myself, I may affirm that no one could possibly have taken up and pursued a subject more independently, or in a more unbiassed manner than I have taken up and pursued mesmerism. I have not drawn my ideas of it from books, but from experience; I have even abstained from reading articles on it, lest I should lose the originality and freshness of personal observation. But I need not insist on this.

My work itself contains internal testimony that our coincidence of opinion is honest, not concerted ; — for, in truth, we differ while we agree. I have not described, because I have never seen, the curious delirium, and coma, which some of your patients have displayed. Not being myself of the medical profession, I have naturally treated mesmerism as a phenomenon of our nature, rather than as a curative means; and the maladies, which you have so successfully combated by the new agency, have of course produced modifications in its action, which the healthy subjects to whom I have confined my practice could not have displayed. However various the degrees of mesmeric developements portrayed by me, the principal features of the state have been similarly characterised throughout, my object being to delineate that species of mesmeric sleepwaking, which, I conceive, may be *induced*, to a certain extent, in any indifferent person. So far we are shown as drawing from separate and independent sources; and this involuntary kind of testimony is favourable to our cause: but, inseparable from this benefit, is a collateral disadvantage, on which I must briefly touch.

The greater part of the London, I may say of the English, world have derived their ideas of mesmerism from *your* experiments, which so many have personally witnessed. Hence the general reader, comparing his preconceptions on the subject with the portrait I have set before him, may surmise that the new science is not in unity with itself, confounding, by a very common mistake, diversity with discrepancy. But, in truth, while there is much that is different in our facts, there is nothing whatever that is contradictory. The subject is large, and cannot but present itself in various points of view to various observers. Even where we most appear to disagree, it must be remembered that the same phenomenon may have more than one phasis, just as the celebrated shield, that was black on one side, was not the less certainly white on the other. Thus, it is very true that a kind of delirium may be developed under mesmeric influence, while, at the same time, it is

capable of eliciting the highest state of moral and intellectual advancement, to which man, in this existence,
can probably attain. This remark is the more necessary to be made, inasmuch as, throughout my work, I
have laboured to prove the mesmeric condition a rise on
our actual mode of being ; and, according to the view
I have taken of the subject, if it be not this, it is nothing. For what does a writer achieve, who does not
contribute, in however small a measure, to the hopes
and welfare of humanity ?

It has struck me that the world, who is very fond
of proving that an author did not write his own book,
may inquire — What share had Doctor Elliotson in
this treatise ? I therefore think it not perhaps altogether useless to state, that neither yourself, nor
any body else, has dictated or suggested to me one
opinion which is therein contained. The only debt
that I have to acknowledge, is on the trifling score of
some advice respecting the terms to be employed in
writing on this particular subject. You observed to
me, that the phraseology of mesmerism could not too
soon be fixed, and rendered precise ; and you suggested the substitution of Mesmeric Sleepwaking for
Induced Somnambulism, on the ground that Somnambulism, strictly speaking, was not always, nor
necessarily, an adjunct of the condition I wished to
describe. In all other respects, the faults or merits
of the publication must be charged on my own head.

For the former, indeed, as probably numerous, few persons would like to become responsible; and on their account (especially should any repetitions disfigure my pages), I shall have to claim indulgence from my readers, in consideration of the circumstances under which the work was composed; one part being despatched from abroad to the printer, while other portions were written at distant intervals, as health and opportunity permitted. I know not, however, whether, on the whole, a residence on the Continent has not been favourable to the consideration of such a subject as mesmerism; for I can assure my countrymen that their own prejudices respecting it are no measure of the reception which it meets with from enlightened foreigners. I have scarcely conversed with one person of education in Germany, who was not able to detail to me some interesting fact relating to mesmerism, which had been personally witnessed and authenticated; and every where abroad, during those travels which in search of health I have undertaken, my information respecting this remarkable phenomenon of our nature has been extended. Opportunities also of mesmerising different individuals (many of them distinguished for rank and science) have been freely and agreeably accorded me. Since I sent to England a list of the persons I had mesmerised, I have experimented on some thirty others at Rome, Naples, &c., and I have still found the

proportion stated in my work of mesmerisable individuals singularly preserved, and the phenomena exhibited perfectly accordant with previous observation.
In the Appendix, some documents relative to these
experiments will be found, some of which are of
high interest; especially those from the pen of Professor Agassis of Neufchâtel, and of Signor Ranieri,
the distinguished historian of Naples.

But I am insensibly converting a letter to you
into an address to the general reader, and I perceive
I have so far extended a dedication as to render a
preface unnecessary.  For this service, at least, let
me thank you.  I have always thought prefaces very
troublesome things to write.  And are they not, in
fact, pieces of falseness throughout? standing first,
though written last — expressing a humility which
the author does not feel — and claiming the especial
attention of everybody, while it is a wonder if they
are read by anybody.  I therefore not unwillingly
take this opportunity of making my introductory bow
to the Public, while at the same time I beg you to
believe me,

My dear Sir,

With sentiments of high respect,

Yours very sincerely,

C. HARE TOWNSHEND.

Innspruck, Nov. 25. 1839.

# CONTENTS.

## BOOK I.

## BOOK II.

### SECT. I.

### SECT. II.

### SECT. III.

## BOOK III.

### SECT. I.

### SECT. II.

### SECT. III.

SECT. IV.

BOOK IV.

SUPPLEMENT.

# MESMERISM.

## BOOK I.

REVIEW OF THE CAUSES THAT HAVE MADE MESMERISM
UNPOPULAR, AND WHICH RENDER IT A SUBJECT
DIFFICULT TO BE TREATED.

HAVING had many opportunities of convincing myself
that man can really influence his fellow in the manner
called mesmeric, I have determined to arrange and
classify the phenomena, relative to this influence, which
have fallen under my observation, in the hope of re-
ducing them to a few simple and general principles.
In thus coming forward, I am fully aware of the obstacles which I have to encounter. The fatal word
Imposture has tainted the subject of my inquiry; and
Ridicule, which is *not* the test of truth, has been
pressed into the service of talent, in order to annihilate
the supposed absurdity before the dread ordeal of a
laugh.

But it is not only the witty who have set up mes-

B

merism as the mark of their fine arrows; men of
science have attacked it, because they could not make
it harmonise with their preconceived notions; and
many of the Galens of our day, instead of wisely
taking it under their patronage, and into their own
hands, have treated it with a desperation of hostility;
— as if, were it allowed to flourish, their glory was
tarnished and their "occupation gone."

What is worse, some of the friends and supporters
of mesmerism have done it more disservice than its
bitterest foes.  Instead of setting before the world
the simplest features of the new discovery, they have
at once produced to view its most astounding marvels,
thus dazzling into blindness the eyes which, by a
more cautious conduct, they might have taught to
see.— Then it cannot be denied that the name of
Mesmerism, or Animal Magnetism, has sometimes
served as a watchword to exploded quackery and
impudent deceit: and who does not know how difficult it is to separate the merits of any doctrine from
the faults of its partisans?  Mesmerism has occasionally been found in company with the vicious and
the designing; and its good repute has suffered
accordingly.

Another circumstance has contributed to the disgrace of this unfortunate subject.  In its palmy days,
when De Maineduc plunged his visionary fingers
into the stomachs of gouty earls and dyspeptic mar-
chionesses, nothing less was expected from the new
remedy than the renovation of human nature, and
an absolute conquest over all diseases.  That it should

subsequently fall into oblivion, will not astonish those who have observed the invariable fate of every fashionable panacea — of Peruvian bark, for example. From thinking that it can perform everything, men doubt whether it is capable of anything; and the more extravagant have been the hopes which it has excited, the more deep is the disgust which disappointment naturally creates.

But mesmerism has been looked upon as worse than false or nugatory. Many, who have believed in its powers, have believed only to tremble. Credulity has done it worse service than incredulity. It has been proscribed as an unholy thing : books have been written upon its dangers; the good Catholic crosses himself when he names it ; and the careful Protestant, even if he hesitates to brand it as diabolical, thinks at least that it may be prostituted to evil purposes. And so it undoubtedly may be. But then the same objection applies to all that is most beneficial upon earth. All great engines are capable of great perversion. This is tacitly allowed in the whole conduct of life. Yet we do not abstain from the use of fire because it can destroy, nor from the medical application of laudanum because it is a poison.

That the world's quarrel against mesmerism should be so very bitter, is hardly accounted for by any of the foregoing considerations. *Primâ facie,* one would say that there were attractions in this despised doctrine, more than sufficient to countervail every objection that might be brought against it.

Fond as we are of the shadowy and the unknown, its very mystery might seem congenial to our nature; and even those vague suspicions concerning its sinfulness which hover darkly round it, might be supposed to have a charm for man — the wilful — who rushes to the forbidden with so keen a zest. But there is a sort of unexplained cdium attached to mesmerism, which quells curiosity and deadens interest. From this odium it appears almost necessary that I should raise my subject, before entering fully upon it; I must secure a hearing before I can plead my cause; I must show that there is in itself no inherent fault, which can justly exclude it from a fair and candid examination. This end, as it appears to me, I shall best attain by exposing the primary causes which have degraded mesmerism; and by showing that these are extrinsic to itself.

Having done this, I propose, also, to touch upon the reasons that make the subject I have chosen not only distasteful to the general reader, but of extreme difficulty to the Author himself.

*First.* The original cause of the ill reception which mesmerism has met with from the world, is undoubtedly to be found in the character of its discoverer — Mesmer, — in his want of candour and philosophic strictness. Had it been introduced to notice by a Newton or an Arago, by one who would have stated his facts honestly, and drawn from them none but legitimate conclusions, the difference of its career may be estimated by all who are aware how much depends upon a propitious beginning. But unfortunately, from the very outset, mesmerism was associated with the soiling calculations of self-interest, and the errors of an over-heated brain. Mesmer wished to make a monopoly of that which should have been the property of all mankind : he sold his secret — he bartered for gold his future fame and the reputation of his darling subject * ; and, losing the light which emanates only from an upright spirit, he became the dupe of his own miracles, so miserably as to surround his really simple and sublime discovery with fictitious terrors and misleading puerilities. The result of this moral and scientific suicide has been the degradation of mesmerism. First associations are, from the very law of our minds, all but indestructible ; and therefore it is that with a few original thinkers alone one can hope to replace the subject on its true and primitive footing — namely, its own merits. The false has been so blended with the true, that it is no wonder that both should be rejected together. The waters come not to us pure, but from a fountain-head that is itself disturbed and sullied ; so that, instead of spreading forth into a lucid mirror, reflecting heaven and earth, and enlivening all around, they stagnate in a thick and blinding

* In justice to the memory of Mesmer, it should be stated that against the fact of his having sold his magnetic secrets for a hundred louis to each candidate for initiation, should be set certain extenuating circumstances, which are related in Mr. Colquhoun's Isis Revelata, vol. i. p. 237.

marsh. Had mesmerism been announced to the world, not as a studied enigma, but in the form of a simple proposition; had all men been invited to test the truth of the principle, and to investigate the laws of its operation; had it been practised in unostentatious privacy instead of crowded assemblies; had there been in the chambers devoted to its service, neither mystical machines nor exciting music, no convulsionaries, no hysterical women; had mesmerism from the first appeared that which it eminently is — a spirit of calmness and of reason; then had it interested the scientific and conciliated the wise; then had it been transmitted to the present age pure and unenveloped by the mists of prejudice. The mere fact that man can produce a kind of slumber in his fellow-man by a few and simple means, is surely not to be confounded with the heap of absurdities attached to it. To say that the one is inextricably and necessarily linked with the other, were want of sense as well as of candour; and, unless we choose to admit a principle which would make even our religion answerable for the sins committed in its name, we must allow that mesmerism is in no way affected either by the errors of its partisans or the prepossessions of its enemies.

*Secondly.* As mesmerism was ill-omened in its birth, so also was it unfortunate in its baptism. Shakspeare's often-quoted query finds here, if no where else, its answer; — and we are forced to reply — There is *much* in a name, except where true love or true philosophy renders the mind insensible to those externals which are all in all with the generality of mankind.

The disservice rendered to mesmerism by its name, is this: — It has turned men from true inquiry, and, like a tub thrown out to the whale, has served as an object of attack, while the real point in debate has remained untouched altogether. We have asked whether such a *power* as mesmerism exists; when we should rather have demanded whether there is a *state* so denominated. It will at once be seen in what material respects the two questions differ. The first presumes, even while it professes to seek,˙ a specific cause for certain phenomena; — the second merely regards the phenomena themselves, and inquires — Do such and such facts exist? Each inquiry should be kept carefully distinct, and yet they have unfortunately been mixed up together; or rather an unhappy priority has been granted to the first, involving the very existence of the second. For it is plain that when we demand, " Is there such a *power* as mes merism ?" the answer may ever be " No;" and then, by a too common injustice, we extend the negative over the whole question, there being but few who will not confound a mistake and the object mistaken in one general anathema. Did we, however, clearly perceive that " Power is nothing more than the relation of one object, or event, as invariably antecedent to another object or event,"* we should perceive that the facts called mesmeric, have as much claim to be

* Brown's Philosophy of the Human Mind.

considered realities, as if indeed there *were* a magnetic power or influence.

Of the error of the mesmerists in bestowing an ill-
judged appellation, the opposite party have taken
ample advantage. They have thrust forward the unlucky cognomen into the very van of discussion, and
have thus compelled an inquiry into the cause of
mesmerism, before the phenomena could be well
considered. Surely it must be conceded that so
singular an inversion of true philosophical investigation cannot but have proved highly detrimental to the
subject of our discussion. In what other matter
have we acted so strangely as to inquire into the
secret cause, before men are well agreed respecting the visible effects ? Do we not, in conducting an
important analysis, first ascertain the phenomena,
their characteristics, and the circumstances under
which they appear ; and then, after long and careful
induction, name — but with caution— some pervading principle into which they may all be harmoniously
resolved ? Not only is it natural thus to commence a
course of reasoning with what is nearest to our apprehensions, but, by so doing, we secure that essential
requisite to an argument — a firm and undebateable
ground, where both he who would convince, and he
who is to be persuaded, may meet as on a neutral ter-
ritory, and, taking a common point for starting, be
advantaged with at least a probability of arriving at
a common conclusion. Facts are this neutral ground;
—facts are this point of vantage ; — for it is to be remarked that respecting outward and visible pheno-

mena there is ordinarily but little difference of opinion, while, in reference to hidden causes, men ever are and must be divided. For instance, every one knows the sensation of light, and may convince himself, if he pleases, that under such and such circumstances, such and such phenomena do occur. So far mankind are agreed : but when we would assign the primary cause of light and its phenomena, we find the war commenced, and opinions far divided — some adopting the undulatory theory, while others fondly cling to the Newtonian doctrine of emitted rays.

It is no wonder, then, that the premature assumption of a specific cause for the phenomena of mesmeric somnambulism should have shocked men in the outset, and have indisposed them, through the medium of their prejudices, for calm investigation. Under this head of offence may be classed, also, such technical expressions as "magnetised water," "magnetised trees," &c. The ideas, which these terms convey, are vague and unpleasing. Who can tell that their meaning simply is, — water, or trees, breathed upon, or touched by the hand, after a certain formula? The vocabulary of every science has been, to the uninitiated, foolishness; — but the nomenclature of mesmerism is worse: it has been a scare-crow even to the wise.

I confess that of all the causes, which have contributed to render mesmerism unpopular, this is to me the most discouraging. To make a dispute verbal is to make it endless, and the erroneous way in which our subject has been considered, has trans-

ferred its merits from things to words — a sad and
barren exchange ! — All hope of an accommodation
between the friends and foes of mesmerism is thus
rendered doubtful. How, indeed, can they calmly
discuss the matter, when the first bare mention of it
is a sort of Slogan or battle-yell, that raises a conflict
sufficient to drown the voice of Reason for ever ?
How shall they decide a truth, whose fate depends
upon the propriety or impropriety of its name ? How
shall they adjudge the real business in hand, seeing
that preliminaries can never be adjusted ? The
subject has indeed fallen into the very Chancery
Court of Philosophy.

That men should be so easily entrapped into pursuing a shadow, while the substance has eluded their
observation, may seem strange ; but, words being the
media of our thoughts, we are naturally so influenced
by them that even the most clear-judging find it difficult
not to be diverted by a verbal inaccuracy from more
important matters. The fate of Hartley's Theory of
Sensation is a proof of this. All the world fastened
so vehemently upon the unlucky term " vibration,"
and were so intent upon proving that the soft and
loose chords, which compose the nervous system,
could never properly be said to vibrate, that the true
merits of a very beautiful hypothesis were overlooked.

*Thirdly.* Nearly connected with the erroneous naming of mesmerism, is the circumstance next in order,
which has thrown a blight over its pretensions, — I
mean the decision, in its disfavour, of the French
Academy, in 1784. Since then, it has been con-

sidered as worse than unexamined; — it has been conceived to have undergone examination, and to have been found worthless. Since then, to revive an interest upon such a condemned subject has been nearly as impossible as to restore life to a corpse by galvanism. Since then, instead of a mystery, it has been regarded as a delusion; and the world has turned away from it with the same sort of flat disappointment which we experience, when, after having trembled at the haunted chambers, dusky veils, and heaving coverlids of a romance, we are chilled into our sober senses by the earthly explanations of its concluding pages. The magician's wand is broken for ever: we like to be frightened, but hate to find ourselves deceived. The attractive ever perishes with the supernatural. But to those who read and inquire for themselves, there is nothing whatever in the decision of the French Academy of Science which can be considered as destructive of mesmerism. What is it that the Academy decided? Few know, — few care to inquire. A general impression has gone abroad, that mesmerism received, in 1784, its *coup de grace;* and there men are content to let the matter rest; for are they not thus saved the trouble of thinking? How pleasant to believe in the eternal banishment of a subject, concerning which, the idle, the self-interested, and the prejudiced have each their several motives for saying, as some honest churchman did of the Athanasian Creed, " I wish we were well rid of it." The blessed consummation is not, however, so easy of attainment; for what is the truth? The French Academy merely decided that there was not sufficient evidence to show that the phenomena called magnetic are caused by the action of a fluid. As to the phenomena themselves the Academy not only allows them, but, as it appears to me, concedes the real question in debate, in terms the most explicit. These are the very words in which the examiners sum up their report. " Ce que nous avous appris, ou, du moins, ce qui nous a été confirmé d'une manière demonstrative et évidente par l'examen dés procédés du magnétisme, c'est que *l'homme peut agir sur l'homme* à tout moment, et presque à volonté, en frappant son imagination; c'est que les gestes, et les signes les plus simples peuvent avoir les plus puissans effets; c'est que l'action de l'homme sur l'imagination peut-être reduite en art, et conduite par une méthode, sur des sujets qui ont la foi."*

Translating the above as literally as possible, we find that the Commissioners have thus expressed themselves : —

" That which we have learnt, or, at least, that which has been proved to us, in a clear and satisfactory manner, by our inquiry into the phenomena of mesmerism, is — that *man can act upon man*, at all times and almost at will, by striking his imagination; — that signs and gestures the most simple may produce the most powerful effects ; that the action of

---

* Exposé des Expériences qui ont été faites pour l'Examen du Magnétisme Animale. Paris, 1784.

man upon the imagination may be reduced to an art, and conducted after a certain method, when exercised upon patients who have faith in the proceedings."

The above passage will show that the French Academy decidedly agrees with the first promoters of mesmerism as to the existence of certain remarkable phenomena : it is only respecting the cause that the two are at issue ; the one asserting that it is a fluid — the other that it is imagination. Whether the latter may not herein be as guilty of a *non-sequitur* as the former, we may hereafter consider. At present I merely remark that, granting the magnetic phenomena to spring from the imagination, we do not thereby condemn them to contempt, or oblivion ; far less do we (as some persons seem to suppose) abrogate them altogether.

Were mesmerism only viewed as illustrative of the mind's influence upon the organisation, it would hold out high claims to notice. What subject, we may ask, is more fertile of important reflections ? Materialists, curiously enough, adduce it to disprove mesmerism, which makes against their views of man as a mere machine — of thought as a mere elaboration of the brain ; — but, in so doing, they open a field of speculation, which they would be sorry to enter upon further than they find convenient. In the very terms of their dissent, they admit of two things of a wholly different nature — that is, the imagination and the organisation ; — the first not only modifying the action of the latter, but actually mastering it altogether. Granting that the imagin-

ation can influence the bodily frame to the extent
allowed by the French Academy in its report on
mesmerism; if what they say be true : — " Nous
avons vu l'imagination exalteé devenue assez puis-
sante pour faire perdre en un instant la parole." —
" We have seen the imagination, when exalted,
become powerful enough to make a person lose the
faculty of speech in a moment : "—if we believe this,
then indeed we admit of a motive force, to which we
can hardly set bounds, and which makes mesmerism,
with all its pretensions, a feeble and unnecessary
agent. The supposition of an existing fluid can add
nothing to miracles like these. By demonstrating
the omnipotence of mind over organisation, the
opponents of mesmerism prove too much. They
bring us to a point where it is only one analogical
step further to inquire, whether the force, which can
so act on the organisation to which it is attached,
may not, peradventure, extend its influence to a
sphere external to itself? For peremptorily answering this question in the negative, we can assign no
reason, but that external offices for such an agent
come not within the pale of our experience ; — and
reasoning like this we cannot but reject as an
effectual bar to all inquiry whatever. We must
remember that human experience is ever on the
increase, and that there exists not one power, even
the most palpable, of which we can affirm that we
have ascertained all the functions. There is not a
discovery of modern times, but teaches us that the
force which performs some things whereof we are

cognisant, may also perform others of which we have, as yet, no notion whatever. Let my reader bear in mind, that I by no means affirm that the imagination is the agent in mesmerism, internally or externally. All I mean to assert is, that the careful French Academy, and their subsequent followers, have gone further than the magnetisers themselves, in attributing power to mind, and in relating wonders to prove that power. Say what we will, the important point in mesmerism, — the influence of man upon his fellow,—was conceded by the French Academy in no doubtful terms ; — *through* the imagination, it is true ; yet still it *is* conceded. The imagination is but considered as the means ; man is still the agent. Are the phenomena themselves altered by being referred to the imagination, instead of a magnetic fluid ? Not in the least. Do we comprehend them better, when so referred ? By no means. We only seem so to do, because in the one case they are referred to a cause of which we cannot but confess our ignorance ; while, in the other, they are attributed to the action of a power which we think (how vainly !) that we comprehend.

*Fourthly.* The fourth cause, which has banished mesmerism from the rank and position to which it is entitled, is the early attempt to assimilate it to the certain sciences in an erroneous manner. In our researches into the discovery of Mesmer, we have, from the beginning, struck into a path which never could lead us to the desired end ; and then, most unreasonably, we have charged upon the subject of our inquiry the fault which existed in ourselves. Because we have found nothing, when we did not seek aright, we have impatiently concluded that there was nothing whatever to be found. Perhaps the error has originated with the Mesmerists themselves. To secure the suffrage of scientific men for their favourite pursuit is evidently desirable. How should they accomplish this? The days are long since past, in which men were content to reason after the vague mode of the Aristotelian philosophy, which leant upon conjecture rather than experiment, and discussed absurdities as gravely as the ordinary phenomena of nature. The world has reached an era, in which facts, attested by the senses, independent of the human will, and invariably reproduced under the same circumstances, can alone engage the attention of the learned. With the scientific men of our day, (and far be it from me to censure this) certainty is the great object. In order, then, to claim the notice and the fostering protection of science, the friends of mesmerism have long endeavoured to identify their presumed agent with physical forces, already ascertained and of invariable action. In their principle, perhaps, they are right — in its application, wrong altogether. Forgetting that mesmerism is a mental and vital, not less than physical phenomenon, and that mind and life are in perpetual opposition to the laws of the material world, they have endeavoured to recognise in mesmerism an operation as constant as that of the galvanic battery, or the electric vial. A certain school of German writers, especially, have theorised on our subject after the false method of explaining one class of phenomena in nature by its fancied resemblance to another. Wishing, perhaps, to avoid the error of the Spiritualists, who solve the problem in debate by the power of the soul alone, they have ransacked the material world for analogies to mesmerism, till the mind itself has been endued with its affinities and its poles. Such attempts as these have done the greatest disservice to the cause we advocate. They submit it to a wrong test. It is as if the laws of light should be applied to a question in acoustics. It is as if we should expect to find in a foreign kingdom the laws and customs of our own. Thus wrongly biassed, we turn away from mesmerism, as provoked at finding it other than we deemed it to be; as the prince in the fairy tale, who found his betrothed, though very charming, not in the least like her portrait, and so sent her back in disgrace. Who has not experienced a thousand times the same feeling? We read, perhaps, a description of some lovely scene; we thence form an image of it in our thoughts. We at length behold it, and are discontented to find it endued, perhaps, with even more beauty than we had imagined; because the beauty is of a different kind. There is rock where we expected smooth turf; there is wildness, where we looked for cultivation; there is a withered oak, where we had in fancy placed a human dwelling. If, in matters of taste, preconceptions like these prepare the way for disappointment; in matters of reason they are not less calculated to awake disgust. A science that is misconceived labours under peculiar disadvantages. Thus, then, till the initial step towards a comprehension of mesmerism be taken anew, there is no hope that it will ever be understood or appreciated. Why unavailingly seek to reduce it to a formula of which it is insusceptible? If we ascribe it to a power already ascertained, why not treat it, at least, as an entirely new function of that power? Why limit it to what we know, when possibly it may be destined to extend the boundaries of our knowledge? Why are we to be trammelled with foregone conclusions? Yet upon these very restrictions, the opponents of mesmerism insist; thus taking away from men the means of investigating the agency in question, by forcing them to set about it in the wrong way.

The case is the more perplexing, inasmuch as this apparent unfairness is grounded on instinctive propensities of our nature, with which it were vain to quarrel, since they are actually a part of man's most valuable inheritance and absolutely essential to the whole conduct of life. That attraction towards sensuous objects, which is fitted to a being that lives in a visible and tangible world, draws us inevitably away from whatever shuns the senses to that which may be seen and handled; and that constitution of our minds, which leads us to rely on the invariableness of nature's sequences, necessarily brings with it a love of the certain and the permanent. With tendencies like these, at once to be cherished and guarded against, it is difficult to deal. Thus reflecting, I am not surprised at the hostile reception which mesmerism has met with in almost every quarter — I see that it could not have been otherwise. Unfortunately too, those men who are the last to be gained over, are generally the best worth gaining ; while those whose habits of mind lead them to embrace the new doctrine without hesitation, are mostly visionaries, who find themselves more at home in the affairs of heaven than of earth, and are ever more disposed to speculate wildly than to submit to the rules of strict and severe analysis.

But, on the other hand, let it not be forgotten that it is the province of man to mount from the known to the unknown, and to reason from that which can be seen to that which is invisible. If one strong instinct teaches us to repose in the objects which lie open to our immediate apprehension, another not less powerful calls us imperatively to rise to the detection of secret causes. And true wisdom consists in the just balance of these two contending propensities. Nor are we invited to ascend beyond the level of more obvious things by the constitution of our minds alone. Between the quenchless desire of our bosoms, and the dispositions of the external world, there is a harmony as perfect as it is beautiful. The simplest of nature's sequences, should we trace it further up the scale, conducts us inevitably to somewhat beyond our visual ken, while at the same time we are encouraged to proceed, by our success in analysing the invisible and ascertaining the obscure. For instance, by imperceptible exhalation and absorption, there is carried on throughout the world a perpetual loss and gain, conversion of substance, and exchange of power, which is scarcely conceivable, yet which man, in many instances, has reduced to rule, and expressed in number. We have measured the undulations of light, and the waves of sound. The microscope has shown us worlds, of which our unassisted senses have no cognisance; and science, aided by the telescope, has traced from stars, unseen by the naked eye, the one great law of attraction prevailing to the very verge of ascertained creation. Ought we, then, in any case, to turn away from facts that indicate a hidden influence, merely because that influence has not yet been tested by the senses? Even our conduct in earthly matters will reprove us here. Of the machinery, wrought by human hands, the results alone are presented to vulgar view, while it is left to the thoughtful and the wise, to explore the finer springs and hidden motory powers. How much more, then, in viewing the mechanism of the universe, should we believe that we discern, as it were, only the hands and the dial-plate, while, behind these, " wheel within wheel involved " is hidden from our gaze, in order to tempt our inquiry, and to develope the higher part of our nature by the stimulus of curiosity?

There is, then, in the nature of things no necessity that mesmerism, because it is recondite, should be unexplored. To investigate it may indeed be difficult; but appears to be rather enjoined than forbidden — enjoined by the tendencies of mind, not less than by the constitution of matter. It may, however, still be urged in reply, that the variableness, which must be granted, in some respects, to characterise mesmerism, is the true disqualification which debars it from the realms of science. How, may it be argued, can that which is confessedly irregular be shaped into a system ; how can phenomena, which contradict each other at every turn, admit of classification ?

I answer that such an objection, if carried out into a principle, would effectually arrest the progress of all human knowledge. It would make us rest satisfied with the apparent and the superficial, when we should be striving after the real and the profound. It would make our own imperfect observation the measure of Nature's regularity. Are we entitled to conclude, in any case, that, because we have not hitherto been able to assign a law to certain operations, they are therefore absolutely without law ? Are we to assert that the orderly dispositions of the universe are deformed by a monstrous exception, or is it not wiser to believe that our own knowledge is in fault, whenever Nature appears inconsistent with herself ? Surely we have enough order around us to suggest that all, which to us seems chance, is " direction which we cannot see;" that all apparent anomalies are but like those discords which, in the most masterly music, prepare the transitions from one noble passage to another, and are actually essential to the general harmony. In many instances, this is not mere conjecture. How much of fancied imperfection and disorder has fled before our investigation ! The motions of comets at first appear to offer an ex-

ception to the exact arrangements of the universe.
They traverse all parts of the heavens. Their
paths have every possible inclination to the plane of
the ecliptic; and, unlike the planets, the motion of
more than half of those, which have appeared, has
been retrograde; — that is, from East to West.*
Yet have we been able to detect the elements of
regularity in the midst of all this seeming confusion,
and to predict with certainty the day, the hour, and
the minute, of a comet's return to our region of the
sky.

Experience also shows that apparently insulated
and lawless phenomena may not only be reduced to a
law, but to a well-known law ; — that many a familiar
agent puts on strange disguises; and that events, with
which, in their mazy channels, we seem to be unacquainted, may be perfectly recognised by us at their
source. Thus galvanism and the magnetic force are
proved by recent discoveries to be only forms of electricity ; showing that a fact may be altered — not in
itself, but in the circumstances that surround it, and
that complexity of development is perfectly consistent with unity of design. Instances like these,
while they encourage us to inquiry, should teach us
to believe that all which is needed to vindicate the
regularity of nature is a more extended observation
on our parts.

Far, then, from granting that mesmerism is de-

* Mrs. Somerville's Connection of the Physical Sciences.
Page 377.

barred from the province of strict analysis by an apparent irregularity, I cannot admit that there exists in it one disqualification, that has not equally been shared, at one time or other, by every science. All that I concede, is that it is in a different stage of its existence, compared with subjects of ascertained knowledge. It must not be likened to optics, or hydrostatics, in their actual development, but to those sciences in their dim and early condition; when rude guesses were accepted as substitutes for truth, and when objects were discerned as mistily as the landscape is —

> " What time the shepherd, blowing of his nails,
> Can neither call it perfect day nor night."

There is indeed this difference — the advances that we have made in knowledge, generally, render it more easy for us to study any subject in particular. Still mesmerism (represented indeed by some as in its dotage) is but in its infancy, and it is a folly to call upon it to exhibit the features of a riper age.

Moreover the uncertainties of mesmerism have been much exaggerated. I may ask, can that be so irregular an influence, which the French Academy, in the height of its learned displeasure, acknowledged was capable of being conducted by art and method?

My own experiments, also, will hereafter show that it exhibits a definite " form and pressure, " and justify the conjecture that it may have its own laws

of certainty, though not precisely those to which we
would refer it.

On the whole it appears, there is no reason to conclude that mesmerism, because once wrongly presented to science, should never be presented to science
at all; — that because its agent has not been ana-
lysed, it is therefore unfit for analysis; or that
because often erroneously *identified* with known forces
it should at no time be found to be *connected* with
them; but —

*Fifthly.* Our impatience of whatever we cannot account for, has prevented, and will, I fear, long prevent mesmerism from being rightly examined, or
even from being classed generally amongst realities.

This cause, at the basis of all the others, which
have, at any time, operated most powerfully against
mesmerism, deserves our particular attention.

This it is, which induced the friends of mesmerism
to bestow on it a name, and refer it to agents, which,
by seeming to indicate its origin, should render it
easier of belief.

This it is, which led the members of the French
Academy of Sciences — not being able to get rid of
it altogether — to banish it to the airy realms of the
imagination. And this it is, which, at the present
day, can alone explain the standing miracle — as
great as that of mesmerism itself — namely, that
facts, which have been witnessed by thousands, should
be rejected as if no testimony to their truth had ever
been offered.

If in inquiring into the deep grounds of this

singularity, I should trespass on my readers' patience, let the following consideration plead my excuse. Every secret cause of the judgments we form, in which the constitution of our own minds is concerned, requires to be thoroughly probed and exhibited in its soundness or unsoundness; for how shall we attain to certain truth, unless sure of the principles on which we seek it? As he, who would take an observation of the sun must first ascertain that his instrument is correct, so must we, before examining nature, duly regulate the organ of our perception.

Whence, then, arises it, that with the world in general, " I comprehend not," is equivalent to " I do not believe " ?

Many of our most important and habitual modes of thinking would seem to revolt against this error. Granting that whatever is beyond the pale of our knowledge is a delusion, we, by parity of reasoning, must deny the existence of ourselves, and, consequently, of all external things: for the soul is an unsolved problem, and man is the great anomaly of creation. Yet thus do we not argue? On the contrary, we not only acknowledge that we are, but cease not to inquire what we are, ever studying the volume of our intellectual being, in order to discover some law, which should be to mind what attraction is to matter. Consenting thus to take our minds as we find them, and to observe and to ascertain rather than to decide and to excommunicate, why is it that, on so many other occasions, we pursue a directly contrary

c

course? For this reason. With regard to ourselves, our own consciousness compels us to begin by recognising the reality of certain phenomena ; but, with regard to other things, we are at liberty to speculate before we examine. Unless restrained by some strong controlling force, our natural tendency is ever to rush onward to causes and conclusions — to simplify and to generalise — and hence to dismiss from our consideration whatever does not harmonise with our own systems. This propensity to resolve the complex and the partial into the simple and ultimate appears to be a part of our original constitution, and is one of those strong impulses which seem made to carry us too far, lest they should fail of carrying us far enough. By it we trace that unity amidst variety, which stamps creation as the work of one hand ; but by it also we are led to invert the true process of reasoning, and to snatch at the higher antecedents of nature's sequences, before we have brought the lower within our grasp. Those comprehensive principles which are, in truth, the latest fruit of experience, we would force to our impatience, and mature at our pleasure. All knowledge being identified with a progression towards primary causes, and primary causes being of necessity the simplest, we fall into the mistake of striving after simplicity that we may reach wisdom — whereas it is only through wisdom that a safe simplicity can be reached. Instead of accommodating ourselves to nature, we would compel nature to accommodate herself to us. Instead of ascending cautiously from the individual to the general, we take our general rule, and force the individual circumstance to bend to it. Instead of adapting our theory to our facts, like a gracefully fitting garment, we clip and torture our facts to make them suit the construction of our theory, or, should they prove especially rebellious, banish them altogether. This is fatal to the very clearness we aim at. To simplify rashly is but to confuse. Until we consent to view things as they are, rather than as we would make them to be, we shall advance only to be led astray. Until we learn to consent for a while to philosophic doubt, we shall never arrive at certainty. For, in truth, we are debarred from that comprehensiveness of principle, which is so attractive to our pride by the deficiencies of our own understanding.* Our laws must necessarily be multiplied because our faculties are bounded. Nature, indeed, may be one; but the aspect she presents to us is varied, and we cannot climb high enough to behold her mighty landscape subject to our gaze. Our ardent inclination to generalise thus struggling with our defect of power,

* " I thought that the first step towards satisfying several inquiries, the mind of man was very apt to run into, was to take a survey of our own understandings, examine our own powers, and see to what things they were adapted. Till that was done, I suspected we began at the wrong end, and in vain sought for satisfaction in a quiet and sure possession of truths, that most concerned us, whilst we let loose our thoughts into the vast ocean of being, as if all that boundless extent were the natural and undoubted possession of our understandings." — *Locke's Essay on the Human Understanding.*

how difficult is the task of impartially examining a
new and extraordinary agent! Impatient of our
ignorance, and making no estimate of our own minds,
we attempt to soar into the full light of truth, without considering whether our pinions may not be
made of such materials as to melt beneath its beams.

From these considerations it appears that the
only course of investigation which is fitted to our
faculties is to observe with patience, and not to be
frightened from our propriety because a phenomenon
appears strange to us, either in itself or its concomitants. In a world, where it must be allowed
that all is wonderful, or that nothing is wonderful,
there is no intrinsic reason why one sequence of
events should be more astonishing than another.
Neither should we be afraid of multiplying laws
(which are, in truth, nothing more than " the expression of the most general circumstances, in which the
phenomena to which they refer have been felt by us
to agree"), as if, by so doing, we offended against
some unknown and mysterious power. In proposing
a new law, we only adopt a convenient, and it may
be, temporary expedient ; we only mean to acknowledge that we have not sufficiently observed a new
fact as yet to perceive its relationship with others,
wherewith we are already acquainted. And, indeed,
if offence there be, it is not of our creating. It is the
agreement or disagreement of natural phenomena
by which the number of our laws is diminished or extended. Above all, we must not forget that it is not
we who are to give laws to the fact, but the fact to

us. We have simply to trace its immediate and re-moter antecedents, keeping in mind that a fact is a virtual unity, and that the nobler province of analysis lies not in ingenuity of dissection but in mounting, step by step, the scale of apparently divergent causes until we re-unite them in one common centre ; — not in distinguishing the parent stem from the branches and the leaves, which together make up one tree, but in tracing back the tree itself to that which is the general origin of all plants — a seed. Thus Gall, as I have heard, developed the brain, instead of cutting it in pieces, and demonstrated that its convolutions were but parts of one beautiful and consistent whole.

But even should we utterly fail in this, — should a phenomenon stand single and alone, and baffle all our attempts to assimilate it to aught around us, we cannot, in true philosophy, pronounce that it exists not. The circumstance that water expands when cooled down below a certain temperature, is utterly at variance with the usual law by which all substances contract under a diminution of heat. Yet we do not deny that the fact is so. Surely, when we consider that every event is but one term of a series, the beginning and the end of which are alike lost in infinity, we should expect to trace back many a phenomenon into an obscurity that should seem impenetrable. Surely, when we reflect on the vastness of the visible creation, and on the still vaster universe of thought; when it seems probable that our system is but one of countless numbers, that are revolving round some unknown centre, in periods which thought cannot measure nor science determine; that these, again, are peopled with various orders of intelligence, —we should rather wonder that we know so much, than that some facts should transcend our knowledge, and acquiesce in our ignorance rather than deny exist-ence to that which we cannot comprehend.

But I may be reminded that mesmerism may be one of those subjects which surpass our intellectual vision; and that, if so, to examine it is superfluous, and to search for its principle a pure waste of time. Undoubtedly. But are we now, or can we be, indeed, at any time, entitled to class mesmerism amongst unfathomable things? If "the proper study of mankind is man," this, as regarding man especially, presents to his faculties a fair field of investigation; and with respect to such legitimate subjects of inquiry, the point at which we should abandon enterprise has not yet been ascertained. While, then, we admit the difficulties that beset us on every side, let us advance, and fear not, and take caution with us in our pursuit of truth; not to check but to steady us in our course. Discovery may be nearer to us than we suppose.* The enigma in which every point seems to contradict the other,

* This, above all, is not the time for despair, when such a man as Elliotson has dedicated his powers of learning and of observation to the service of mesmerism; treating the subject (as far as I can judge from the scanty account of Dr. Elliotson's experiments which has reached me abroad) in a spirit which seems to render any remarks of mine useless.

may be rendered clear, perhaps, by one word alone; and may set us wondering that we did not sooner perceive its simple solution. To me, at least, nature has just read a beautiful lesson, inculcating doubt in ourselves, trust in her, and patience to wait upon her revelations. All day the mountains were covered with a mist so uniform that, to a stranger's eye, they might have seemed like clouds, or altogether have escaped his notice; — but at this moment there is light and clearness; and where almost a blank was, are now the glorious Alps, snow-covered, lit up with sunshine — seeming to link earth with heaven.

Having now detailed, at too much length, perhaps, the principal causes that have rendered mesmerism obnoxious to the world, it now only remains for me to touch briefly on one or two minor causes of offence. Mesmerism is one of nature's great resources in the cure of maladies; and it is not, therefore, wonderful, if some of its most striking effects should have been developed rather in the ailing and the delicate than in the healthy and robust. Hence the world, always ready to build up error on truth, has connected it, in idea, with weakness of mind as well as of body, and has classed it amongst those idle imaginings which beset the fanciful invalid. But what is the fact? Mesmerism does, indeed, act more peculiarly on the nervous system; and, on that account, affects, in an especial manner, persons whose nervous system is finely organised. But we must not confound sensitiveness with imbecility. The universal temperament of genius gives the lie to such

c 4

an error; and it would be plainly ridiculous to say
that the timid and susceptible author of an elegy in
a country churchyard, or Rousseau, or Pascal*, who
were both nervous even to hypochondriacism, were
weak in intellect, because they were strong in sensibility. Besides, before we identify mesmerism with
weakness of any kind, it should be shown that none
but the feeble are susceptible of its influence. Now,
as far as my experience goes, I can affirm, that not
only does a certain degree of intelligence appear
requisite for the favourable manifestation of the
mesmeric phenomena, but that persons in perfect
health have frequently exhibited them. It may also
be asserted, that fear and nervous agitation are
wholly incompatible with their genuine development. These may, indeed, accompany a spurious
sort of mesmeric affection, but are wholly distinct
from the power with which they co-exist, and to
which they are invariably hurtful. They are the
corruptions of the true faith, and not the faith itself.
In fine, sensibility, and not weakness, is the real
condition on which mesmerism depends.

But it is not only an apparent identity with
weakness which has degraded this unfortunate
subject; — it has suffered still more from a re-
pulsive connection with disease. The discoverer of
mesmerism was a physician, and its extraordinary

---

* Pascal, after having once been nearly upset into a river
was, for a long time, tormented with the idea of being continually on the edge of a precipice.

curative powers have naturally placed it in medical
hands; so that of the existing works upon it there
are few that are not written by members of the
medical profession, and fewer still that do not bear
immediate reference to the treatment of maladies.
This alone is sufficient to exclude it from general
interest, and to lower its pretensions to literary
consideration. Medical books are read but by a
few, — ordinarily, indeed, by those alone whose
vocation compels them to the perusal. To the
greater part of mankind they are as closely shut
as the dissecting room. It might, however, be
supposed that, as all men are liable to pain and
sickness, all would be interested in that which, at
one time, claimed to be a panacea. But whatever
may be said of the sufferings of humanity, the mass
of health is greater than the mass of disease; and
most persons instinctively dislike to have their
well-being disturbed by hearing of remedies which
they do not require, or of ailments with which they
cannot sympathise. There is something also in
medical details, which is peculiarly quenching to the
imagination, and consequently distasteful to the
man of letters and refinement. It is not, then,
extraordinary that mesmerism, treated medically,
should have been restricted to a narrow sphere; but
is this the fault of mesmerism itself? Surely not.
On no subject is it less permitted to

" Give up to party what's meant for mankind."

The doors of this temple should be thrown widely
c 5

open to the world. As mental, it regards the meta-physician and the moralist; as physical, the physiologist and the man of science; and I would call upon each and all, separating it from its ridiculous concomitants, to contribute to raise it to that intellectual grandeur of which it is so eminently susceptible.

Such are the causes which have condemned mesmerism to lie beating, like a wreck, on the shore of substantial knowledge. The vessel, in itself, was beautiful and well built, but adverse currents turned it from its course, and by many a storm its sails were rent, and its noble frame was shattered.

But it is not so much the causes which I have separately enumerated as their combination which has marked the discovery of Mesmer with peculiar odium, and rendered it a by word and a reproach. All the circumstances which are unfavourable to mesmerism end in one fatal word — contempt. Everything tends to raise a laugh at its expense; — and against a laugh who shall have the courage to contend? This is the last possible degradation. Men love the mysterious and the proscribed, but shrink from the ridiculous; they can bear to be thought wicked, but not to be deemed fools; they will endure to be hated, but not to be despised. Now mesmerism has become not merely a persecuted but a ridiculous faith. There is no pomp of circumstance about it to uphold the proselyte who is called upon to defend it to the death. The glory of martyrdom for its sake is done away. There is no dignity in suffering in such a cause.

Thus, then, the advocate of mesmerism comes
before a prejudiced tribunal, and labours under
disabilities which it were just to bear in mind when
we feel inclined to measure the merits of his subject
by the reception with which it meets.  We should,
however, take but a partial view of the case, were
we to attribute the writer's want of success entirely
to the partialities of those whom he addresses.  Before we can have a true notion of the difficulties of
his undertaking, the causes of failure, personal to
himself, must also be added to the account.  He has
to battle not only with the minds of others but with
his own — to fight his way through obstacles which
arise from within as well as from without, and to force
himself to calmness where all conspires to agitate and
to excite.  However difficult it may be for him to
obtain a hearing; to treat his subject impartially
and so as best to conciliate an adversary is perhaps
a still harder task.  He is his own worst enemy, and,
amidst every ambush laid for him, has most to
beware of the snares of his own spirit.  For what is
a man's state of mind, when, for the first time, he
withdraws the curtain which separates the region of
mesmerism from the scenery of common life ?  He is
bewildered with rushing thoughts and wondrous
speculations.  He has beheld, but scarcely knows
what it is he has beheld.  A world of magic has
opened on his gaze; and, should he be a person of
imagination, who, like Schiller's idealist, is tired of
rough reality, he finds there all that can intoxicate and
enchain, and keep him lingering on enchanted

c 6

ground when he should be struggling onward to the domain of truth. Thus fascinated, thus enthralled, it is no wonder that, if called upon to abdicate his paradise, he should exclaim, like Milton's Eve when sentenced to quit Eden —

> " From thee
> How shall I part, and whither wander down
> Into a lower world, to this obscure
> And wild ?  How shall we breathe in other air
> Less pure, accustomed to immortal fruits ? "

There is, moreover, in the sensations of him who finds that he is capable of exercising the mesmeric influence that peculiar charm which ever waits upon the development of a new faculty. Even the swimmer, who learns at length to surmount the boisterous surf, or to stem the adverse stream, will revel in the consciousness of awakened power. How much more must the mental enthusiast riot in the display of energies so long concealed, so wondrously developed! Self-love adds her flattering lure to the attractions of novelty ; — the pride of exerting an influence over others awakens in his breast. It is he himself who is the author of his own enjoyment ; and the fairy scenes appear to him fairer still, because they are of his own creating. Unexpectedness, too, that principal ingredient of pleasure, yet more entrances and bewilders the astonished novice, who perceives such mighty effects resulting from his employment of a few and simple means. He feels that he is " greater than he knows," and he advances into the yet un-

conquered province that lies before him, with all those
alternations of rapture and surprise which agitate yet
please the explorer of strange regions. He trembles —
he hesitates — he catches a glimpse of a new prospect. Still is he tempted onward — ever onward
fresh wonders still opening around him, and fresh
complacency awakening in his heart : for mesmerism
is not one of those pursuits of which a man soon tires.
Founded on human nature, it presents a perpetual
variety, like that of humanity itself. Its successful
votary runs the risk, therefore, of turning his thoughts
upon it not only too vividly, but too exclusively.
Absorbed by a mighty interest, he is too apt to pass
into that mood wherein, as to a lover, all nature is
to us but a reflection of the beloved object; when, if
we turn a page, we think that it bears reference to
the subject of our ceaseless contemplation ; — when
every sound teems with appropriate oracles, and
every sight with omens addressed to us alone — all
figuring the image of our own thoughts. Of this there
is the greater danger, because there exists, in fact, so
much of unconscious mesmerism in life, that it is
not surprising if, like Alonzo in The Tempest, we
should exclaim —

> " Methought the billows spoke and told me of it;
> The winds did sing it to me, and the thunder —
> That deep and dreadful organ-pipe — pronounced
> Its name."

But, to an enthusiast in particular, how much
there is about him and within him to recall and

strengthen the first impression, and to speak to him
perpetually of that which has kindled his imagination !
That mysterious bliss, wherewith the very atmosphere
of those we love seems to be impregnated ; — that attraction of an unknown face, which beams upon us
from the passing crowd, and which we would give
worlds to see again; — the sleep of the infant that
is "rocked by the beating of its mother's heart ; " —
all teems with an influence, potent yet invisible, and
which we may call mesmerism, if we will. But, alas !
feelings are not proofs; and, on these occasions, how
much we feel that we can by no means prove — how
much we seem to understand of which we can
render no account ! It will be easily seen how unfavourable is this vagueness of emotion — this universality of reference — to the strictness of philosophy
and the particularity of rigid research. We have at
hand too ready a solution of every difficulty. We
are tempted to generalise before we have performed
the work of analysis, and to allow hypothesis to
supersede investigation. On a subject so wide and
so fertile, analogies offer themselves in crowds, and
being easier than inductions, are more readily
accepted by our effervescing imagination. So much
of nature's inner processes is apparently laid open to
our view, that we seem to be on the point of snatching the last word of her secret, and of breaking down
the barriers that separate the visible from the unseen,
the finite from the unbounded.

Thus excited, how shall we subside into the calmness, wherein alone the mind can compare, can

select, or reject? How shall we exchange the wand
of the magician for the disenchanting rod of truth?
How, from the very heights of contemplation, shall
we stoop to the observation of petty facts — to the
detection of fallacies — to the routine of ordinary
examination? Deeming that we know all things by
intuition, how can we condescend to reason? Rather
are we tempted to frame some vast and all-embracing
theory — to rush at every subject, and consequently
to fail in all. It is at least worthy of remark, that the
writers on mesmerism, who have treated it otherwise
than medically, have been too much of that school,
which philosophises " de omnibus rebus, et quibus-
dam aliis." They have passed the flaming bounds of
the universe, when they ought to have been considering the peculiar influence which one human being,
under certain circumstances, exerts upon another.

Mesmerism, then, is no matter of indifference, on
which a man may argue with a clear head, because
he argues with a quiet heart. It is a subject which
not only calls into action, but has a decided influence
on all the most essential principles of the human
mind; disturbing the emotions, and through them
the exercise of the intellectual powers. Self-love,
wonder, and desire of novelty, we have already seen
enlisted in its service, and troubled by its apparent
miracles. But deeply as it may engage these
feelings, still greater is the effect which it produces
upon the principle of belief — a part of our mental
constitution which so largely concerns our modes of
action — our present, and our future well-being, that

whatever tends to modify it is of vital consequence, and should be treated with all the caution which the importance of the occasion demands. With how much self-watchfulness, then, should we approach a subject which is calculated to work an entire re- volution in the very regulator of our being — to substitute a rash credulity for a sound and safe belief? This is the great snare of his own mind, against which a writer on mesmerism has perpetu- ally to be on his guard. He has beheld phenomena which are beyond the pale of his former experience; and, impelled as we all are to reason from that which we ourselves have seen to that of which we are personally ignorant, he is tempted to ask himself, — " If this be true (as I perceive it is), why not also a thousand other things, of which I have hitherto doubted?" Thus is he led to trust rather to analogy than to observation; and his willingness to accept a fact upon due evidence is turned into a disposition to credit all things upon no evidence whatever. The barrier of prudent hesitation is cast down, and his mind is left a prey to the invasion of every idle fancy. That he was once even of a sceptical spirit will profit him nothing. Rather, with that propensity to rush into extremes which is a part of human imperfection, the more incredulous he has hitherto been, the greater will be his rebound to credulity. The more he has hitherto refused to put faith in the testimony of others, the more entirely will he submit himself to that of his own senses; and finding that he is forced to believe what once he deemed incredible, he feels that he has scarcely a right to refuse assent to the wildest dreams of a diseased imagination. The wise admonition, — " Believe not every spirit,"— is disregarded ; — nay, there is danger that he should become in love with marvels, and not only welcome them, when they are offered to his consideration, but go out of his way to seek them, even where they exist not ; elevating every trifle into importance, till he falls into the absurdity of perpetual wonder. Thus it is to be feared that he who adventures his bark of discovery into the great deep of mesmerism will not, Columbus-like, steer ever towards one anticipated point, deterred neither by portents nor by passions in mutiny ; but that he will enter upon a course resembling rather the fabled voyage of Ulysses, haunted by prodigies and embarrassed by delays; and it will be well if imagination, like another Circe, does not seduce from their allegiance, and transform those faculties which ought to guide and serve him under the stern control of reason.

But, even supposing that a man has strength of mind sufficient to elude these perils, and to discipline his fancy to the proper standard of belief, stating only that which he has ascertained by the most rigid observation, there is still a difficulty in his way, which relates mutually to himself and to those whom he addresses. However reasonable he may be, and however unprejudiced his reader, still each party is, as regards the subject in question, in a very different state of experience. It is true that, in a degree, this remark applies to every case where information is given and received; but then, in all other cases, there is a groundwork to go upon, common both to the instructor and the instructed. The experience of the former is indeed more enlarged than that of the latter: yet it is ever to the experience of the pupil that the teacher appeals; and the basis of the new edifice of knowledge which he undertakes to erect consists in a few simple principles which are generally recognised by all mankind. For instance, an ignorant man, when told of the enormous distances at which the stars are placed, relatively to the earth and to each other, may at first feel inclined to doubt the facts: but let the size of the heavenly bodies be stated to him; let him be reminded that all objects appear less in proportion to their distance from us; and, that which he knows already helping him to that he does not know, the matter is at once brought within the scope of his comprehension. But, as regards mesmerism, we are excluded from any such appeal to universal experience. On this subject, he who undertakes to persuade, and he who consents to listen, are not only in different, but in opposite states of mind. There is no common point from which they may start together, — there is no centre of belief where they may meet and be reconciled. The pupil, at the very outset, is called upon to assent to a series of propositions not only startling but repugnant to his experience. That a man can throw another into a kind of sleep by certain gestures; that the person thus apparently deprived of consciousness can answer questions rationally, and

discern objects, though his eyes are fast shut; —
these are the extraordinary statements which alarm
him into opposition before the discussion is begun,
and excite either his incredulity or his scorn. By
them he is thrown into a perplexing dilemma.
Unexplained as are the facts of mesmerism, or (what
is worse) inefficiently explained, it seems as if he had
only to choose between rejecting them as absurdities,
or admitting them as miracles. In either case, an
inquiry into their causes seems a pure waste of time.
If nullities, they sprang from nothing; if miracles,
they cannot be accounted for. Moreover, to admit
them as miracles is repugnant to human feeling in
general. Men love not to have their faith taxed
more than is absolutely necessary; and, in this case,
there is no religious motive which can compel faith :
on the contrary, many a person of weak but zealous
piety fears to lessen the miracles recorded in
scripture, by allowing that any thing whatever,
which appears to interrupt the course of nature,
can, at this day, be wrought by mortal power. With
these feelings, though once, perhaps, experienced by
himself, the mesmeriser finds it difficult duly to
sympathise. He has left them far behind. To him
the facts of mesmerism are no miracles, — for one
plain reason : from the moment a thing takes its
station amongst every-day occurrences it ceases to
be a wonder. Let any unwonted term in any
sequence of events with which we are already acquainted interpose itself sufficiently often, and it
will itself become a part of the series which it

appeared, at first, to interrupt. In brief, our
familiarity with a fact is the measure of our opinion
concerning it. Drawing lightning from the clouds
was once a feat that appeared almost necromantic;
but custom has deprived it of its marvel, and we
place conductors on our houses as a matter of course.
Had the sun always begun the day in the west, such
would appear to us his natural mode of rising. Now
the mesmeriser has probably been in the constant
habit of witnessing those phenomena, at the bare
mention of which others are alarmed or revolted;
and so rivetted together in his thoughts are certain
trains of physical or mental changes, that *not* to
behold them would be the surprise; *not* to find the
usual result consequent upon the usual causes would
be the miracle.

Thus, then, the writer on mesmerism and his
reader are, as it were, at antagonist poles of belief;
and it is as difficult for the one to divest himself of
his present knowledge, and to revert to his ancient
incredulity or wonder, as for the other cordially to
assent to facts which belie the whole tenor of his
experience. To find the resolution of these discords
is no easy task. Let the author approach his subject
as guardedly as he may, still he cannot but state
facts that are astonishing; and the manner of his
discourse will avail him little while the matter of it
is so exceptionable. To those who disbelieve the
phenomena whereof he treats, an inquiry into their
causes is but an impertinence. The very calmness
with which he propounds his heretical opinions, and

which results from a deep conviction of their truth, must appear to others the worst symptom of incurable error; — his impartiality will seem prejudice; his utmost caution madness. Let him be ever so sound in mind, he will encounter more than one Felix who will be ready to exclaim, "Thou art beside thyself."

An opposition so great between the writer on our subject and his reader, would seem to render the task of the former not only difficult but hopeless. On more mature reflection, however, we shall perceive that though there is much to make him lower his expectations of success, there is enough to encourage him not to abandon them altogether. Amongst those to whom facts interesting and new are stated, there will ever be some persons who will be led to think and to inquire for themselves. Thus the mesmeriser only loses the suffrages of the prejudiced who refuse, or the careless who are incompetent, to examine the question, and the very causes that contribute to narrow the circle of his auditors tend only to render it more select.

In conclusion : — The result of the preceding deliberations seems to be this. We should lay aside all prejudice, connected either with the origin, name, or injudicious exposition of mesmerism, and try the subject, wholly and impartially, upon its own merits.

Unalarmed by the apparent strangeness and incongruity of the phenomena to be investigated, we should call to mind how frequently "appearances of external nature, puzzling at first sight, and seemingly irreconcileable with one another, have all been solved and harmonised by a reference to some one pervading principle," and should thus be led to surmise that the irregularity and variations of the mesmeric world may be found, upon mature observation, less inexplicable than a careless spectator could imagine. Even should this hope be long deferred, we are not, on that account, to deny the reality of well-attested facts. Are these things so? is the one great question which we have to ask; and to separate this from all its accidental accompaniments is the first step towards its satisfactory solution.

# BOOK II.

## SECT. I.

### MESMERIC SOMNAMBULISM, OR, MORE PROPERLY, SLEEPWAKING.*

IN entering upon the examination of a new subject, we should strive to place it in the clearest and least disputable point of view. To this end, I shall for the present consider mesmerism simply as a distinct species of somnambulism, into which it has been asserted that man has the capacity of passing, through the influence of his fellow-man. Were this all — did none other of the extraordinary allegations respecting mesmerism exist, — this alone appears to me a subject of deep interest, and worthy of being submitted to that patient investigation, which should either confirm it as a fact, or banish it as an illusion. Mesmeric sleepwaking — if proved to be a reality — will claim to be considered as one of the conditions of man, and, as such, must concern all mankind — but especially the philosopher, who, in order to investigate our nature truly, should explore it in its

* Dr. Elliotson, in the chapter on mesmerism in his PHYSI-OLOGY, has adopted the term *sleepwaking*, because walking is but one result of the combination of the waking with the sleeping state, and because in this state persons may not walk, or may even be unable to walk.

weakness and its strength, its integrity and its disorder. Sleep, fever, irritation of the nervous system, have each their several trains of physical and mental phenomena, which are anxiously investigated by him who would know man as he is. Even drunkenness has had its anatomists, and with what fearful interest we strive to fathom the abyss of madness, as if we could evoke from thence the secret of our complicate and wonderful existence ! There is good sense in this. From the derangement of a machine we are often led to ascertain the uses of its several parts. Now, though mesmeric sleepwaking can scarcely be called a derangement of the animal economy, still it is an aberration from man's normal state ; and, so considered, cannot fail to be replete with instruction to the careful observer. Unless we peruse this page, our study of the great volume of human nature must be incomplete.

Surely likewise, independently of the philosophical view of the question, all who share this suffering mortal frame — all who, during long and tedious hours, have vainly courted sleep, that perverse power which, in the poet's beautiful words, is " still last to come where it is wanted most, " — all who mourn the past, or fear the future, cannot but be interested in a discussion respecting the reality of a power in man to withdraw for a while the senses of his fellow-mortals from this world of troubles.

In thus considering mesmerism as nothing more than a state to be investigated like any other state into which man has the capacity of passing, I shall

probably surprise those who have been accustomed
to regard the matter either as too mysterious to
claim affinity with aught on earth, or as too fanci-
fully wild to be brought under the laws of rational
research. It may be doubted also, whether, with
some persons, our subject may not lose its charms,
when robbed of its controversial title, and whether
the plain view I have taken of the question may not
equally offend the argumentative opposer who seeks
to show his eloquence in dispute, and the warm par-
tisan who would invest his favourite science with an
awful and unknown dignity. But the earnest and
impartial pupil of nature will be content to take his
stand with me on the sure ground of experimental
proof, and, even should he become my opponent, will
attack the true subject in debate instead of the man
of straw, which has so long and so often been warred
with under the name of mesmerism.

The proofs, which I shall offer, regarding this peculiar phasis of our mortal state, will be drawn,
almost exclusively, from the proceeds of my own
observation. To make use of the materials accumulated by others, in order to construct a theory
of one's own, has always appeared to me inconsistent
with the earnestness of true inquiry. That deep
conviction of the reality of his subject, which is an
author's life and soul, is the result of personal experience alone. Besides, it should be the aim, as I
conceive, of every one who undertakes a cause, to add
to the stock of real information thereupon. Thus
only can the boundaries of human knowledge be en-

larged; — for what new fact can be elicited by those, who copy everlastingly from the old? Moreover, a person who observes for himself may perchance remark what has hitherto escaped the notice of even clearer eyes than his. Let twenty men witness a transaction, and the attention of each will probably be directed to a different feature of that transaction. By the combined result of such individual scrutiny, we come to a perception of the whole truth. Each carefully observed fact is an heirloom to mankind; and he who verifies a new phenomenon does more, perhaps, for science than he who constructs a theory.

Before, however, bringing forward the records of my own personal experience, I must warn my reader that if he imagines I have anything miraculous to relate he will be disappointed. It is true that I might, were I so inclined, give a zest to my pages by recording circumstances which were passing strange; but I have avoided these, from reasons which partly regard myself, partly those whom I address.

With respect to the latter, I think that persons in general would only be revolted by particulars which they would scarcely credit on any testimony; and thus I should rather injure than promote the cause I have undertaken to defend.

Secondly, as concerning myself. For my own sake I shun those tracts of thought, which, if not belonging to superstition, at least closely border upon her domains. When once we pass the bounds of the definite and the probable, we can scarcely settle with our own minds what degree of credit we ought to attach to the mystic circumstances that surround us, or on what basis we should place them. Of every fact of this nature there are ordinarily two explanations to be given — a rational, and a marvellous. The marvellous would lead us too far; the rational does not content us. As, for instance, a sleepwaker describes to me what her family, in a distant house, are doing at a particular hour. I inquire subsequently into this, and find that she has been correct in every point. Now there are two interpretations of this seeming miracle : — we may either suppose that the sleepwaker has really seen what she stated ; or that, from her knowledge of what the occupations of her family generally are, at a certain hour, she was enabled to divine, almost certainly, the truth.

Again : a sleepwaker tells me — " My brother, who has been to the Havannah, and of whom we have not heard for some months, is returned, and is at this moment anchored off Flushing. He is just coming from watch, and on such a day he will be here." This is verified to the letter ; but again there are two methods of explaining wherefore : — either the sleepwaker was gifted with extraordinary knowledge and prescience, or, from expecting her brother about a certain time, she calculated on probabilities, perhaps, more acutely than in her natural state ; and this, combined with that sort of coincidence which occurs more frequently than some may suppose, is

sufficient to account for the verification of her assertions and her prophecy.  This may be true ; but I
confess that such reasoning as the latter does not
always satisfy me, and that, in my opinion, after
subtracting from the account all of the mavellous that
we possibly can, there will yet remain a  residuum of
something strange and perplexing.  For this reason,
I turn my attention to facts that appear  to me  more
important, as well  as  more certain ; conceiving  that
to dwell on wonders, that are, at best, doubtful, is far
from healthy to the mind.   I  have therefore always
discouraged  my  sleepwakers from  making  to  me
extraordinary  revelations, and I have endeavoured to
confine  both  them  and  myself  within  a  walk —
narrow, indeed,  but safe ;  less interesting,  perhaps,
but certainly more direct.

All  facts  connected  with  the  senses,  or  which
illustrate the  close  affinity  between  the  mesmeriser
and his patient (indicative as they are of some medium
of communication existing between them) seem to me
unexceptionable in  their tendency, and strictly relative to the influence we have to  consider.   To these
then I  have  chiefly limited  myself at  the  hazard of
offending by over-caution, rather than by superfluous
zeal..  Hence my narrative will  have  an humble air,
which  may  contrast  oddly,  perhaps,  with  the  high
terms in which I have  spoken of  the pretensions of
mesmerism in my first book.   I  may  be  asked  the
*cui bono* of certain trifling details, and may be told
that  the  experiments  I  have  related  are  neither
beautiful  nor  useful.   To  what  end,  it  may  be

demanded, do these sleepwakers perform things asleep which were better done awake? I can imagine a critic saying—" Never was the ' parturiunt montes,' more admirably exemplified than in the work before us. We are told, in the first place that the mesmeriser withdraws a curtain that conceals a magic world, and we are then presented with nothing better than a Bartholomew show of farthing rushlights. We first read of awakened powers and intellectual influences that unlock the mysteries of nature, and, then, proceeding to the experiments, we meet with trash little better than the following : — ' Are you asleep?' 'Yes !' — 'Do you see me?' — ' No !' &c. &c." This is all easily said; but I beseech my reader to remember the vast disproportion between the experiments themselves and the deductions to be drawn from them: they are but mean steps to a mighty temple ; yet they must be mounted before we can enter the sanctuary itself. Is this preliminary humiliation in approaching a science peculiar to mesmerism? Let us turn to Newton's optics, where we shall find such small particularities as the following : — " I made in a piece of lead a small hole with a pin, whose breadth was the forty-second part of an inch; for twenty-one of those pins laid together took up the breadth of half an inch." We are neither amused nor prejudiced by this, because we know to what mighty consequences the apparent trifles led. But in mesmerism unfortunately we look not forward thus ; and unfortunately also there is no existing subject which more requires men to look beyond externals. The object experimented on is man; and, in the very nature of things, whatever has to do with man contains some of the ludicrous. Human life and the pages of Shakspeare will show how near akin in us are tears and laughter, sublimity and littleness. Our postures, our gestures, our peculiarities of countenance may all become matter of mirth to each other of us. When their best friend falls down, the first impulse with most persons is to laugh. It is no wonder, then, that in mesmerism we experience a sense of the comic, which the inanimate objects that are operated upon in other sciences cannot excite. No one feels disposed to ridicule a crystal, or to smile at a mineral; but there will be, perhaps, moments in mesmerism, when "to be grave exceeds all powers of face." The motions of the mesmeriser especially must seem ridiculous to those who see him gesticulate, and know not to what end. He is using those means which experience has shown him to be most efficacious in calling forth the phenomena he desires to produce;— but the uninitiated cannot be aware of this, and an action that appears to want motive is essentially ludicrous. Any one who has stopped his ears to the music, while people are dancing, will remember how fantastic the scene of gaiety appeared to him. This, and more, I can allow. I have as lively a sense of the ludicrous as any man, and could, myself, most easily travestie mesmerism, and exhibit it in a ridiculous point of view; for ridicule is always easy. I can also laugh, where a grave subject is goodhumouredly

bantered; as, for instance, where Hood writes a lively
and pleasant article — into which one can plainly see
no malice enters — on what he facetiously calls som-
bamboozleism; and represents a Yorkshire grazier
as much disappointed, when he finds that mesmerism
is not a new mode of fattening animals and has
nothing to do with mangel wurzel. But there is
something in solemn and malignant ridicule which
is not so easy to be endured; — though, indeed, we
should remember that whether scoffers laugh at
mesmerism or not can matter little. The subject is
not for them, nor for any who cannot look
through the trifling to the important, without
being diverted from their object by things which
may make them smile at the moment but ponder
deeply afterwards; — as, for example, the mesmeriser makes faces, and the sleepwaker, though seeing
him not, imitates them exactly. Some may stop
short at the ludicrous part of this; but, for him who
views the experiment as proving an unseen medium
of communication between the mesmeriser and his
patient, the fact is full of an absorbing interest which
checks all lighter thoughts. Thus some have mocked
at the simplicity of diction or of events that characterises the writings of Wordsworth, while others have
seen that the poet has only adopted nature's mode of
teaching us philosophy. This, at least, must be
granted. The merest trifles are interesting that
suggest to us an action in man independent of his
present organisation. Now mesmerism teems with
more than slight indications of this; and we should

treasure up such glimmerings of futurity — however
faint, and however presented to us — as inestimable
proofs that we possess a germ of being which God
permits us to behold partially unfolded here, in order
to confirm our faith as to its fuller development
hereafter.

This being premised, I proceed to give an account
of the first experiments in mesmerism which it fell in
my way to witness ; conceiving that such a narrative
will form not only the simplest but the best commencement to the series of proofs I have to offer,
respecting the state and characteristics of mesmeric
sleepwaking.

If to have been an unbeliever in the very existence
of the state in question can add weight to my testimony, my reader, should he be also an heretic on the
subject, may be assured that his incredulity in this
respect can scarcely be greater than mine was up to
the winter of 1836. That at the time I mention I
should be both ignorant and prejudiced on the score
of mesmerism will not surprise those who are
aware of its long proscription in England, and the
want of information upon it which till very lately
prevailed there.

In the course of a residence at Antwerp, a valued
friend detailed to me some extraordinary results of
mesmerism, to which he had been eye-witness.
I could not altogether discredit the evidence of one
whom I knew to be both observant and incapable of
falsehood ; but I took refuge in the supposition that he
had been ingeniously deceived. Reflecting, however,

that to condemn before I had examined was as unjust to others as it was unsatisfactory to myself, I accepted readily the proposition of my friend to introduce me to an acquaintance of his in Antwerp who had learned the practice of the mesmeric art from a German physician. We waited together on Mr. K——, the mesmeriser, (an agreeable and well-informed person) and stated to him that the object of our visit was to prevail on him to exhibit to us a specimen of his mysterious talent.    To this he at first replied that he was rather seeking to abjure a renown that had become troublesome, half the world viewing him as a conjuror, and the other half as a getter up of strange comedies ; but," he kindly added, " if you will promise me a strictly private meeting, I will, this evening, do all in my power to convince you that mesmerism is no delusion. This being agreed upon, together with a stipulation that the members of my own family should be present on the occasion, I, to remove all doubt of complicity from every mind, proposed that Mr. K—— should mesmerise a person who should be a perfect stranger to him. To this he readily acceded; and now the only difficulty was to find a subject for our experiment.    At length we thought of a young person, in the middling class of life, who had often done fine work for the ladies of our family, and of whose character we had the most favourable knowledge.    Her mother was Irish ; her father, who had been dead some time, had been a Belgian, and she spoke English, Flemish, and French, with perfect facility. Her widowed parent was chiefly supported by her industry, and, in the midst of trying circumstances, her temper was gay and cheerful, and her health excellent. That she had never seen Mr. K—— we were sure; and of her probity and incapacity for feigning we had every reason to be convinced. With our request, conveyed to her through one of the ladies of our family, for whom she had conceived a warm affection, she complied without hesitation. Not being of a nervous, though of an exciteable temperament, she had no fears whatever about what she was to undergo. On the contrary, she had rather a desire to know what the sensation of being mesmerised might be. Of the phenomena which were to be developed in the mesmeric state she knew absolutely nothing. Thus all deceptive imitation of them on her part was rendered impossible.

About nine o'clock in the evening our party assembled for what in foreign phrase is called " une séance magnétique." Anna M——, our *mesmerisée*, was already with us. Mr. K—— arrived soon after, and was introduced to his young patient, whose name we had purposely avoided mentioning to him in the morning — not that we feared imposition on either hand, but that we were determined, by every precaution, to prevent any one from alleging that imposition had been practised. Utterly unknown as the parties were to each other, a game played by two confederates was plainly out of the question. Almost immediately after the entrance of Mr. K—— we proceeded to the business of the evening. By his directions, Mademoiselle M—— placed herself in an arm-chair at one end of the apartment, while he occupied a seat directly facing hers. He then took each of her hands in one of his, and sat in such a manner as that the knees and feet of both should be in contact. In this position he remained for some time motionless, attentively regarding her with eyes as unwinking as the lidless orbs which Coleridge has attributed to the Genius of Destruction. We had been told previously to keep utter silence, and none of our circle — composed of some five or six persons — felt inclined to transgress the order. To me, novice as I was at that time in such matters, it was a moment of absorbing interest. That which I had heard mocked at as foolishness — that which I myself had doubted as a dream, was perhaps about to be brought home to my conviction, and established for ever in my mind as a reality. Should the present trial prove successful, how much of my past experience must be remodelled and reversed!

Convinced as I have since been to what valuable conclusions the phenomena of mesmerism may conduct the inquirer, never, perhaps, have I been more impressed with the importance of its pretensions than at that moment when my doubts of their validity were either to be strengthened or removed. Concentrating my attention upon the motionless pair, I observed that Mademoiselle M—— seemed at her ease, and occasionally smiled, or glanced at the assembled party; but her eyes, as if by a charm, always reverted to those of her mesmeriser, and at length seemed unable to turn away from them

Then a heaviness, as of sleep, seemed to weigh down her eyelids, and to pervade the expression of her countenance; her head drooped on one side; her breathing became regular; at length her eyes closed entirely, and, to all appearance, she was calmly asleep in just seven minutes from the time when Mr. K—— first commenced his operations. I should have observed that, as soon as the first symptom of drowsiness was manifested, the mesmeriser had withdrawn his hands from those of Mademoiselle M—— and had commenced what are called the " mesmeric passes," * conducting his fingers slowly downward, without contact, along the arms of the patient. For about five minutes Mademoiselle M—— continued to repose tranquilly, when suddenly she began to heave deep sighs, and to turn and toss in her chair. She then called out, " Je me trouve malade ! Je m'étouffe ! " and, rising in a wild manner, she continued to repeat, " Je m'étouffe ! " evidently labouring under an oppression of the breath. But all this time her eyes remained fast shut, and, at the command of her mesmeriser, she took his arm, and walked — still with her eyes shut — to the table. Mr. K—— then said, " Voulez-vous que je vous éveille ? " " Oui, oui," she exclaimed, " Je m'etouffe." Upon this, Mr. K—— again operated with his hands, but in a different set of movements, and, taking out his handkerchief,

---

* The term means nothing more than such motions as I have described.

agitated the air around the patient, who forthwith opened her eyes, and stared about the room like a person awaking from sleep. No traces of her indisposition, however, appeared to remain, and, soon shaking off all drowsiness, she was able to converse and laugh as cheerfully as usual. On being asked what she remembered of her sensations, she said that she had only a general idea of having felt unwell and oppressed; that she had wished to open her eyes, but could not. They felt as if lead were on them. Of having walked to the table she had no recollection. Notwithstanding her having suffered, she was desirous of being again mesmerised, and sat down fearlessly to make a second trial. This time it was longer before her eyes closed, and she never seemed to be reduced to more than a state of half unconsciousness. When the mesmeriser asked her if she slept, she answered in the tone of utter drowsiness, " Je dors, et je ne dors pas." This lasted some time, when Mr. K—— declared that he was afraid of fatiguing his patient (and probably his spectators too), and that he should disperse the mesmeric fluid. To do so, however, seemed not so easy a matter as the first time when he awoke the sleepwaker. With difficulty she appeared to rouse herself, and even after having spoken a few words to us, and risen from her chair, she suddenly relapsed into a state of torpor, and fell prostrate to the ground, as if perfectly insensible. Mr. K—— entreating us not to be alarmed, raised her up, placed her in a chair, and supported her head with his hand. It was then that I distinctly recognised one of the asserted phenomena of mesmerism. The head of Mademoiselle M—— followed everywhere, with unerring certainty, the hand of her mesmeriser, and seemed irresistibly attracted to it, as iron to the loadstone. At length Mr. K—— succeeded in thoroughly awaking his patient, who, on being interrogated respecting her past sensations, said that she retained a recollection of her state of semi-consciousness, during which she much desired to have been able to sleep wholly; but of her having fallen to the ground, or of what had passed subsequently, she remembered nothing whatever. To other inquiries she replied that the drowsy sensation which first stole over her was rather of an agreeable nature, and that it was preceded by a slight tingling, which ran down her arms in the direction of the mesmeriser's fingers. Moreover, she assured us that the oppression she had at one time felt was not fanciful, but real — not mental, but bodily, and was accompanied by a peculiar pain in the region of the heart, which, however, ceased immediately on the dispersion of the mesmeric sleep. These statements were the rather to be relied upon, inasmuch as the girl's character was neither timid nor imaginative.

## Second Sitting.

On this occasion Mademoiselle M—— sat down, as she assured me afterwards, with a resolution to resist the mesmeric drowsiness to the utmost, and with a strong impression that she sohuld not sleep.

This want of co-operation on her part seemed to render the mesmeriser's task more difficult, yet in about twenty minutes she was brought into that state of imperfect slumber into which she had fallen on the first evening. As then, so now, she affirmed, "Je dors, et je ne dors pas;" while all the time she seemed incapable of opening her eyes. At other times she expressed an impatience of her semi-conscious state, and exclaimed, "Faites-moi dormir tout-a-fait;" but the nagain she murmured; "Non, vous ne le pouvez pas." In this half torpid state she remained so long, that we feared no other result would be obtained; and Mr. K—— himself turned round his head towards us, and shook it, as much as to say that he despaired of success. At length, however, appearing to rally his powers, he laid his hands upon the shoulders of his patient, who almost instantly heaved a deep sigh, and fell back in her chair with every symptom of the profoundest slumber. Soon after this, I observed that her head, as on the last occasion, followed every motion of the mesmer-iser's hands. Mr. K—— now asked her if she slept, and she replied softly, but distinctly, "Oui, je dors." He then asked her if she would quit her chair, and walk a little, but to this proposition she strongly objected, declaring that she wished to rest, and that she thought she should fall were she to attempt to get up. Mr. K——, however, assured her that such would not be the case, and rising from his chair, and going to a short distance, he made a motion with his arms, as if he would draw her towards him. Sighing deeply, she then got up, and moved towards him, but tottered, and exclaimed, " Oh je vais tomber." Upon this the mesmeriser seized her hand, and said, in a tone of decision, " Non, vous ne tomberez pas." In effect, she afterwards stood firmly, and even walked about the room, holding the arm of Mr. K—— but still retaining the expression of a sleeping person, though the extreme torpor which weighed down her head before she left her chair seemed, in a great measure, removed. To all appearance her eyes were closely shut, and there was even a sealed look about the lids which was very remarkable. Notwithstanding this, she appeared to be very sensible to the light, and never approached the lamp without complaining of the glare. One of our party having expressed a doubt that she slept, and this having been explained to her by the mesmeriser (for she did not seem to attend to any one else), she exclaimed, with an air of surprise, " Yes, certainly, I am asleep." During the time (upwards of an hour) that she remained thus abstracted, or entranced, I noted the following phenomena : —

1st. *Attraction towards the mesmeriser — mental and physical.* — She could not endure Mr. K—— to leave her for an instant; and there was a nervous anxiety in the manner in which she often exclaimed, " Mais, il ne faut pas me quitter." The most apparently uncomfortable posture seemed agreeable to her so long as she could rest her head on the mesmeriser's hand. If another hand

was substituted, however adroitly, she was immediately aware of the change, and betrayed uneasiness. On one occasion I took the mesmeriser's
place at a moment when he stood behind her. After
a few seconds, an expression of utter discontent and
repugnance came over her countenance, and she removed her head from my support. At this moment
Mr. K—— was stealing away noiselessly to another
part of the room, and I observed that her face turned
towards him, and followed his every motion. When
he made a gesture as if he would beckon her towards
him, she got up immediately, and walked to where
he was, threading her way, exactly as he had done,
between the chairs, tables, &c., without stumbling or
coming in contact with any obstacle.

2dly. *A knowledge of what the mesmeriser ate or drank,
indicating community of sensation with him.* — This
phenomenon appearing to me to constitute one of the
most interesting features of the mesmeric state, I
took particular pains to verify it. Mr. K—— stood
behind the sleepwaker, but in contact with her (for
otherwise the experiment did not seem to answer) one
hand slightly touching her shoulder. In this position
of the parties, we handed to Mr. K—— from behind
(so as to prevent any possibility of the sleepwaker
seeing what we were about, even had her eyes been
open) different things to drink or eat. While the
mesmeriser was taking these, Mademoiselle M——
imitated precisely all the motions of tasting and swal-
lowing, just as if she herself had taken the substances
into her mouth. On being asked what she tasted,

she stated correctly what the mesmeriser was occupied with at the moment.  In this way she recognised tea, wine, and water, and distinguished
between them, change them as often as we would.
The accuracy with which she described a peculiar
kind of roll, which we gave Mr. K—— to eat, was
extraordinary.  " Ce n'est pas du pain ; ce n'est pas
précisement du biscuit.  C'est quelquechose que les
matelots mangent."  The last assertion was perfectly
correct, for at Antwerp this kind of roll is a favourite
article of diet with the sailors.

3dly. *An increased quickness of perception.*—Privately
and with' precaution, I changed rings with the mesmeriser.  The moment the sleepwaker subsequently
took his hand, she exclaimed, " Vous avez quelque
chose, qui n'est pas de vous ! "  " A qui est ce donc ? "
asked Mr. K——: she immediately named the right
owner.  I should remark that, in her waking state,
she assured me she did not know whether even I
wore a ring or not.  Another proof of the acuteness of her perceptions was the accuracy with which
she distinguished her mesmeriser's hand from every
other.  We made the experiment together with
her mesmeriser, giving her all our hands successively many times, in a different order, but she
invariably recognised the hand of Mr. K—— at
the first touch, exclaiming " Ah c'est de lui."

4thly. *A development of the power of vision.* — I am
aware that I here approach dangerous ground.  Our
very nature rises up in arms against whatever seems incompatible with our personal experience; and that men

should, under any circumstances, see otherwise than in the appointed way of vision appears so entirely to contradict the very purposes of our organisation, that we shrink from the bare mention of such an anomaly. Accordingly, what is called *clair-voyance* has been ever the great stumbling-block to the unbeliever in mesmerism. To decide that mesmeric sleepwakers do really possess the faculty of seeing through closed lids, or other obstacles, would in this place be premature. I waive that question until I can place it upon the sure ground of proof. But allowing to pass for the present an hypothesis which has been suggested to me; namely, that mesmerised persons, being in a state of nervous excitement, see, like those afflicted with a peculiar irritability of the retina, with less light and through a smaller aperture between the lids, than others; still, even the fact that the sensibility of vision can be thus increased by apparently insignificant means, appears to me highly interesting and worthy of investigation. This I can safely affirm — that there was something new and astonishing in the manner in which Mademoiselle M—— distinguished objects. The closest observation of the six members of the party, who witnessed the experiments I am now detailing, could detect not the slightest opening of her eye-lids; nevertheless she gave undoubted proofs of possessing a considerable degree of vision, for she recognised and named every person present, when brought up to her in any order. The exercise, however, of whatever new mode of vision she had acquired, appeared to her difficult and fatiguing; and, when an object was held before her, she usually declared it to be too small for her to know what it was. When urged to look at any thing, she expressed the greatest repugnance to do so; and it was only at the reiterated command of the mesmeriser that she aroused herself to the necessary effort. At those times, her whole deportment was that of a person who wished to rest, yet who by some external force was compelled into exertion. No exorcised spirit could have done its work more grudgingly; and, like the enchantress evoked by Odin, she continually entreated to be left to repose. Sometimes addressing those around her, when asked to look at any object, she exclaimed, "You, who see so much better than I do, why do not *you* rather tell *me* what it is. I beg of you to tell me." Twice, however, she gave singular proofs of correct vision. Some music-paper was put into her hand, and she was asked what was *written* on it. She replied, "There is nothing written on it; it is music paper." The mesmeriser gave her his watch, and asked her the hour. After the usual reluctance and some delay, and moving of her fingers over the watch-glass, in the direction of the hands, she named the hour and minute with precision.

During this sitting, there was no return of the oppression from which Mademoiselle M—— had suffered on first being mesmerised. Both in countenance and manner she remained perfectly calm.

The experiments being at an end, she was asked if she wished to awake, when she answered in the

affirmative, but at the same time declared her inability to do so unless the mesmeriser took off the
heaviness from her eyes, as in the preceding sitting.
Upon this Mr. K—— made some lateral motions
with his hands, across her face, as if he were removing
something from before it, and in about a minute the
sleepwaker opened her eyes, rubbed them, and stared
about, like a person awaking from heavy slumber.
Her first question was, "How long have I been
asleep?" She then complained of considerable
heaviness in her limbs, but this was soon removed,
as it seemed, by some mesmeric "passes" which Mr.
K—— made over them. When interrogated as to
what she remembered of her past state, she said that
the last thing which she distinctly recollected was the
mesmeriser's laying his hands on her shoulders —
that they felt to her burning hot, and as heavy as
lead, and seemed to weigh her down into unconsciousness. All subsequent to this was a blank, with
the exception of a confused notion, as of something
in a dream that many hands had touched her own.

### Third Sitting.

This time, Mr. K—— brought to us, as his
mesmeric patient, a young man of about seventeen
years of age, whom I shall call Theodore, the son of
most respectable persons in Antwerp. Mr. K——
had already mesmerised him several times, and had
excited our curiosity respecting him by detailing to
us extraordinary proofs of his powers of vision in the
state of sleepwaking. From the moment of his

entering the room, this young man prepossessed every one in his favour. His quiet, and even timid manners, and his ingenuous countenance, made us feel that we might rely upon, as genuine, any phenomena he might exhibit. He was in a state of perfect health.

In about ten minutes from the time he was submitted, in the usual manner, to the influence of Mr. K—— he gave tokens of having passed into the mesmeric state, his head following every motion of the mesmeriser's hand. Slight shudderings, from time to time, seemed to pass over his frame, and were occasionally repeated during the whole time of his sleepwaking. This unusual state was accompanied by a marked change in his demeanour. He seemed to have lost his shyness, and conversed with Mr. K—— in an unembarrassed manner, as if no longer conscious of the presence of strangers. He became more animated, frequently laughed aloud, and even went the length of committing a pleasantry. When led up to a very pretty girl in company, the mesmeriser (in order to try his powers of vision) said, "Savez vous qui est ce jeune *homme?*" to which Theodore replied directly, "Je voudrais bien voir un jeune *homme* aussi joli que cela."

The principal phenomena which I noted in the case of Mademoiselle M—— were all developed in this sleepwaker, with the slight exception of his more patiently suffering the mesmeriser to go to a short distance from him. On the former, however, trying to steal from the apartment as noiselessly as possible, he rushed to him and dragged him back with considerable agitation, which was only calmed by the mesmeriser caressing him, as it were, with the hand. On all occasions he testified the strongest attachment and affection to Mr. K—— frequently leaning his head upon the shoulder of the latter, and running to him, when placed at a little distance, on the slightest motion of his hand, with such simple and natural gestures as a child uses when, half playfully, it runs for shelter to its mother's side.

His physical attraction towards the mesmeriser was as strongly marked as the mental. As he was sitting down, Mr. K—— approached his hand to the leg of the patient, and then slowly drew it upward. The limb, as if compelled to follow, rose also, and, as long as Mr. K—— held his hand over it, remained in a raised position.

A very important proof that this physical attraction was wholly independent of the patient's imagination was the following : — Theodore was standing with his back to the mesmeriser, while some one engaged his attention in front, and the mesmeriser going to a good distance, so as to render any cognisance of his gestures by feeling impossible, gently beckoned towards him the patient, who actually ran backwards for some steps, then, turning quickly round, hastened to the mesmeriser. Community of sensation was manifested not only by a recognition of what the mesmeriser ate or drank, but in another way, which seemed to indicate that the sense of feeling, as well as of taste, was placed in some mysterious communication with the corresponding function of the mesmeriser. Mr. K—— having left Theodore standing apart for a few instants, went to the fire and warmed himself. During this time the patient rubbed his hands and spread them out, as if over the flame, precisely as the mesmeriser did, and with a look of evident satisfaction and enjoyment. Mr. K—— said, " Que faites vous la Théodore ? " " Je m'échauffe " was the answer.

Our young patient also exhibited other phenomena, which I have since found to be generally characteristic of mesmeric sleepwaking, and which I shall call —

*Occasional Community of Motion with the Mesmeriser. — Isolation from all others than the Mesmeriser.*

In illustration of the first, I might adduce the fact, that there was no person present who did not remark that, in walking, Theodore took the step and air of Mr. K—— as if his body were a machine directed by the mesmeriser. But this may be objected to as a proof not sufficiently rigorous. The correctness, however, of the following experiment cannot be contested.

The mesmeriser was placed with his back to the patient. In this position, the former made sundry grimaces and contortions of visage, which were exactly and simultaneously imitated by the latter.

*Secondly.* Upon first passing into the mesmeric state, Theodore seemed absolutely insensible to every other than the mesmeriser's voice. Some of our party went close to him, and shouted his name, but he gave no tokens of hearing us, until Mr. K——, taking our hands, made us touch those of Theodore and his own at the same time. This he called putting us " en rapport " with the patient. After this, Theodore seemed to hear our voices equally with that of the mesmeriser, but by no means to pay an equal attention to them.

With regard to development of vision, the eyes of the patient appeared to be firmly shut during the whole sitting, and yet he gave the following proofs of accurate sight : —

Without being guided by our voices (for in making the experiment we kept carefully silent) he distinguished between the different persons present and the colours of their dresses. He also named with accuracy various objects on the table, such as a miniature picture, a drawing by Mr. K——, &c.

When the mesmeriser left him, and ran quickly amongst the chairs, tables, &c., of the apartment, he followed him, running also, and taking the same turns, without once coming in contact with any thing that stood in his way.

He told the hour accurately by Mr. K——'s watch.

He played several games at dominos with the different members of our family, as readily as if his eyes had been perfectly open. On these occasions the lights were placed in front of him, and he arranged his dominos on the table with their backs to the candles in such a manner that, when I placed

E

my head in the same position as his own, I could scarcely, through the shade, distinguish one from the other. Yet he took them up unerringly, never hesitated in his play, generally won the game, and announced the sum of the spots on such of his dominos as remained over at the end, before his adversaries could count theirs. One of our party, a lady, who had been extremely incredulous on the subject of mesmerism, stooped down so as to look under his eyelids all the time he played, and declared herself convinced and satisfied that his eyes were perfectly closed. It was not always, however, that Theodore could be prevailed upon to exercise his power of vision. Some words, written by the mesmeriser, of a tolerable size, being shown him, he declared (as Mademoiselle M—— did on another occasion) that it was too small for him to distinguish.

Towards the conclusion of the sitting, the patient seemed much fatigued, and, going to the sofa, arranged a pillow for himself comfortably under his head, after which he appeared to pass into a state more akin to natural sleep than his late sleepwaking. Mr. K—— allowed him to repose in this manner for a short time, and then awoke him by the usual formula. A very few motions of the hand were sufficient to restore him to full consciousness, and to his usual character. The fatigue of which he had so lately complained seemed wholly to have passed away, together with the memory of all that he had been doing for the last hour.

I must now pause to set before my reader my own

state of mind respecting the facts I had witnessed. I perceived that important deductions might be drawn from them, and that they bore upon disputed questions of the highest interest to man, connected with the three great mysteries of being — life, death, and immortality. On these grounds I was resolved to enter upon a consistent course of inquiry concerning them : though, as yet, while all was new and wonderful to my apprehensions, I could scarcely do more than observe and verify phenomena. It was, however, necessary that my views, though for the present bounded, should be distinct. I had already asked respecting mesmeric sleepwaking — does it exist ? and to this question the cases which had fallen under my notice, and which were above suspicion, seemed to answer decidedly in the affirmative. But it was essential still further to inquire — does it exist so generally as to be pronounced a part — though a rarely developed part — of the human constitution ? In order to determine this, it was requisite to observe how far individuals of different ages, stations, and temperaments, were capable of mesmeric sleepwaking. I resolved, therefore, by experiments on as extensive a scale as possible, to ascertain whether the state in question were too commonly exhibited to be exceptional or idiosyncratic. Again, the two cases that I had witnessed coincided in characteristics. But could this coincidence be accidental ? It might still be asked, were the phenomena they displayed uncertain, mutable, such as might never occur again ; — or were they orderly, invariable, the growth of fixed causes, which, being present, implied their presence also ?  In fine, was mesmeric sleepwaking, not only a state, but entitled to rank as a distinct state, clearly and permanently characterised ; and, as such, set apart from all other abnormal conditions of man ?  On its pretensions to be so considered, rested, I conceived, its claims to notice and peculiar investigation.  To decide this point was, therefore, one of my chief objects ; and respecting it I was determined to seek that certainty which can only be attained by a careful comparison of facts occurring under the same circumstances.  To sum up my intentions, — I desired to show that man, through external human influence, is capable of a species of sleepwaking differing from the common, not only inasmuch as it is otherwise produced, but as it displays quite other characteristics when produced.

Thus the inquiry on which I desired to enter was twofold, for I had to consider —

1st. Sleepwaking, as mesmerically induced.

2d. As possessing, when so induced, its own class of
        phenomena.

Before relating with what success I pursued the above objects, it is necessary that I should say a few words as to the plan to which I shall adhere in the detail of my " proofs by experiments." Each branch of my inquiry ought, perhaps, to be treated separately ; but, in that case, I should have, in the first instance, only to draw up a barren statement of the number of different persons, in whom I had seen sleepwaking induced ; and then I should have to return to these same cases for an elucidation of the constancy of the phenomena.

This would be at once cumbrous and useless, and I therefore take the simpler course of detailing my mesmeric cases just as they occurred, strengthening my cause at once by their number and by their identity of feature. That I should thus let the two arguments run parallel has this advantage also, — I present a portrait of mesmeric sleepwaking from the outset — a very needful preliminary. So at least will they who differ from me in opinion be sure that they and I are speaking of the same thing.

It may be also that to paint mesmerism as it is will remove objections to its validity more surely than a method more elaborate. It should never be forgotten that in all states there is a spurious and a true ; and of our unfortunate subject it may especially be affirmed that there exist many counterfeit presentations. It has been drawn by its enemies, and even in early times by its friends, as a convulsed and decrepid hag, with Superstition on her right hand and Spleen upon her left ; but the influence whereof I treat is of a tranquil brow, and has Truth and Reason for her handmaids. Should I become tedious by presenting again and again the same features to view, I have at least thus much to urge in my defence : each successive copy of the original type strengthens the evidence that the thing is ; while, if the picture once drawn be faithfully adhered to, there will be every reason to conclude that the thing is such as represented, and no other. Facts, unaccompanied by deductions, may,

E 3

indeed, be wearisome ; but, at present, the great
point is to portray phenomena as they occurred ;
for a discussion upon mesmerism is not like a discussion upon any known science, which men are at least
agreed to treat as a reality.    The inquiry is not
merely into its nature but its very existence.

I must further observe that, with regard to the
great characteristics of mesmerism, I have never had
occasion to change my opinion from the beginning.
I trusted to my impressions then, on the ground that
the first view of a subject often seizes its character the
best, and that a fresh eye will fix upon the salient
points of phenomena, to which habit may afterwards
blunt the perceptions.    Thus reasoning, I have not
since been convicted of erroneous views.    The production of a bodily sleep through another's influence,
while the mind continued waking — the community
of motion and sensation with the mesmeriser — the
development of the perceptions, — these were the
phenomena which, from the first, struck me in
mesmerism as the most important, the most replete
with interest; and they seem so still.    I have beheld
them often enough repeated to fix in my mind a conviction of their truth ; and, believing, as I do, that
they may be brought home to every man's business
and bosom, I am anxious to impart to others the
belief which I possess myself.

Thus, then, it is my intention plainly to narrate
that which I have witnessed, leaving it in general to
my reader to refer to their separate tracts of inquiry
the phenomena which I shall detail, and to mark their

correspondence with others previously observed.
Nor shall I confine myself altogether to a description of the completed state of mesmerism, since I conceive it to be interesting in its dawn also ; and it is instructive to see in certain cases, the same phenomena,
indicated in their rise, which were, in others, developed during their progress.  At any rate the begin-
nings of mesmeric sleepwaking, even if they go
no further, tend to prove that there is an influence
which one human being can exert over another, —
hitherto but partially observed, yet more general than
is usually supposed.  By such cases my first lemma
is advantaged, even if the second is not benefited
by it.

It may be even that the imperfect cases of mesmerism present the strongest testimony to its truth.  As
a force confessedly acting on the nervous system, the
strong and the insusceptible should be altogether
subtracted from its influence, so that the circumstance
of such persons being subdued at all is more convincing than the entire subjugation of the weaker and
more sensitive.  That a healthy, unimaginative man
should be reduced to a sort of spell-bound slumber
in five or ten minutes, by the agency of another,
under circumstances the most unfavourable to sleep,
strikes the mind, perhaps, as much as any of the
mightier miracles of mesmerism.  It is a fact, of
which the senses are clearly cognisant, and it has
made its converts accordingly.  It is true that, in the
physical sciences, the detail of an imperfect experiment is altogether futile ; but it is not so in respect

of the investigation of mind. Relative to this, the merest trifle is important, the slightest indication interesting. Besides, every state has a beginning, a middle, and an end; and the first is surely best examined when it remains stationary, instead of merging rapidly in the others. Let my reader, then, keep in mind that I divide mesmeric sleepwaking into imperfect and perfect; the former being altogether in harmony with and elucidating the latter.

In all the subsequent experiments, I was myself the operator, having, through the kindness of Mr. K——, received such instructions as put it in my own power to practise the mesmeric art; for it is plain that the course, on which I was desirous of entering, required that I should be independent of the extraneous assistance which I could by no means always have at hand, and demanded the full force of the most careful and personal experience.

CASE I.

The first person whom I succeeded in mesmerising was Theodore. The phenomena of the third sitting were repeated — with this additional proof of the patient's community of sensation with the mesmeriser: —

A member of our party, having heard it asserted that mesmerised persons could feel any injury that was inflicted on the mesmeriser, standing behind me, pulled my hair, without warning, and without a possibility of my patient being made aware of the circumstance in any usual way; — for, first, his eyes were closed; secondly, his head was bent down and drooping, as frequently happens in the mesmeric sleepwaking; thirdly, the position in which I stood completely screened from him the person who tried the experiment. Nevertheless, Theodore immediately winced as if he himself had felt the injury, and put up his hand to that part of the back of my head where my hair had been pulled, throwing his arm round my neck, as if to defend me from any other attacks that might be made upon me.

I observed that when he exerted himself to play at dominos he was more particularly affected by those slight convulsive twitchings which I had remarked on a former occasion.

CASE II.

A. L——, a Belgian gentleman of our acquaintance, who thoroughly laughed at mesmerism, consented, mockingly, one evening, to let me mesmerise him. He was twenty-three years of age — in good health; — in character imaginative, but of a scoffing turn, and lively even to boisterousness. On first sitting down, while I held his hands in the usual mesmeric manner, he laughed much and made ridiculous faces, opening his eyes, and pretending, in his turn, to mesmerise *me*. Shortly, however, the same drowsy expression which I had remarked in Mademoiselle M—— stole over his countenance. He became serious, and seemed to struggle against the torpor which was evidently coming over him. At length, as if disliking to be vanquished, he got up from his chair with evident effort, and walked about

E 5

the room, but soon, complaining of an irresistible drowsiness, lay down, as if he did not at all know what he was doing, on the sofa, in the presence of ladies, before whom, under other circumstances, he never would have dreamed of committing such a breach of etiquette. I came to him, and, curious to know how far my influence over him might extend, made the usual mesmeric passes in front of him. Upon this, his upper eyelids, with a motion almost as slow and imperceptible as that of the hour hand of a watch, began to fall, while his eyeballs seemed drawn upwards. After a time, the lids became stationary and remained a little open, showing only the whites of the eye. His head began to follow the motions of my hand. I now asked him if he slept, to which he replied, in a very low voice, " Not altogether ; I wish I could." " Do you suffer in any way ? " I inquired. " Yes, I have a pain in the back of my head."

Upon this, I made mesmeric passes over the part affected, and soon after asked —

Q. — Do you feel better now ?

A. — Yes, the pain is quite gone.

Q. — Do you wish to be awakened ?

A. — No, I wish to sleep more soundly.

Q. — Why do you not open your eyes wider ?

A. — I cannot; it is as if lead were on them (Ils sont comme plombés).

Q. — Can you see any thing ?

A. — I see you.

Q. — Nothing else ?

A. — Nothing whatever.

I now made a motion as if I were going away, when suddenly the patient seized my hands, one in each of his, and exclaimed, " You must not go away, on any account." After this, he grasped my hands so tightly that it would have been no easy matter for me to liberate them. Fearing to irritate him, I remained quiet, and as long as I did so he lay perfectly motionless, but always with his eyes a little open. His complexion, which was naturally pale, was changed to a ghastly hue, and his whole appearance was so death-like that the spectators in the room became alarmed, and I own that I was myself uneasy. To my questions, however, he replied that he did not wish to wake, — that he was very well, but that I must not leave him. Thus circumstanced, I scarcely knew what course to take. That he should be awakened was desirable, for it was late ; but every effort that I made to free my hands only made him grasp them more firmly. My position was sufficiently ridiculous. I had raised a spirit which I could not quell, and the work of my own hands had become as unmanageable as the creation of Frankenstein. At length, after matters had remained in this state a full hour, I so strongly insisted upon his awaking, that he reluctantly permitted me to take the necessary measures for arousing him. I then made before the patient the motions which are usually employed for dispersing the mesmeric slumber ; and it was curious to observe how quickly, by means (as it seemed) of these few and simple gestures, the torpor

E 6

passed away. At the same time, the natural colour and appearance returned to the countenance of my patient, who got up, rubbed his eyes, looked much confused at having slept in the company of ladies (a circumstance that shocked his foreign politeness), and, with many apologies, bade us hastily good night. The next day, when I called to inquire after him, he seemed to be half ashamed of his mesmeric exploits, but owned that the way in which he had been affected was at least a convincing proof to him of the influence he had denied. A drowsiness, he said, and slight confusion of head were still felt by him; and these, he declared, were relieved by my moving my hand across his forehead. When I asked him what he remembered of his mesmeric state, he said that it seemed to him as if he had always retained in it a sort of twilight consciousness, and that he recollected having been very anxious to exchange these " twinklings of oblivion " for a deeper slumber. Of my having spoken to him, however, or of his having answered me, he remembered nothing; and when I told him how forcibly he had kept me near him, he was much surprised.

Shortly after making the foregoing experiments, I returned to England, it being the spring of 1837. Like a person full of a new subject, I discoursed, amongst my friends, upon what I had witnessed of the mesmeric influence. The result may be imagined. Those who knew me well were forced to believe that I had really beheld the phenomena I described; but here they were content to stop. It was all very extraordinary, they declared, and they would rather not think about it.  Some swore —

"'Twas strange — 'twas passing strange."

And others that —

"'Twas pitiful — 'twas wondrous pitiful."

A few, more sagacious than the rest, surmised that it savoured of the black art; and the ladies thought, indeed, that "the naughty man *might* have something to do with it."  One or two confessed that they were too old for the introduction of new ideas into their brains, — and I fully believed them when they said so.  More distant acquaintances listened to my story with a look of polite incredulity; — others yawned.

The practice of the Baron Dupotet, and the efforts of Mr. Mayo and Dr. Lardner in promulgating Dr. Elliotson's experiments, had not yet familiarised the London world with, at least, the name of mesmerism; — so, in the great thinking, unthinking metropolis, I met with scarce any disciples of the new creed.  The only experiment, therefore, which I had any opportunity of making was upon a young cousin of my own, — a lad of about fifteen.  He, however, proved at that time insensible to the mesmeric influence; but he took an interest in the subject of mesmerism, and actually tried to mesmerise a younger brother of his, who was not more than twelve years old.  The results of his attempt are sufficiently curious to tempt me to swerve from my usual rule of only recording that which I have personally observed. Besides, the actors in the little drama are so well known to me, that it almost appears as if I had been eye-witness to the scene, the circumstances of which were related to me by the mother of the two boys.

The elder, in her presence, sat down, partly in joke, to mesmerise his brother, who was wholly ignorant of the process and effects of mesmerism; and, taking the hands of his patient, went through the forms of which he had seen me make use when mesmerising. Be it remarked that the time was midday, when the light, the noise in the streets, and the natural liveliness of youth, would particularly indispose a boy to slumber. Nevertheless, in about ten minutes, a manifest effect was produced. The patient's eyes closed as by a spell, and his head followed the mesmeriser's hand in the usual manner. Charles, the elder brother, now got up from his chair, when Edward rose also. The former raised his right arm; the latter immediately raised his; — his left arm — the same result. Charles walked forward — Edward advanced also. Charles stopped, and Edward stood still on the instant; — in short, all the gestures of the one were faithfully imitated by the other. But the mother became now alarmed, and insisted upon the young mesmeriser awaking his patient. This was soon effected, and Edward, opening his eyes, and staring about like one just startled from slumber, expressed the greatest as- tonishment at finding himself standing in the middle of the room, — declaring, at the same time, that his unconsciousness had been complete.

In one point of view the above relation lays claim to a peculiar interest. The effects of a new influence, when manifested by the young, are especially to be relied upon; they can be neither prompted nor impeded by those prepossessions which are the growth of maturer years; nor can they be attributed to those nervous imaginings which spring from infirmity. Youth, moreover, is the season of candour, when all is bold, healthy, unsophisticated; and the phenomena which it exhibits are valuable in proportion as they are exempt from the suspicion which attaches to the doings of the older inhabitants of this crafty world.

## SECT. II.

### SHOWING THE CLAIMS OF MESMERIC SLEEPWAKING TO BE CONSIDERED A PECULIAR CONDITION OF MAN.

In the spring of 1837, I went to Cambridge, on a visit to some friends. That I should have succeeded in interesting them, and others of the University, upon the subject nearest my thoughts, will not surprise those who have observed that, where science is most studied, new statements respecting natural phenomena obtain the most patient bearing. I did, in fact, at Cambridge, meet with many persons really desirous of witnessing experiments in mesmerism, and even of submitting themselves to the mesmeric influence.

#### CASE III.

In my first essay of mesmerism, at Cambridge, I experimented — before seven or eight persons — upon a servant, whom I had brought from Belgium, a man of about 27 years of age, in good health, lively and quick-tempered; able also to speak English. I had attempted to mesmerise him once before, but, at that time, he had advanced no further than imperfect sleepwaking, during which he retained his consciousness, but could neither move nor open his eyes. On the present occasion, after a quarter of an hour's mesmerisation, he closed his eyes, but seemed restless,

and, in answer to a question I made him, declared
that it was impossible for him to sleep on account of
noises in the street and the constraint caused by the
presence of so many persons. Scarcely, however, had
he said this, when I perceived that his head followed
my hand; and, continuing the mesmeric passes for
about five minutes more, I again asked him if he felt
disposed to sleep. To my surprise, (for I had
not expected a perfect result so soon) he replied, —
" I am asleep." " Can you rise from your chair?"
I inquired. " Yes, but I had rather stay quiet."
When, however, I got up and removed to a short
distance from the patient, he also rose and moved a
step or two towards me, but tottered and would have
fallen, had I not hastened to his assistance. He
leaned heavily against me, and seemed unable to
stand without my support; I therefore placed him
in a chair, when he complained of a pain at the chest,
and begged me earnestly to wake him. " Can you
not awake of yourself?" I asked; to which he
replied, " By no means — but I beg of you to
awake me." Reluctant to do so without exhibiting
to the party present some of the marked phenomena
of mesmerism, I soothed my patient as well as I could,
and persuaded him to remain a little while longer in
the state of sleepwaking. As he, however, manifestly suffered, I was unwilling to fatigue him by
many experiments. The following were the principal : — wine, water, and coffee were handed to me,
and I tasted them, successively, in such a way as to
prevent the patient from perceiving, by any usual

means, what the liquids were. He, however, correctly named them in their order. The order was then changed, and the results of the experiment were the same.

Flowers were given me to smell: I was holding the patient by one hand, at the time, but turning altogether away from him to a table, over which I bent, so as to interpose myself between him and any thing that might be handed to me. He, however, when I smelt to the flowers, imitated the action, and on my asking him what he perceived, replied, without hesitation " flowers." Upon this, one of the party silently changed the flowers for a bottle of *eau de cologne*, when he observed, " That is not the same smell; it is *eau de cologne*." With the manner of conducting this experiment, and its results, all who were present declared themselves perfectly satisfied.

The patient, on being awakened, declared that he remembered nothing of what had passed since the moment when he complained of the noise in the street. His mesmeric sleepwaking had been complete.

CASE IV.

T. S., an undergraduate of Trinity College, consented to try how far I could influence him mesmerically. He possessed abilities of no common order, was extremely lively, by no means of a wonder-seeking nature, and altogether seemed the last person in the world who could be worked upon by nervous apprehensions. He sat down to be mesmerised, laughingly, and with perfect calmness. In about ten minutes his eyelids drooped, and closed gradually; his head followed my hand; his features became fixed and rigid; his colour fled, and a dead stillness came over his countenance: the change was the more striking on account of the usually animated and mobile character of his physiognomy. It painfully resembled the alteration caused by death. Nevertheless, on being interrogated as to his state, he declared that he was well, with the exception of a slight *pain in the back of his head* *, which was shortly relieved by the mesmeric passes. To other questions he replied that he did not sleep, but was unable to move, or to open his eyes. As, after long mesmerisation, I saw no probability of his advancing further into sleepwaking, I awoke him in the usual manner, when his countenance resumed its accustomed expression. On attempting, however, to rise, he complained of feeling extremely weak and dizzy, and, for a short time, he was unable to stand with any firmness. He was convinced that he had experienced the effects of an influence independent of the imagination.

CASE V.

C. M., a bachelor of Trinity, aged about twenty-five, submitted to be mesmerised; he was clever, thoughtful, and rather grave in character. On first sitting down, he seemed very wakeful, and declared that it would be strange, if, under circumstances the

---

* See Case ii.

most unfavourable to sleep, he could be charmed into
any thing like drowsiness.  In about a quarter of an
hour, during which a perfect silence was by no means
observed, some force seemed to close his eyes as
irresistibly as if they had been touched by Mercury's
wand of slumbrous notoriety, and he passed into a
state of imperfect sleepwaking, in which his head
followed my hand and he answered my questions.
On my removing to a short distance from him, he
leant forward towards me, and would have fallen,
had I not resumed my former position.  I awoke
him in about half an hour; he had not been in any
way, unpleasantly affected, and retained a vague
recollection of an agreeable state of reverie, similar to
that which precedes slumber.

CASE VI.

On this occasion some ten or twelve persons were
assembled to witness my experiments; and amongst
them was a man of luminous intellect and varied
acquirements, whom I was naturally desirous of
interesting in the question of mesmerism.  He was
decidedly sceptical on the subject, but I knew that
his was a mind which, if once fairly convinced,
would be firmest in faith and foremost in investigation.  I was therefore vexed when, after long mesmerisation, my servant, whom, from his having once
passed into perfect sleepwaking, I had chosen as
the subject of our experiments, remained uninfluenced.  Two or three of the party, tired of waiting
for a result that came not, went away, and the person

whom I was chiefly anxious to convince, and whom I will call V——, having an engagement, was about to follow their example, when I urged him to stay a very short time longer, while I tried another patient, D. C., an undergraduate of Trinity, who wished, as he said, to try the effects of mesmerism, in order to prove their nothingness. After this, I need not add that he was very incredulous on the subject; every circumstance seemed to diminish the probability of my success. The man I had agreed to mesmerise was in the strength of three and twenty years of age, six feet in height, and muscular in proportion. The stillness of the meeting, once broken, could not be restored. Persons were talking, and moving about the room, and my recent failure had thrown an air of ridicule about the proceedings of the evening, which, if the mesmeric influence were dependent on imagination, would have been sufficient to annihilate it at once. V—— looked hopeless of seeing any thing remarkable, and had taken up a book.

The following facts then occurred: —

I had not held the hands of D. C. more than five minutes, when I remarked a dizzy look about his eyes, which is peculiarly indicative of the incipient stage of mesmeric sleepwaking. Encouraged by this success, I had recourse to the mesmeric passes, when by degrees the eyes of the patient closed, and shortly after the head followed every motion of my hand. V——, at this moment looking up from his book, was surprised to see what had been effected. I beckoned him to come near, and, by reiterated trials, convinced him that my hand had an attractive power over the patient.

I now spoke to D. C., and asked him if he were asleep ; to which he replied, " Not precisely." I then, at V——'s request said, " How do you feel." " Very strangely," he said, " as I never felt before." Shortly afterwards, V—— himself spoke to the patient, and I called him by his name, but he seemed to pay no attention to the circumstance. When V—— happened to lay his hand upon his shoulder, the effect of the touch was like that of an electric shock. The patient's whole body quivered, his features were convulsed, his countenance became deadly pale, and he seemed to gasp for breath, like a person who has been suddenly immersed in cold water. Nevertheless, he did not awake, and the affection, whatever it was, seemed entirely physical (as if his mind had no longer its usual partnership with the body). He himself appeared to be unaware of it; and, when I asked him what had disturbed him — what was the matter with him, he said that he felt nothing whatever. Still, however, he continued to tremble, until by the application of my hand to his forehead, and by mesmeric passes from the head downwards, I restored him to tranquillity. V——, whom, since he touched him, the patient seemed to hear equally with myself, now recommenced speaking, and asked D. C. if he knew him. He replied in the affirmative, and named him. " Do you see at all ? " V—— inquired. " Not much; I see a red light about *so* large ; " and D. C. made a circular motion with his hands to express the size of

the light. I then asked him, " Do you see *me* ? " to
which he answered " Yes, I see you always." " Well,"
I said, " I wish you now to get up from your chair."
" Oh no, no ! I prefer staying where I am ; but *you*
must stay with me." " No ! " I replied, " I am
going : " whereupon, he seized both my hands, and
exclaimed, " You must, on no account, leave me."
I, however, rose up, when the patient also rose; I
walked forward, and he walked also, but unsteadily,
and leaning upon me. As he seemed to dislike
remaining in an upright position, I placed him in a
chair in a part of the room opposite to where he had
been. I then proceded to show V—— some of the
characteristic phenomena of mesmerism. I drank
wine, water, coffee, with the usual precautions, and
the patient distinguished them all, going through the
motions of tasting simultaneously with myself. There
were various articles of food upon the table, amongst
others some sandwiches. These last were given me,
and he said directly, " You are eating sandwich."
This may seem strange — it is so to myself; but I
state the fact as it occurred. Some snuff that I took
from the chimney-piece behind me, turning my back,
and suppressing, as much as possible, the usual indications of smelling, was named directly ; this being
exchanged for flowers, the result was equally satisfactory.

Remembering with what acuteness of perception
Mademoiselle M—— had distinguished between
objects that did or did not belong to her mesmerist,
I asked for the pocket-handkerchiefs of all the party,

twisted up my own in the midst of them, and laid them on the knee of D. C. He was immediately affected with slight shuddering ; and tossed away very quickly all the pocket-handkerchiefs but mine. The experiment, repeated, gave the same results ; but the second time he grasped and firmly held my handkerchief, until the time of his awaking.

At a yard's distance his hand rose up to meet mine, as iron flies to the magnet. From the approach of other hands he recoiled.

These experiments concluded, V—— said to me — " I am now satisfied that all you asserted to me respecting the mesmeric state is correctly true. I do not know that we are liable to elicit any new fact by keeping D. C. any longer asleep. He looks ill and suffering, and I think you had better wake him."

Upon this, I asked the patient if he desired to wake, and he replied " Yes ; I feel much fatigued."

I awoke him, by the usual mesmeric passes, when he expressed the utmost astonishment at finding himself in a different part of the room from that where he had first been mesmerised. The last half-hour had been a blank to him, with this single exception — he thought that he remembered hearing V——'s voice, asking him if he knew him. Faithful, however, to his character, he refused to believe that he had exhibited the phenomena to which we bore witness.

The next day, as may be supposed, I talked over with V—— the circumstances of our mesmeric evening. I found him entirely persuaded that mesmeric sleepwaking was a distinct state, worthy to be

investigated as a part of man. His concluding
words were these : —

"I thank you for having enlarged my experience
by facts perfectly new to it. I have seen a something which is not sleep, which is not delirium, but a
*tertium quid*, for which, as yet, we have no accurate
name. The thing is most interesting, but should, I
see, be exercised with caution. The effect of my
touching D. C. was fearful. You have verified many
phenomena. It has occurred to me that you might
try, when you have an opportunity, whether the sense
of touch in the mesmerised sympathises with that in
the mesmeriser, as do the senses of smell and taste.
Try whether, *through you*, the mesmerised person
may have a cognisance of form — whether he can
ascertain by means of your perceptions if you are
handling square objects or round — rough or
smooth." To this suggestion I promised to attend.

## CASE VII.

On this occasion my patient was G. T——, aged
19, the son of a gentleman at whose house in
the country I spent a few days, after quitting Cambridge. He was strong and healthy, and possessed
good sound sense, with but a trifling infusion of
imagination. After a quarter of an hour's mesmerisation he passed into imperfect sleepwaking, in
which state his corporeal powers were held in complete and singular enthralment, while those of his
mind remained unaltered. Unlike my other sleep-

F

wakers, he was utterly incapable of speaking during the mesmeric trance, though, as he told me
subsequently, he heard every question I put to him,
and was desirous of answering me. He was also
unable to open his eyes, or to move, except so far as
regarded the involuntary tendency of his head towards my hand. Another time, being curious to ascertain how far the influence (whatever it might be)
which I exercised over the patient was dependent upon
physical causes, I made the experiment of mesmerising him in the open air, on a fine June day, when
the weather was warm, but refreshed by partial
breezes. The following were the differences between
this mesmerisation and the last : —

The patient did not close his eyes so soon as on
the preceding occasion, nor pass into the imperfect
crisis by so regular a progression.

In the beginning of his sleep, the power to move
and to open his eyes seemed slightly restored whenever the gust came by.

He told me afterwards that when I commenced
mesmerising the action of the air produced a decided
difference in his sensations. When the wind swept
across his forehead it was, he said, as if there were a
disturbance of some drowsy weight that was gathering over it. By degrees, however, he became insensible to this apparent remission of the sleeping
influence. Nothing, he declared, could be more singular than the feeling of utter inability to speak, move,
or open his eyes. He heard me command him, urge
him, in every way, to answer my questions, but his

organs of speech seemed to be as completely paralysed as those of vision. In fact, the only external
indication that he had given of his desire to reply
to me was the slightest possible motion of his lips.

CASE VIII.

The family with whom I was staying having but
few prejudices on the subject of mesmerism, I was
permitted— in consideration, I suppose, of my years
and gravity — to mesmerise the sister of my preceding patient, a young lady of most amiable temper,
cheerful, and by no means of a nervous temperament.
Of the experiment to be tried she had no apprehension, as she had already seen her brother support mesmerisation without injury ; and it should be especially
kept in mind that her ideas of the mesmeric state
were limited to what she herself had witnessed of it;
for I had expressly abstained from describing the
higher phenomena of sleepwaking to herself or to
any member of her family, in order that their minds
might be uninfluenced on the subject. Under these
circumstances, to ascribe the curious results which I
am about to detail to imagination or to imitation
would be absurd.

I sat down to mesmerise my patient about nine
o'clock in the evening. In about seven minutes her
eyes closed, and in three minutes more her head
followed my hand. I asked several times, " Do you
sleep? " but no answer was given, and I began to
fear that this incapacity of speech under mesmerism
was a family failing. At length, however, on a

F 2

repetition of my question, Miss T——— answered in a very low voice, " Yes." From this time she replied to all I said, and displayed the higher characteristics of mesmeric sleepwaking in a perfection which I never saw evinced upon a first essay. The effects of mesmerisation upon the patient seemed to be altogether of an agreeable nature. She felt no pain of any kind, and her colour did not vary, or, if any thing, was slightly heightened. Her appearance only differed from the natural in these respects; — her eyes were entirely closed, and her head, drooping as if with the heaviness of slumber, swayed dizzily from side to side, except when I strongly commanded her attention, and desired her to raise her head, at which times she seemed with an effort to resume the power over her organisation. In this state, the phenomena manifested by Miss T——— may be ranged under the usual heads, as follows : —

1st. *Attraction towards the mesmeriser.* — Though indisposed at first, like others in the mesmeric sleep, to motion, Miss T———, when I arose from my chair, rose also. Without touching her, I moved onwards, and she seemed compelled to follow me. At one time I ran quickly to a distance from her, when she remained standing where I had left her, but bent towards me, and wavered as if she would fall. I made a motion with my arms as if I would draw her to me, when, recovering herself, she slowly, and, as it were, reluctantly, came to where I stood.

When her hands were lying on her lap, if I held mine above them they were attracted upward, more quickly as they drew nearer my own. It was much like the effect which a piece of rubbed sealing-wax has on a bit of paper. In proportion to the attraction towards her mesmeriser seemed the patient's repulsion from others. If any but myself presented their hands to her she drew back her own with a slow but singularly repugnant shrinking, her fingers becoming rigid and curved inwards. This experiment was often repeated with the same results. Remembering how painfully one of my patients had been affected by an unadvised touch, I was careful not to allow any one to come into actual contact with Miss T——, until I had established what mesmerisers call a " rapport" between the parties, and which is effected by the mesmeriser taking, at the same moment, the patient's hand and that of a third person.

2dly. *Community of sensation with the mesmeriser.* — Wine, water, and a stand of liqueurs were on a side-table. Placing the patient as far from me as was consistent with my still touching her hand, turning also away from her, I tasted alternately water and wine, between which she distinguished correctly, moving her lips as if she herself were drinking. Some brandy and water, already mixed, was in a tumbler on the table ; for this I suddenly changed the first mentioned liquids, when the moment I had taken it into my mouth the patient made wry faces, and exclaimed, " Nasty, nasty." " Why do not you like it ? " I asked. " Because it is too strong." " Have you ever tasted it before ? " " Yes, once, when I had a sore throat." " You know, then, what it is ?" " Yes ; it is brandy and water."

A member of the party gave me, from behind, some wild flowers, which the young people had gathered that day. "Do you know what I am smelling?" I inquired. "Something that comes from the fields," was the reply.

The flowers were silently changed for snuff, when the patient again exhibited symptoms of dislike. On being asked the reason, she said, "I smell tobacco."

It now occurred to me to try the experiment which V—— had suggested with respect to the cognisance of form. With every precaution, I took up from a table, on which were many other articles, a small square box, and passed my finger over the edges; I, at the same time, asked the patient, "Can you tell me what sort of thing I am touching?" Upon this Miss T—— made motions with her fingers, as if she had the object under her own hand, and replied, "It is something with edges, like a box." I next took into my hand a chess-man, some parts of which were carved in points, and felt alternately the carved and uncarved portions of the piece. Between these she discriminated correctly, saying, "Now it feels rough; now smooth;" and always before speaking she went through the same motions of touching with myself.

One of the patient's sisters, without giving notice of her intention, went out of the room and fetched an egg, which she concealed carefully in her hand and gave me from behind. The patient now said, "I feel something smooth and round." Being urged to tell what it was, she said, "I think it is a ball." Stretching out my hand behind me, I whispered to one of the party, to hurt me in some way; I was pricked with a pin, when my patient started and shook her hand as if she felt the injury. On being asked what she felt, she answered, "As if they pricked my hand."

At another time, when I inquired if she suffered in any way, she replied, "No, only my feet are very cold." My own feet being exceedingly cold at the moment, I suspected that her feeling was sympathetic : I conjectured rightly, for, awaking soon after, she assured me that her feet were perfectly warm.

3dly. *Isolation from every one but the mesmeriser.* — Even after having been placed "en rapport" with all present, the patient seemed incapable of hearing any voice but mine, unless the person who spoke were in actual contact with me and with herself at the same time. On one occasion, when I was asking her if she knew what some object was that I held before her, her father told her very loudly that it was a wine-glass (which it actually was), but the patient did not profit by the intelligence, for on being again questioned she said impatiently, "I do not know; I cannot tell." Her name shouted close to her ear by different members of her family seemed to make no impression whatever upon her organs of hearing; — while, on the contrary, she attended to the slightest word that I addressed to her.

4thly. *Development of vision.* — The patient's eyes were, to all appearance, firmly shut during the whole period of the mesmeric sleep. When I first asked her if she saw any thing, she answered in the

F 4

negative; but added, after a pause, "I see *you*."
After this. she seemed by degrees to acquire a more
general vision. She recognised, on being led up to
them, the several members of her family, consisting
of nine persons, and could not be deceived by any
change in their position. At length, I held a book
before her on a level with her eyes, when she,
without hesitation named what the object was.
Upon this I opened the book at the title-page, and
pressed her to try if she could distinguish any words.
She at first begged me not to trouble her to do this;
but at length, as if making a great effort, she made
out the word "London," at the bottom of the page.
This was the first time I had seen any sleepwaker
read any written or printed word. The experiment,
however, though curious, was not strictly enough
conducted to serve as the basis of any theory on the
mysterious subject of vision through the closed lids·
The book, at the moment when the word was distinguished, was resting upon the patient's lap, below
the level of her eyes. Those, however, who were
observant of the circumstance, could detect not the
smallest opening of the eyelids.

Holding my watch before the patient, in the usual
line of vision, I asked her the hour, when she correctly
told that it was five minutes to eleven.

Directly after this last experiment, I awoke my
patient, who testified great surprise at finding herself
on a chair instead of the sofa, where she last remembered to have been seated. Mr. T—— now
came up to her and said, "H——, do you know

what o'clock it is?" to which she replied, " I suppose
about ten, or perhaps half-past nine. I am sure I
cannot tell." Mr. T——, then said, " Why, it is
not five minutes since you told us the hour."
Upon this Miss T—— seemed much astonished,
but declared her utter ignorance of the circumstance, and indeed of all that had occurred during
her sleepwaking. The last thing which she recollected was hearing me ask whether she slept,
when she desired to answer me but could not. A
moment after, she became unconscious. Her feelings
after the mesmerisation was over were those of plea-
surable excitement, and she felt none of the heaviness which sometimes succeeds sleepwaking. The
next morning she declared that she had not been
able to close her eyes all night, but had felt no want
of sleep, nor subsequently did she experience any of
the lassitude usually resulting from a wakeful night.

On a second mesmerisation Miss T —— confirmed by repetition the results of most of the preceding experiments. A cambric pocket-handkerchief
tied over her eyes made no difference whatever in
her powers of vision. She did not even seem to be
aware when I put on the bandage, or when I took it
off. While bandaged, she read the words " Edin-
burgh Guide," and " Travels in Russia," from the
title-pages of two books, which I took up at random
from amongst others lying about the room, and held
before her, in about the usual position a person holds
a book in reading. The effort to distinguish these
words seemed to be great, and subsequently Miss

T—— refused to make any more exertions. I urged her to look at my watch, but she remained silent, and soon after large tears forced their way through her closed lids. Having tranquillised my patient by such motions of the hands as I have ever found beneficial and soothing to mesmerised persons, I awoke her, when she at once resumed her cheerfulness, and began laughingly to ask her sisters what she had been doing, and how long she had slept. She was not in the least aware that she had been crying, and I begged that she might not be told of it.

I might here, perhaps, close my series of proofs relating to mesmeric sleepwaking and its phenomena; but, as regards this argument, numbers are strength, and each additional instance of the mesmeric sleep, accompanied by the same characteristics, is a step towards that firm and decisive conclusion on the subject, which can only be attained, on any, by a repetition of the same fact under the same circumstances. At the risk, therefore, of wearying my readers, I add the following cases, which occurred after I had returned to the Continent in 1837 :—

CASE IX.

J. G——, of Antwerp, aged 22, not very strong, but having no particular ailment, was brought by me into imperfect mesmeric sleepwaking. His head followed my hand, and he was unable to move or open his eyes, though he replied to my questions. He felt a drowsiness during the whole of the day succeeding to the mesmerisation.

CASE X.

W——, of Aix-la-Chapelle, aged 45, in delicate health, of a calm, reflective mind, and sceptical as well as uninformed on the subject of mesmerism, was thrown into perfect sleepwaking at the first sitting. He displayed the usual characteristics of the state. There were circumstances relating to this patient which I shall have occasion hereafter to detail, and which are therefore omitted here.

CASE XI.

T. B——, (English) aged nine, an extremely lively boy, and full of the restlessness natural to his age, consented to be mesmerised, but observed that he thought he never should be able to sit still while I tried my experiments. Notwithstanding this, and the time of day (two o'clock), his eyes closed after about ten minutes' mesmerisation, and shortly after he gave every token of having passed into perfect sleepwaking. He answered all my questions, heard no other voice than mine, was attracted towards me in the usual manner, and, with his eyes to all appearance shut, saw me, and indicated the position of my hands and the number of fingers which I held up before him. In about half an hour he begged me to awake him, and, on returning to the natural state, declared that his unconsciousness had been complete.

CASE XII.

Mrs. O—— a married English lady, clever and
F 6

firm-minded, was thrown by me into imperfect sleep-waking in about a quarter of an hour. In this state her countenance was pale, her head moved towards my hand involuntarily, and she was unable to get up, or to open her eyes, until I had de-mesmerised her.

## CASE XIII.

R. V——, a young Belgian, of the medical profession, was so far susceptible of the effects of mesmerism as to display the usual symptoms of imperfect sleepwaking. Having an engagement on the evening that he was mesmerised, and rather ashamed of having slept, he hurried away before the mesmeric drowsiness had been entirely dispersed. The consequence was that he nearly fell down in the street, and was obliged to have recourse to the support of a fellow student. On the following day, he still felt a confusion in his head. This is not the only occasion on which I have remarked that the sensations which attend and follow an imperfect crisis are by no means so agreeable as those which result from complete sleepwaking. In the one case, there is an apparent disturbance of the system, accompanied by feelings of drowsiness and fatigue — in the other, the frame seems to be invigorated, and a wakefulness and vivacity are the after symptoms.

## CASE XIV.

At the time when this occurred, I was staying at a country house in Switzerland, and my friends, the Countess of —— and Lord —— were passing a few days at the town which was nearest our residence. Having heard from me some interesting details respecting mesmerism, they were desirous of witnessing its phenomena; but, as no patient was forthcoming, it seemed unlikely that the wishes of my noble friends would be gratified. However, late one evening, the Countess sent her carriage and a note, pressing me to come to her immediately, as she had met with some friends amongst whom I should find one at least willing to be mesmerised. On arriving at the town of —— I was introduced, for the first time, to a most agreeable English family, the friends of whom the Countess had spoken; and after some conversation, in the course of which I ascertained that the newly arrived visiters had but a very vague notion of what they were to witness or to experience, it was decided that I should mesmerise Miss ——, a young lady, who might have seen some eighteen summers, of a lively disposition, and not in the least timid or nervous with regard to the experiment about to be made. Lady —— took down a kind of *procès verbal*, during the proceedings of the evening, and from this I extract the following details : —

"*Mr. T——* began to mesmerise Miss —— at ten o'clock. Five minutes after her eyes closed, her head followed the mesmeriser's hand wherever he moved it. The mesmeriser then asked — ˎ

Are you asleep ?

Patient. — No; I am not asleep.

Mesmeriser. — Do you like being mesmerised ?

P. — No. (On repetition) Yes.

The apparently mesmerised person here started from her seat, opened her eyes, and said " I'm not in the least mesmerised." All present thought she had been feigning, and were preparing to join in a laugh against the mesmeriser, when, to the surprise of every one, the patient fell back into her chair, her eyes at the same time closing with a sleepy expression. The mesmeriser continued his passes. The head, as before, followed the hand.

The mesmeriser then said —

Do you feel sleepy ?

P. — Yes.

M. — Why then do not you sleep ?

P. — I do.

M. — Are you quite asleep ?

The mesmeriser here drank some wine, turning his head away, and asked —

Do you taste any thing?

P. — No, but I smell wine. (The wine was, however, too far off from the patient to have been distinguished by her, in any way, under ordinary circumstances.)

M. — Do you know where my hand is ?

P. — (her hand going directly to where the mesmeriser's was) Yes.

M. — Can you see me ?

P. — Yes, but not very distinctly. (Her eyes appeared to be closely shut.)

M. — (smelling to a nosegay which he took from the table) What is this ?

P. — Flowers.

The mesmeriser now urged the patient to rise from

her chair, but she expressed the strongest reluctance to move: at the same time she entreated the mesmeriser not to leave her. He, however, went to a little distance, when Miss —— seemed to be forced to get up and follow him. Supported by the mesmeriser, she walked half across the room, but begged that she might be permitted to sit down again. The mesmeriser then placed her in a chair in the middle of the apartment, and held up various objects (not belonging to him) before her eyes. These she declared she could not see, but whenever the mesmeriser held before her any thing which belonged to himself, she named the object directly. In this way she told rightly when the mesmeriser successively presented his pocket-handkerchief, purse, and watch, also a letter which he took from his pocket. On being asked if she could distinguish the hour on the mesmeriser's watch, she held it before her closed eyes, and said, "It is a quarter to something." In effect it was a quarter to eleven.

The mesmeriser now asked —

What am I tasting?

P. — (seeming to swallow something) Cold water. (Right.)

M. — What am I eating?

P. — (seeming also to eat) Something like bread, but sweeter. (It was cake.)

M. — Do you like it?

P. — Yes.

M. —What am I smelling?

P. — Eau de Cologne. (Right.)

Miss ——'s father and mother now approached her, and would have taken her hand, but she drew it back with a shudder. To the mesmeriser's hand, on the contrary, her own seemed always attracted, and rose up to meet it when held at a considerable distance. She distinguished between her father's and the mesmeriser's handkerchiefs, by throwing the former away, and retaining the latter in her possession. She also refused to part with a small piece of coin on which the mesmeriser had breathed, and which she held fast grasped in her hand.

The patient being now asked if she was tired, replied that she did indeed feel much fatigued, and seemed pleased to rest her head, leaning forward, upon the mesmeriser's hand : at this moment she seemed to sleep very deeply. Her countenance was extremely pale. She seemed to hear no one speak but the mesmeriser, but always answered his questions. On his asking if she felt well and comfortable, she replied, " Oh, so comfortable ! " nevertheless the position of the patient, as she bent forward, with her neck stretched out, was the most uneasy that could well be imagined.

M.—Would you not be more at your ease on the sofa ?

P.—Oh no ! I am so well here ! But you are not going away ?

The mesmeriser here rose and went to the sofa, when Miss —— seemed obliged to do so likewise.

M.—Are you not better here.

P.—Yes, so—sitting quiet.

The mesmeriser now asked his patient if she would awake, but she said she would rather remain as she was, and begged of him to allow her to sleep some time longer. At length, however, the mesmeriser said, " I must wake you," and passed his hands rapidly over the patient's eyes, when they partially opened, but she shut them again directly, exclaiming, " I had rather sleep." The mesmeriser however continued to use the same motions as before, when the patient again opened her eyes, and to the question " Are you quite awake?" answered in the affirmative. She then looked about with an air of surprise and asked how she came to be sitting on the sofa? Then she held up her hand, which still remained closed upon the piece of money that the mesmeriser had given her, and exclaimed, " What is this?" When it was explained to her that she had refused to part with the coin, she seemed much surprised, and, when informed of all that had passed, would scarcely believe what we told her. The only circumstance that she remembered of her sleepwaking was having seen, at a moment when her eyes were partly open, the upper part of Lady S——'s face. She declared the only disagreeable thing in mesmerism was the awaking, and compared it to what she should imagine of the feelings of a drowned person in returning to life."

Fearing lest my reader should now exclaim with Macbeth —

" What! will the line stretch out to the crack
Of doom? ———— ———— I'll see no more!"—

I here close my detailed cases of mesmeric sleepwaking, merely stating in conclusion that, in addition
to those already particularised, I have mesmerised,
between January 1837, and December 1838, nine
other individuals, seven of whom passed into the
perfect, and two into the imperfect crisis. These all
displayed phenomena coincident with the above,
though in some cases higher and more interesting in
their degree of development. Of course as I proceeded in my career of investigation, my views
cleared, and my experiments became extended. It
was impossible to be so long conversant with one set
of facts without hazarding some conjectures as to their
efficient causes. I theorised, and endeavoured to
prove or disprove the theories that I constructed by
bringing them to the test of actual experiment. How
far I succeeded in this labour it is my intention to
relate hereafter. My intention, up to this point, has
simply been to answer the objections raised against
mesmerism on account of its limited manifestation,
and variable as well as hypochondriac character. I
have wished to show that mesmeric sleepwaking is a
state, and a peculiar state, into which man generally has the capacity of passing ; and surely I have
said enough to remove from it the reproach of being
limited to a few nervous and fanciful persons, chiefly
of the weaker sex. Out of three and twenty individuals in whom I induced sleepwaking, more or
less perfectly, six only were women, one only a decided invalid. Let me, however, hasten to anticipate
an objection which may have been forming in my

reader's mind. I ought, it may be urged, to have
noticed my cases of failure as well as of success.
This objection shall not long be valid — I have
already stated that a cousin of mine could not be influenced by me mesmerically. The trial was however scarcely serious or protracted enough to be
considered as decisive one way or the other. At
Cambridge two persons experienced no effects from
mesmerisation. The one was determined to resist
the influence, and to that end was solving an abstruse
mathematical problem all the time that I was mesmerising him. The other disliked being mesmerised
and was afraid of it ; — and fear I have always found
to be in mesmerism a most disturbing force. Subsequently among those whom I have essayed to mesmerise, I have met with but four persons who manifested either no symptoms of being affected, or those
so slight and equivocal that they may not be relied
upon. Of these, two were ladies, who ought (according to the received notion of mesmerism) to have
been more easily influenced than persons of the other
sex, especially as they allowed me a fair and sufficient
trial, while, on the contrary, the two other individuals
alluded to jumped up from their seats after a mesmerisation of a few minutes, one of them exclaiming,
" I feel nothing ; and now believe in mesmerism
less than ever !"

However, allowing these cases to pass as absolute
failures, it appears that in the space of less than two
years, the number of persons mesmerised by one
single individual was in the proportion of twenty-

three to eight.    Striking off even the imperfect cases,
there will remain fourteen persons out of thirty-one
in whom sleepwaking was fully developed with all its
attendant train of characteristic phenomena, — these
too, not being selected by myself as likely subjects
for mesmerism, but offering themselves accidentally :
and surely they who consider how difficult it is, first,
to prevail on persons to submit to mesmerisation at
all, and, secondly, to secure an adequate trial, will
wonder that the cases are so many rather than so
few.    The great argument, therefore, against mesmerism, of infrequency and irregularity, falls to the ground.
Nor can any one rationally demand a *universality* of
mesmeric sleepwaking, before he will admit that it is
one of the states into which man generally has the
capacity of passing.    The exceptions forbid not the
existence of the rule.    All persons are not, it is to be
hoped, mad; yet we pronounce madness to be an
affliction to which any man whatever is liable.

> " Perhaps some doctor of tremendous paunch,
> Burly and big — a black abyss of drink," —

may see all his fellow topers under the table, himself
sober the while ; yet this by no means affects the proposition, that drunkenness is a state into which man
generally has the capacity of passing.

But there yet remain two important questions to
ask and to answer, respecting the phenomena which
I have detailed.

1st.  Have I been deceived ?

2d.  Have I been mistaken ?

The first implies that my patients feigned; the second that my senses were incompetent to discern the objects presented to them.

The wilful and determined opposers of mesmerism (carefully to be separated from the class of honest objectors) are very ready, in judging of mesmeric statements, to have recourse to one or both of the above suppositions, not perceiving that, in so doing, they decide abstract questions which can never be limited to the single subject of mesmerism, but which bear largely upon the whole conduct of life.  In their blind zeal to overthrow the particular object of their dislike, they would strike away universally the two great props of humanity, namely, — belief in our neighbour, whereby we greatly regulate the moral part of our nature ;  and confidence in the sufficiency of our senses, whereby alone we judge of any thing external to ourselves.  Allow that a large proportion of persons deceive, (as we must do, if we suppose all whom we mesmerise to be feigning,) and what becomes of our faith in the founders of our holy religion, even in the credibility of human testimony on which we ground our present actions and our future hopes ? Again, grant that we are deluded by our senses to the extent we must be when we are mistaken in a multitude of cases, and what remains in the universe but dreams and uncertainties?  The splendid labours of Galileo and Herschel, based as they are on the competency of the human senses, are but as a vapour, and the discoveries of Newton, sublime as they appear, must be as illusory as the organs through which they were transmitted.

But it may be replied, — We are not to distrust our senses generally, but only in those particular cases, when they appear to bear testimony incompatible with former experience. This manner of viewing the question is plausible, and has accordingly been much adopted by those who would make short work of mesmerism and banish it at a blow. On these grounds, some have even gone so far as to say — Should we even see the imputed wonders of mesmerism with our own eyes, we should not credit our senses. Whether such determination be as wise as it is magnanimous, remains to be proved. There is unfortunately just that mixture of good in it which makes evil dangerous. Every reasonable man will grant that we should examine with becoming caution any appearances that may seem to contradict our former experience ; — but to reject, when we should only distrust — to be positive when we ought to be merely sceptical — is certainly not the part of reason. The mistake seems to have originated, as so many other misconceptions on important points have done, not so much in a false idea as in the misplacement of a truth — in taking that as a basis which is only superstructure. The maxim, "Doubt all that does not harmonise with previous knowledge," is just in itself, but, in this case, becomes, by its position, erroneous. It is excellent as a monitor, but insufficient as a guide — an admirable servant, but a miserable master. Were the ultra ante-mesmerists consistent,

and should they carry through their principle, what
groundwork would they have for any belief what-
soever? I may be answered, "Experience!" Yes;
but should it be forgotten that experience itself is the
fruit of the senses? What becomes of the product,
if the parent itself be found so very fallible? If we
grant that we are the fools of our senses in one important branch of their observation, why not in all?
The absurdity is this: we are to trust our general
experience, but allow that it may err in the particular,
as if the general were not composed of the particular!
It is as if we were to talk of a good form of government under which every individual was grossly
wronged. Now, trust in the testimony of our senses,
or rather of our consciousness, is not only the groundwork of all we believe, but is the cement of each fact
that makes up the whole fabric of our certain knowledge. Of the external world we can ascertain nothing but through the faculties by which we observe
it; and of the external world we can believe nothing
but by our trust in the accuracy of those faculties —
our organ of information is either competent or incompetent. If it be competent, there is no reason why
we should distrust it, *cæteris paribus*, in one case more
than another. If incompetent, there is no reason why
we should trust it, *cæteris paribus*, at one time more
than another. There is no alternative: either we
must place confidence in our sole medium of knowledge, or take refuge in universal scepticism. Grant
that every anomaly presented to us be a reason for temporary doubt; still, when it is often enough presented

to us, our doubts as to its existence cannot but cease.
Allow that to accept facts on insufficient grounds
is folly; still, to reject them when properly demonstrated is equally absurd : nay, the greater danger
waits upon unfounded hesitation. A too great rash-
ness may correct itself; but distrust in the evidence
of our senses is a radical defect, which, as it annihilates
all things, so does it preclude the possibility of a remedy ; — not that any wise man would attempt to deny
that, though our senses are competent, we ourselves
may be in fault; and that, through a careless use of
our means of knowledge, we may sometimes err in
our estimate of things. But though this may happen
once or twice, the most slovenly observer can scarcely
be deceived as to the actuality of a series of facts
which occur before him again and again. Moreover,
in witnessing aught so remarkable as an exhibition of
the mesmeric phenomena, the gazer's mind is in the
attitude of roused attention ; a state the most opposite imaginable to the careless mood, in which we are
apt to be deceived. They who would represent
mesmerism as the cheat of our own eyes should remember, that the worst mistakes of our senses are
but transient, even momentary, and that their very
errors tend to correct themselves. For instance,
when alone in my apartment at the witching time of
night, I may suppose I see a face gazing at me from
out the window-curtains. But my immediate impulse is to look again, and to bend a stricter scrutiny
upon the delusive appearance, which I discover
forthwith to have arisen from certain fortuitous

arrangements of light and shade. It might, indeed, have happened that, overcome by sudden panic, I should have fled from the room, while still under a delusion, and have proclaimed ever after I had seen a ghost. But such conduct is, happily, the exception to, and not the rule of, that of mankind in general.

Besides, there is plainly no analogy whatever between a momentary and fear-engendered delusion of this sort, and a persistence in error, with regard to objects which we behold very frequently and in the calmest frame of mind. Other apparent inaccuracies there are in the action of the senses : but these are well known, and deceive no one ; and exceptions, when known and invariable, become themselves a part of the original law from which they seem to deviate. Allowances are made for them ; and on every subject to which they relate our judgments are quite as exact as if no such limitations were enjoined. Thus, when we apparently see a flash of light from bringing into contact zinc and silver in our mouths, we soon come to know that there is no such thing; and when a juggler by the rapidity of his movements bewilders our apprehensions, we admire his dexterity, but are not deceived by it in the least.

But the phenomena of mesmerism come not within the category of these transient illusions. The patients who exhibit them are not jugglers, but persons of good faith, unpractised in sleight of hand or dexterity of any kind. Even supposing them to be adepts in deception, they manifestly cannot play a part of which they know nothing; and I have shown

G

that those whom I have mesmerised were generally unaware what was to be felt or done in mesmeric sleepwaking.  Am I to entertain a thought so monstrous, as that a number of respectable individuals whom I casually met at different times and in different places were united in a conspiracy to deceive me ?  I cannot think so, even if I would.  He, who observes a state frequently, attains at length to a certain experience in its minute symptoms, which, like Ithuriel's spear, detects falsehood at once.  Now, in mesmerism, there are a thousand such trifling but decisive tests of authenticity..  To instance only clockwork lowering of the eyelids and the remarkable manner of their closing — it may be affirmed that these phenomena are not imitable ; for they are out of the reach of human muscular power, as long as it remains under its usual conditions.

Perceiving, then, that imposition on the part of my mesmeric patients is not only in the highest degree unlikely, but actually impossible, and confiding in the testimony of my own senses, I am assured that I have witnessed a state of man which is peculiar and distinctive, 1st, as induced ; 2dly, as possessing its own characteristics when induced.

Nor are these conclusions at all invalidated by the common argument so often adduced by the ante-mesmerists, that the whole of what are called the mesmeric conditions are brought about through the agency of the patient's imagination, and are therefore self-induced. This is plainly quibbling ; for, granting that imagination be the proximate cause, or immediate antecedent of mesmeric sleepwaking, still we require a moving *cause*, or a prior antecedent whereby the imagination itself is set in action; and this moving cause is indisputably external, and to be referred to certain looks or gestures or mere proximity of the mesmeriser. Equally futile is the attempt to nullify the fact of external influence by assertions "that some individuals on whom the effects of mesmerism have *repeatedly* been produced may exhibit like effects, without being mesmerised afresh, provided they are led to believe that the process is carried on as usual and that they are under its influence." This proves nothing more than the usual law of our nature, whereby trains of sensations grow into habits, and occur in certain series when the idea that is linked with them can be raised in the mind. But then they must always owe their origin in the first place to some exciting cause, which remains the true basis of their production; since but for it they would never have existed. Thus the peculiar sensation called tickling was in the first instance made known to us through the contact of something external (it being a curious fact that no one can tickle himself so as to fall into that convulsive laughter which characterises the sensation); yet if any one presents his finger to susceptible persons, as if with an intention of tickling, the feeling will be excited, and those on whom the experiment is tried will shrink and laugh, as if actually touched by the finger. The *idea*, then, of the primary cause of a sensation confessedly precedes the sensation itself; and, when mesmeric patients obey this law, they appear to me, instead of disproving the influence which primarily affected them, to establish it altogether. In supposing the actions of the mesmeriser to be present, the patient owns them as the primary cause of his sensations; and the reproduction of the accompanying phenomena, under certain convictions, is a perpetual recognition of that force which first set the hidden springs in play

In mesmerism, then, the influence of man is always the proper antecedent: — the state of mesmeric sleepwaking the proper consequent. Will any one declare that external causes have nothing whatever to do with the production of the mesmeric state? Yet this he must affirm before he can consistently class mesmerism with self-originating states of mind or body.

In vain, therefore, is the mesmeric sleep likened to, or identified with, natural sleepwaking, hypochondriasis, catalepsy, &c., for it differs in one most important particular, from any of these states: it is consequent upon external influence; it is *induced*, and that (whatever intermediate machinery may be set in motion) by the agency of man. We should consider also how utterly distinct from constitutional disorders like the above is mesmeric sleepwaking, which may be produced, almost at will, in all kinds of temperaments. Catalepsy, which it in some respects resembles, is a disease; but the mesmeric sleep is not a disease. Surely every event has its adequate cause, and, if mesmeric patients are not sleepwakers either from malady or by constitution,

how is it that they become so? It has been shown
that not only the healthy and the strong are susceptible of the mesmeric slumber, but individuals advanced in life, who cannot be supposed to have that
restlessness of the blood which sometimes afflicts
younger persons with sleepwaking. How then can
we confound mesmeric with natural sleepwaking?
The causes which are explanatory of the one have
nothing in common with the other. Besides, the
sleepwaking we are considering is guided and
wielded by external agency; — and this alone,
exclusive of other considerations which we have before suggested, is a sufficient answer to those who
allege that the so-called mesmeric state can be produced apart from the mesmeriser; since, when
genuine, it presupposes the exhibition of phenomena
that spring from the mesmeriser's actual presence.

Another distinguishing feature of mesmeric sleepwaking which stamps it with an individuality the
most remarkable, is (as far as my own observation of
the state extends) the extreme clearness and truth of
its perceptions. Mesmeric patients act upon real
impressions, and in perfect conformity with external
circumstances. They retain all their sense of locality,
all their cognition of time, and their knowledge of
the persons who may be around them. Now natural
sleepwakers are generally acting under some delusion. They will get astride upon a chair and fancy
that they have mounted a horse, and, advancing still
further into hallucination, will whip and spur their
imaginary steed, supposing that they are performing

twelve miles within the hour.   Like Lady Macbeth, they will see what is not, and, their waking thoughts continuing to influence their slumbers, they will suppose themselves still engaged in deeds which have been long numbered with the shadows past.   I do not say that this is always the case ; — but facts will bear me out in the assertion that auto-sleepwaking can never be *characterised* as an unillusive state.

Again, mesmeric sleepwaking, in some respects, resembles the exaltation produced by opium ; but we have only to read the celebrated Confessions of an Opium-eater to be convinced that the votaries of that drug live in a fantastic and ideal world.

With natural sleep mesmeric sleepwaking can least of all be identified.   That it is as different from this as natural sleep itself is from waking may be judged from the following circumstance.   E. A., returning with me and two other persons from an evening party, fell asleep in the carriage.   I made the mesmeric passes over him without contact and in silence.   After about the usual time required for mesmerising him when awake, he began to stir and testify uneasiness.   Soon after, he spoke and upbraided me for taking him at a disadvantage and for changing his natural sleep into the mesmeric.   There was as marked a change in his condition as if he had actually shaken off slumber.   A lady*, whose daughter was sometimes singularly affected by a

* See note page 152. (relating to Madlle. Estelle l'H.) in Book II.

species of auto-sleepwaking, presenting many of the features of the mesmeric state, told me that the patient would occasionally go to bed, while in what she called the " crise," fall into a natural sleep, rest well, yet rise the next morning, still, as it were, self-mesmerised. Chardel also, a French writer on mesmerism, gives an interesting account of two sisters whom he mesmerised, as a physician, with the hope of checking a tendency to consumption which they had both evinced. One evening, being in sleepwaking, they, as if prompted by a natural instinct, entreated their mesmeriser to leave them in that state, only so far demesmerising them as to enable them to open their eyes and to be committed to their own self-guidance. Day after day they renewed their petition — for day after day they felt health returning under the mesmeric influence. In other respects they pursued their usual habits, and their mesmeric existence had its alternate periods of sleep and of waking, as regular as those of the natural life. At the end of three months, their cure appearing to be complete, M. Chardel conducted the sisters, accompanied by their mother, to a beautiful spot in the country, where he restored them to a knowledge of themselves. He describes, in lively terms, their surprise and joy on returning to consciousness. It was winter when they entered the mesmeric state — it was now spring. The ground was then covered with snow, but now with flowers. They were then looking forward to an early grave; but now the feeling of renewed health tinged every thing with hope and

life — almost doubting if they did not dream, they
threw themselves into their mother's arms, gathered
flowers, and smelt to them, and endeavoured, by the
exercise of their senses, to convince themselves that it
was all a blessed reality.      Not a circumstance of the
three past months survived in their memory.

And this leads me further to remark that the absolute forgetfulness which severs the mysterious state of
mesmeric sleepwaking from the cognisance of ordinary
consciousness sets a peculiar seal upon this very remarkable condition of man.      The drunkard has his
glimpses of remembrance; the opium-eater can re-
count his visions; the natural sleepwaker may
sometimes recall the dreams that influenced his sleeping actions : but he who emerges from the mesmeric
state, when it is true and perfect, has shaken hands
with memory on the threshold.      It is true that
reverie shares this complete oblivion ; but then
reverie is a state of illusion and of incoherent thinking.  Darwin, who has treated of it at length, characterises it as such, and mentions a young lady (in
whom reverie had become a malady), who, when
under a paroxysm of this complaint, would converse
with imaginary visitors.      On one occasion, while thus
afflicted, she took off her shoe, looked at it, and said
" A little longer and a little wider, even this would
make me a coffin ! "

With reverie, then, mesmeric sleepwaking cannot
be identified, and in every point of view we seem
justified in considering it as a state apart from all
others — a distinct and peculiar condition of man,

I do not, however, forget that these deductions are founded on the proceeds of my individual observation, nor that it is essential that the experiments which are their base should be capable of repetition, in order that each man may, if he so pleases, bring my accuracy to the test of his own senses. Convinced that personal observation, if sincere, will establish all that I have advanced, I urge every one to inquire and to judge for himself. And this peculiar advantage attends upon mesmerism, — the proof of it, by experiment, is in the hands of all. The mesmeriser makes no mysterious monopoly of his art, but contends that the power to mesmerise and the capacity of being mesmerised, however modified by circumstances, may be developed in every human being. Moreover, for that repetition of the mesmeric phenomena which is necessary to produce a conviction of their reality there is not needed, as in chemistry or optics, an assemblage of substances, or a complicated apparatus which few can command The follies of the baquet and of the magnetic rod have perished with Mesmer, and at the present day the mesmeriser's eye and hand are his sole "conjuration and mighty magic."

But here it may be asked, " Why then, if the proof be so easy, is the world so hard of belief?" The answer is obvious: because there are few who will take the trouble to prove facts, which, from causes already numbered, men had rather not believe. Besides, though I do not hesitate to say that all who are really desirous to obtain proofs of mesmerism may, with comparative ease, obtain them; still, in searching for these, a virtue is required, which is by no means a general attribute of mankind; namely, perseverance. Nothing, indeed, is easier than for a man, who dislikes being troubled with the new ideas which the subject of mesmerism introduces into the mind, to make a hasty trial of the alleged power, and, failing to elicit or to experience any of its effects, to say, " I have proved it, and found it nought." But he who adopts this facile mode of dismissing a troublesome question can scarcely be called an inquirer, nor can his verdict have much weight. To examine a phenomenon is not sufficient : we must examine it also adequately ; and if it be recognised as a principle that a fact is not to be admitted till after a repeated scrutiny, neither, in common justice, should it be rejected till after the same deliberate investigation. But whether the difficulties attendant upon a verification of those phenomena which I have witnessed and carefully noted have been under or overstated, I must, as a principle, inculcate a dependence upon the observer who makes a subject his study, rather than on the captious opponent or indifferent idler who approaches the same theme with a want of temper or a want of patience, which equally disqualify a man from judging rightly. Putting myself wholly out of the question, and speaking as if I were only justifying, and claiming belief for, the first discoverer of mesmerism, I cannot but assert that there was a much greater chance of his being in the right than all who have subsequently proclaimed him in the wrong. In every case, the boldness and acuteness which lead a man to take new views of a subject and to combat ancient prejudices are in his favour. There is an innate consciousness attendant upon correct observation which upholds him and encourages him to war with the world. The sudden light that has struck him is like a light from heaven and brings conviction to his breast. His eyes are not deluded, but sharpened by desire. No other person can have that intimate zeal — that interest in the inquiry which urges him onward. His all is at stake, and consequently no one is so strict with him as he is with himself. Who can compare the cursory and partial views of the great body of mankind with his? We may call him a visionary or a heretic, but, remembering the slow but certain triumph of Harvey and Galileo, we should be cautious how we brand him with epithets which may only recoil upon ourselves. How abundant are the proofs that all which is most opposed to the passions and prejudices of men is most likely to be true!

Again, in the history of all observational science there are precedents to show that the accuracy of the most genuine experiments has been called in question by those who were unable to repeat them. Of this the reception which the Newtonian doctrine of light and colours met with from the world is a familiar instance, the more to be noted on account of Newton's reputation as a strict investigator. Abjuring all hypothesis and cautiously pursuing the method of induction which Bacon had substituted for the worthless dogmatism of the schools, the great philosopher one might have thought, could have had nothing to apprehend from the petulance of opposition. His theorems were facts which every man might verify to himself by the aid of his own senses. And yet in many cases Newton appealed in vain to the common testimony of our common nature. The correctness of his experiments was doubted by many scientific men both of his and other countries; a host of enemies impeached the soundness of his conclusions; and even in a later age, after the general voice had ratified his theory, the celebrated Goethe, quitting his domain of high imagination, undertook to expose its errors in a work which will ever remain a monument of perverted genius.

A certain dependence, then, upon the first observer of new phenomena is inculcated both by reason and by experience, and appears to be requisite, when we would ourselves examine the phenomena, to guard us from impatience and to support us through deceptive appearances.

But, I would ask, is this the course pursued by the opponents of mesmerism? Certainly not: they bring forward the defective trial of an experiment as a complete refutation of the experiment itself, and select, with curious partiality, the one failure amongst innumerable instances of success, as an illustration of the nullity of mesmerism. Yet what would be said of a man who, going to see some well-known chemical or other effect exhibited, and finding it fail through unforeseen circumstances, were to return home convinced that no such thing had ever existed? What, again, would be thought of this man's state of intellect, were he to write a book to prove that certain phenomena, because he had not witnessed them, had never been seen by thousands who could support them by their testimony? Yet, cannot even such an original sceptic as this be considered a whit more absurd than they who fondly fancy that by *exparte* statements and scattered histories of failure, collected from men as prejudiced as themselves, they can put down mesmerism? Do they forget that all which appears to be a mass of negative evidence is but air in the scale when weighed against one positive fact?

" It is a very obvious principle," says Chalmers, "although often forgotten in the pride of prejudice, and of controversy, that what has been seen by one pair of human eyes is of force to countervail all that has been reasoned or guessed at by a thousand human understandings."

Let, then, body after body of learned men deny the phenomena of mesmerism and logically disprove their existence, an appeal may ever and at any moment be made to the " proof by experiment;" and, even should experiment itself fail a thousand times, the success of the thousandth and first trial would justify further examination. Till the authority of observation can be wholly set aside, the subject of our inquiry can never be said to have undergone its final ostracism.

## SECT. III.

SHOWING CERTAIN OF THE PHYSICAL AND METAPHY-
SICAL CONDITIONS OF MESMERIC SLEEPWAKING.

HITHERTO I have exhibited to my reader a succession of patients, the greater part of whom I had
no opportunity of mesmerising again.  I may have
been considered as trying an experiment upon a
great scale, as to the mesmerisibility of mankind in
general.  Such a mode of proceeding has this advantage: — persons, who have been frequently mesmerised
may be charged with having been well practised in
their parts ;  but they who exhibit for the first time
the mesmeric characteristics, without even having
heard them previously described, stand aloof from
all suspicions of the kind.  There is neither habit
nor imitation nor duplicity to be charged upon
them ; and therefore the phenomena they display
may be regarded as eminently genuine.  But, on the
other hand, it should be considered that such phenomena, if more to be relied on, are also humbler in
degree than those belonging to a more advanced stage
of sleepwaking : and this, I think, may be owned
without prejudice to mesmerism ;  for, as I have
before observed, we cannot liken it to those purely
mechanical agencies which affect persons as forcibly

the first time as in all succeeding trials. Many of
the phenomena are of such a kind as to be obviously
capable of developement; and in demanding a certain license on this account we ask no more for mesmerism than is conceded in every case where man's
intellect is concerned. Let it be always kept in mind
that mesmeric sleepwaking is more than an exhibition
of involuntary motions or mere physical changes.
It is a *state* which appears to possess its own laws of
perception and of action ; and, in this point of view,
a mesmerised person may be considered as learning
a new language in which he cannot express himself
with eloquence or with ease until he has mastered
its idioms and possessed himself of its copiousness.
When we reflect upon the difficulty we all find in
acting in any unusual way, we cannot but perceive
that time must be requisite to the developement of a
mode of existence which seems to be abnormal in the
highest degree.

It is evident, then, that, in order to study the phenomena of mesmeric sleepwaking in their maturer developement, it was necessary for me to mesmerise the
same person frequently. This I have done in more
than one instance, and the following has been the
result of my observations : —

In accordance with what has been suggested above,
it is to be remarked that each successive time a person is mesmerised he becomes more easy of mesmerisation, and more at home in his new capacities.
Hence without effort he exhibits many phenomena
which may be called spontaneous, in distinction to

those which result more clearly from the peculiar relationship in which the mesmeriser, *pro tempore*, stands to his patient. The two classes of phenomena should be accurately distinguished, and their difference borne in mind. At present, I have chiefly given instances of the latter ; as was natural, considering that my earlier patients were more dependent upon myself than those whom I subsequently treated, and that in consequence my attention was chiefly turned to the mesmeric processes necessary for producing or directing their sleepwaking. It was not till later, and till the task of guiding my patients could be, in a degree, remitted, that I was able to add some remarks, (which indeed appear essential to complete our view of mesmerism as a separate condition of man), upon the state of the bodily organs, and of the mental faculties, in the mesmeric sleepwaker. Beginning with the former, and dividing the corporeal functions into the sensitive, the vital, and the motory, we first inquire — What is the state of the actual apparatus of the senses during mesmeric sleepwaking ? And first,

## *Of the Eye.*

This is a part of the body which is earliest affected under that which we may call, without theory, the human influence. One of the first tokens a person gives of passing into mesmeric sleepwaking is a look of stupor in the eyes, and an apparent lack of power in the eyelids to perform the usual office of nictation. The upper lid, as I have before observed, falls very gradually over the organ of sight, and sometimes, ceasing to move altogether, remains suspended, as it were, in such a manner as that the eye appears to be three quarters closed. When this continues to be the case, the patient rarely falls into sleepwaking, the exceptions occurring commonly amongst those who sleep with their eyes partly open and who keep them in the same state during the whole period of their sleepwaking. The patient, when interrogated as to his sensations, will declare, (retaining his consciousness at the time) that he feels prickings in his eyelids, and, as far as outward actions can be proof, the same uneasy sensation remains or recurs at intervals during the sleepwaking. For all the sleepwakers that I have seen rub their eyes frequently, so that any one would think they were about to awake; but their slumber remains undisturbed, and the action alluded to seems simply automatic, as when in natural sleep we brush off a fly from the face without being conscious of the movement. I once asked a sleepwaker why she rubbed her eyes, and she replied, " Did I do so ? I suppose it is a trick I have " (" je suppose que c'est une habitude"). I have before remarked that many persons in semi-sleepwaking retain their consciousness, yet are wholly unable to open their eyes. A scientific man, accustomed to investigation, being in this state, assured me that he experienced a gradual paralysis of the nerves connected with the eye, and with the motive powers of the eye-lid, and that at length his utmost efforts of volition were insufficient to make the orbicular muscles obey him as usual. Sleepwaking being complete, the eye is gene-

rally first closed, yet not exactly as in sleep.   The
following differences may be noted : —1st. The place
where the eye-lids meet, in a natural way, is much
below the level of the transverse diameter of the eye ;
but I have observed that in sleepwaking the line of
contact is often thrown so high above its ordinary
situation as to coincide with what Haller calls the
*æquator oculi.*   2dly. There is a compressed look
about the lids, as if they were rather held down by
force than quietly and naturally closed. 3dly. The
ball of the eye is in frequent and violent motion,
which, in those who have prominent eyes especially,
can be plainly discerned beneath the skin of the eyelids.   All these circumstances, it should be remembered, are actually distinctive of the mesmeric sleep ;
for many persons who form a wrong notion concerning it suppose, before *assisting* (as the French say)
at a *sèance* of mesmerism, that they are to see a something precisely like natural slumber, and exclaim, when
the reality is before them, " This must be all a trick,"
only because their erroneous ideas are contradicted.
With regard to the internal state of the eye during
mesmeric sleepwaking, it is, of course, difficult to
judge.   No force, short of that which would seriously
injure the sleepwaker, can wrench asunder the eyelids.   I have tried this and made others try it, and
the resistance to such efforts was so great that, to be
appreciated, it must be felt.   But that which violence
cannot accomplish, the command of the mesmeriser
can ordinarily effect,  though that he should use this
power with great discretion the following occurrence

will show. I once asked Mademoiselle Anna M——,
(whom, after returning to the Continent in 1837, I
frequently mesmerised) whether she could open her
eyes, which, at the time, were fast shut in mesmeric
sleepwaking. Her reply was, "I can do it, if you
insist that I should; but I warn you that you will see
something very disagreeable, and, besides, the effort
will do me no good." Notwithstanding this warning,
I commanded my *mesmerisée* to open her eyes. She
did so. It was really a fearful sight. The eyeballs were turned up, and converged towards the
nose, the white of the eye was bloodshot, and the
whole organ quivering and convulsed. As may be
supposed, I did not long keep my sleepwaker in what
appeared to me a painful state, and, at my command,
she again closed her eyes. On awaking her shortly
after, what was my alarm to hear her declare that
she saw nothing whatever! She appeared to be
much agitated, and my thoughts may be conceived,
as long as I had the least reason to fear that my rash
experiment might end in her being blind for life.
The exigence of the moment, however, roused me to
exertion, and the idea occurred to me that the best
thing I could do was to re-mesmerise my patient as
quickly as possible. This I did. As soon as she
had passed into the mesmeric state, she became calm
and begged me not to be alarmed, as the blindness
she had just experienced was but the momentary
effect of the fatigue to her eyes, caused by the effort
to open them. She then requested me to breathe
on her eyes, and to lay my hand on them, after which,

she assured me that I might awake her without any
further apprehension.    In effect, the transient but
fearful attack of blindness was passed away.    After
this, as may be supposed, I was not very ready to bid
my sleepwakers open their eyes.    I have also found
in them all, when consulted on the subject, an extreme dislike to the idea.    I have, however, reason
to think that the position of the eyeball is, in most
sleepwakers, the same as it appeared in Mademoiselle
M——.    I once asked one of my patients, while in
the waking state, if he knew in what position the eyes
were during sleepwaking, and he replied, " I do not
at all know, but I should suppose just the same as
now.   Why should there be any difference ? "  Having
thus ascertained that he had no preconceptions on the
subject, I again asked him, when in sleepwaking,
" In what position are your eyes ? "    " I will show
you," he replied ; and without a moment's hesitation,
seized a pencil and a paper that lay on the table,
and drew an eye, with the ball turned up to one
corner, and only partly seen.    [I should observe
that this sleepwaker had an extraordinary power
of vision in the mesmeric state, to which I shall
hereafter have occasion  more particularly to allude.]

Another time, I had an opportunity of making further remarks on the state of the eye, under mesmeric
sleepwaking, through the kindness of Mr. Berckmans,
architect of the province of Antwerp — a man whom
I have pleasure in naming, as one who utilises mesmerism by employing it (and successfully) in the relief

of maladies. This gentleman permitted me to see one of his sleepwakers, who, from a natural predisposition, often kept her eyes open during sleepwaking. During twenty minutes or half an hour that I saw her thus, I could most truly affirm that, though her eyes were open, their "sense was shut." A dull film seemed to overspread them; the pupil was dilated, and did not contract with light. A candle brought near, or a hand waved suddenly and quickly before the patient's eyes, produced no perceptible alteration or motion either in the lid or in the apparatus of vision.

Altogether, there seems to be every reason to conclude that the eye in mesmeric sleepwaking is either so disordered or so paralysed in its functions as to cease to convey impressions to the mind — in any mode at least that can be termed ordinary.

*Secondly.* I now proceed to show that a similar derangement of function seems to take place as regards the sense of hearing.

In proportion as persons sink deeper into mesmeric sleepwaking, their external senses seem blunted, one by one, and so far there is certainly a relation between the mesmeric and the natural sleep.

The eye, as we have seen, yields first to the slumbrous influence. Long after this organ has ceased to act, the hearing retains all its acuteness, and the sleepwaker is able to indicate what sounds are going on around; but at length the "porches of the ear" are closed as well as the "curtain of the eye," and the patient, though still alive to feeling, is dead to

every sound save that of the mesmeriser's voice. I
have proved this times innumerable — so frequently,
indeed, that it is better to give the general results of
the experiments I have witnessed than to state one
in particular. Often have the members of my family,
or visitors, who perhaps were but little inclined to
believe in mesmerism, tried to awaken Mademoiselle
M—— or to startle her by sudden noises. Logs
of wood have been dashed against the floor; plates
have been suddenly broken; her name has been
shouted out, close to her ear; in vain. Other persons
present have shown that they were startled — but
not the sleepwaker. Once or twice, indeed, on such
occasions, when asked if she heard any thing, she has
replied, " No, I heard nothing; but I thought, just
now, something pushed against my chair;" a mode
of expression which deserves to be remarked, as
analogous to that used by deaf persons to describe the
sensations given them by the concussion of the air
produced by great sounds. I once met a young
lady, perfectly deaf and dumb from her birth, who
was, in this way, remarkably sensitive to the undulations of the air. I have frequently seen her start
when a door was opened, or when any thing fell suddenly, and the account she gave to me of this, in
writing, was that she felt as if some one had pushed
against her. So susceptible indeed was she to aërial
vibrations that she could distinguish a certain measure and rhythm in harmonious chords which gave
her a marked degree of pleasure. She would take a
stick; and, putting one end of it in her mouth, would

place the other in contact with the piano, while any one was playing on it. Discords struck upon the instrument made her shudder, and convulsed her features with all that pantomimic exaggeration so usual in the dumb ; but soft and pleasing sequences of sound soothed her and brought a satisfied smile over her countenance.

It has appeared to me that the mesmerised possess similar perceptions of sound apart from the natural sense of hearing, and that, like the young lady above alluded to, they require certain conductors, in order to make them apprehend a regular series of aërial vibrations. Be it however remarked that the degree of this isolation from sounds, considered as sounds, depends on the intensity of the mesmeric sleep ; for it should ever be kept in mind that mesmeric sleepwaking has its shades and graduations, varying from consciousness fully retained to its faintest twilight or utter extinction. A due recollection of this truth will prevent many mistakes and unfounded expectations relative to our subject.

In the case of Mademoiselle M——, as being a perfect .sleepwaker, the insulation from all sounds to which I did not serve as conductor was complete. This phenomenon, as indeed most of the others I have mentioned, was rather accidentally offered to my remark than looked for or expected. A lady, present when Mademoiselle M—— had been mesmerised by me, went up to the sleepwaker and spoke to her for some time. She however did not seem to hear what was said, and indeed held her head

down with every sign of inattention. By mere accident
the speaker touched me. Then suddenly the sleepwaker lifted up her head, and assumed the expression
of listening. A slight movement drew me away from
contact with the speaker, when the sleepwaker said —
" I hear nothing now." She then herself took the
lady's hand, placed it in mine, and laid her own upon
both so as to be in contact with each, saying, " Now
I hear well." We subsequently found that it was
under such an arrangement as this that the sleepwaker could best hear any one speak, though did a
person talking merely touch me she manifested some
perceptions of sound. On one occasion, when a
musician was singing and playing during her mesmeric trance, she, seeing him at the piano, expressed
a desire to hear the performance, and, to accomplish
this, requested me to lay one hand on the musician's
shoulder, while she herself laid a hand upon mine.
So placed, she heard the singing well, but the piano
indistinctly.

Others of my sleepwakers manifested similar phenomena, some even the first time of being mesmerised, when they knew nothing of the characteristics
of mesmeric sleepwaking. The scientific person to
whom I have once before alluded, and whose testimony is valuable, inasmuch as his habits of mind led
him ever to separate illusion from truth, assured me,
when in the mesmeric state, that he could hear no
sound whatever except my voice. I made another
person speak who was in the room on that occasion,
and the sleepwaker was unaware that any thing had
been said.

Another patient (E. A——, to whom I shall have occasion to allude hereafter) said, when I was singing, " You should ask Mr. V——" (a musician who was in the room) "to accompany you. " I did so; but, though Mr. V—— made a loud accompaniment to my voice, E. A. kept calling out, " Why does he not play ? "

3dly. *State of the nerves of touch.* — Every one has heard of the insensibility of sleepwakers to external stimuli. Even in the present crisis of hostility to mesmerism, the world seems agreed to go the length of believing, with Mr. Edwin Lee, that, " in the states of partial torpor, not unfrequently occurring in nervous persons," (so does the cautious doctor define mesmeric sleepwaking,) "the individuals are, to a certain extent, insensible to mechanical stimulants. " This being granted, the evidence which I can bring forward to confirm the alleged phenomenon is, fortunately, but little needed; and, in truth, I was too little inclined to try such cruelties upon my sleepwakers, as plunging in pins and applying moxas, to be able to offer numerous proofs upon the subject. Such experiments, however unfelt at the time, cannot but produce effects of which the patients will become sensible afterwards; and, when they are smarting under the pain occasioned by the burning away of half an inch of skin, they will be but ill consoled by the glory of having suffered in the cause of science. On these grounds, I have refused either to torment my sleepwakers or to allow them to be tormented. However, a friend of

H

mine was once determined to try an experiment of
this kind; and, without giving me any notice of his
intentions, he suddenly plunged a large pin into the
back of the hand of Anna M——, so that the blood
came when it was withdrawn. She neither moved
her hand nor started, nor seemed in any way sensible of the injury. About half an hour after, being
still in the mesmeric state, she looked at her
hand, as if by accident, and, seeing the congealed
blood on it, asked, " What in the world is this?"
At another time, being lightly touched with the end
of a feather about the lips and nostrils, (a proceeding
which few can endure,) she gave no token of sensation, and once, while I was mesmerising her, she said,
" I feel now quite comfortable," at the very moment when some one, fond of experiment, was pulling
her ear till it became scarlet. [This inflaming of the
cuticle seems to show, by the way, that the skin, in
sleepwaking, loses none of its functions, though it no
longer transmits sensations to the mind.]    Other
little circumstances, naturally presenting themselves,
have confirmed to me the physical insensibility of
sleepwakers.    Anna M—— had once a swollen and
very painful finger.    In her waking state she complained bitterly of the torment she endured from this,
and was constantly breathing on the disordered part.
Once, in sleepwaking, she seemed to forget her
finger altogether; and, on being interrogated, declared that she felt no pain.

Another of my sleepwakers (E. A——) watched
his sensations, while he was entering into the mesmeric
state, up to the last moment, when consciousness be-

came extinguished. He told me that he could distinctly perceive a gradual deadening of sensation; that at length he no longer felt his limbs; but that all his life and feeling seemed to rally towards the brain. In the mesmeric state, he continued to assure me that he was corporeally insensible; and, anxious, as it seemed, to try experiments upon himself, he would bite his own hand till he drew blood, without, as he affirmed, exciting sensation.

4thly. *State of the organs of taste.* — That these also share the insensibility of the other senses I have had every reason to believe. Anna M——— could never distinguish one substance, that was placed, with precaution, in her mouth from another. I have told her that cheese was very good orange, or water wine; and she, trusting to my veracity, has implicitly believed me: her faith in my assertions being uncorrected by the exercise of her usual faculties, and preponderating manifestly over these. I have tried upon her, when in sleepwaking, the well-known experiment of placing a piece of zinc and a piece of silver, the one above, the other below, the tongue, and then bringing them suddenly in contact; but no metallic taste was perceived by her. The same experiment, repeated in her waking state, produced its usual result.

5thly. *The smell seems equally unaffected by external stimulus.* — I have held, for a considerable time, strong ammonia, which made my own eyes water, even at a distance, close under the nose of a sleepwaker, while the mouth was shut and respiration carried on only through the nostrils. The breathing proceeded as regularly as before; there was no watering of the eye : in short, the patient (a sister of Theodore) gave no token whatever of sensibility in the olfactory nerves, or those of common sensation or touch.

Having finished our review of the sensitive organs, we now proceed to consider the state of those which may be called vital, viz. the respiratory, the circulatory, and the digestive.

In those mesmerised persons who have fallen under my observation, the respiration has been slower and more regular than usual, resembling what it is in sleep. Occasionally, when the sleepwaker has been permitted to repose, either in a sitting or recumbent posture, the breathing has become louder, almost approaching to stertorous. Any one seeing the mesmerised at such times would pronounce them to be really asleep; but the proof that this cannot be the case is, that they still hear and answer the mesmeriser's voice.

I have found, also, that, in mesmeric sleepwaking, the heart beats more slowly and evenly than in the normal state, and that the pulse indicates a corresponding change in the circulation.

On one occasion, two medical men, by no means favourable to mesmerism, were present when I was about to mesmerise Mademoiselle M——. It was during the festivities of the Antwerp carnival; and my fair patient, having been at a ball for two nights running, was fatigued, and feverish — complaining much of pains and stiffness in her limbs. Before beginning to mesmerise, I requested the medical men to feel her pulse. On doing this, their exclamation was, "You ought to be in your bed, Mademoiselle, for you have a great deal of fever." In fact, her pulse was 120, her skin hot and dry, and her tongue white. I then threw my patient into sleepwaking; and after she had continued in that state, reposing on a chair, for about a quarter of an hour, I again asked the doctors to feel her pulse. Each took a wrist, as before, with due professional gravity, and with a stop-watch counted the pulsations. "Marvellous!" they both exclaimed. "The pulse beats quite other than when we last felt it. The sharp wiry rebound of fever is gone, and the pulsations, besides being soft, full, and regular, are not more than 80 in a minute." After this I made my patient rise from her chair, and, during the hour that she remained in sleepwaking, she was in almost constant motion; yet she neither complained of, nor betrayed in her gestures, the fatigue and stiffness from which she had previously been suffering. On awaking, the doctors again inspected her state, and assured me that every symptom of malady had disappeared. The patient's pulse had sunk to 74 in the minute; her skin was cool; her tongue no longer indicative of fever. When asked for an account of her own personal sensations, she declared that she felt so renovated, and so entirely free from pain and uneasiness, that she should have thought that the sleep of two nights had been put together in the single hour of her sleepwaking. Hearing this,

H 3

the doctors were inclined to exclaim, with him of the fraternity in Macbeth, "A great perturbation in nature ! To receive at once the benefit of sleep, and do the effects of watching."

*The Effect of the Mesmeric Sleep on the Digestive Organs seems equally reparatory, and indicative of increased Power.*

E. A.——— a youth aged fifteen, whom I frequently mesmerised, and whose general health was excellent, was suffering one day from an accidental attack of indigestion, accompanied by slight sickness. During an hour's sleepwaking his uneasy feelings were suspended, and, when he awoke, were found to be completely removed. At another time — not with a view to any particular experiment, but in the way of a pleasantry — I kept the same sleepwaker in the mesmeric state for a longer period than usual, so as to make him eat his supper with our family party, while still in sleepwaking. Our evening meal was brought in ; and the patient, at my request, (for otherwise he did not seem to care about taking anything,) ate whatever was given him ; and as our object was to make him *feel*, on awaking, that he had had his supper, and to wonder how this had come to pass, we supplied him plentifully with food, so that he actually made a fuller meal than was usual with him at the same hour. We then had the supper things removed ; and, in about a quarter of an hour afterwards, I awoke the patient. His first question was, " Have I been long asleep ?" " How long should you suppose?" we inquired. "I cannot imagine," he replied; "but I hope we shall soon have supper, for I am very hungry!" At this we all laughed, and assured him that he had just made a capital meal — a piece of information which he at first refused to believe. The bell was rung, and the servant called in to depose to the fact, before he could credit it; and then he appeared by no means disposed to rest contented with what seemed in his idea a refection as visionary as that wherewith the Barmecide in the Arabian Nights regaled the beggar. He begged to have if it were only a crust of bread, to appease the cravings of hunger; and, the wherewithal being afforded, soon made a second repast as substantial as that which we had forced upon him during his sleepwaking.*

* At the time that the above phenomenon presented itself, I had not met with an interesting account of an extraordinary cure by mesmerism, written by M. Despine, physician at Aix-en-Savoie.

Subsequently, on reading the Doctor's pamphlet, I was struck with the following passages and their relation to the subject we have been considering :—

" Estelle a l'appétence de tout ce qui lui convenait, et de ce qu'elle aimait dans ses premiers ans. De plus, elle en mange, pendant sa crise, *avec abondance et impunément*, sans jamais en sentir le moindre mal-aise ; pendant que, dans son état naturel, elle ne saurait s'écarter de son régime végétal habituel, sans en éprouver des crampes, &c.

" On dirait que notre malade semblait avoir deux estomacs — l'un pour l'état de crise, l'autre pour celui de veille.

" Mais ce qui paraîtra plus singulier encore, c'est que les alimens, pris en *abondance* dans la crise, ne paraissaient pas, le

H 4

Again, before the above occurrence had taken place, I mesmerised the same person near the hour of luncheon; he having previously observed that he was very hungry, and therefore hoped I should wake him soon, in order that he might eat his accustomed meal. When in mesmeric sleepwaking, he complained no more of hunger; and, indeed, assured me that he felt nothing whatever of the powerful appetite which had so recently tormented him. He had no desire to awake; and, as I was engaged in trying upon him some interesting experiments relating to vision, I did not think of demesmerising him till late in the day — about half an hour before dinner. The moment he was restored to the normal state, he declared that he was almost mad with hunger; and, when I told him how long he had been in sleepwaking, was quite angry that I had not waked him sooner. Though told that dinner would soon be on the table, he begged to eat something immediately.

---

moins du monde, la rassasier pour le temps de veille, et *vice versâ*."

Since reading the above, I have had an opportunity of seeing the young person mentioned — a most interesting little girl, of thirteen years of age, living at Neufchâtel, in Switzerland. Her mother certified to me all the above particulars, and told me that mesmerism alone had cured her daughter Estelle of paralysis and spine complaint, after all other means had been tried in vain. Of the degree of the patient's previous illness of course I can only speak from the testimony of others; but this I can certify, that, when I saw Mademoiselle Estelle L'H———, she was apparently in perfect health and enjoying the activity natural to her age.

This and the preceding anecdotes seem to indicate that the digestive functions are by no means disordered or suspended during mesmeric sleepwaking; but, on the contrary, that they act more powerfully: while, on the other hand, they do not, as usual, convey intimations of their condition to the mind; for, in the first instance, we have seen indigestion removed during sleepwaking — a proof that the apparatus of digestion is strengthened in the mesmeric slumber. Secondly, we have been made aware of digestion having been rapidly performed in the mesmeric state: a full meal having been so quickly assimilated to the juices of the body as to cause the sensation of hunger to recur almost immediately after it was taken. Thirdly, we have been shown hunger progressing as usual, but unfelt, during the whole period of the mesmeric sleep.

I may add that the increased energy given to the digestive functions during mesmeric sleepwaking has seemed, in some cases, to confer a lasting benefit. Mademoiselle M——, though in other respects strong and healthy, suffered, when I first saw her, from indigestion, which manifested itself in painful fits of colic. Subsequently she thus wrote to me:—" With regard to my complaints of the stomach, I have not felt the least symptoms since my last mesmerising. My malady has been too severe, and the relief I have experienced too effectual, for me to doubt of the good effects of mesmerism."

Again, that the human influence is reparatory of the vital functions generally, I had a proof in the case

of a gentleman (already mentioned in Sect. II.)
who allowed me to mesmerise him when he was
suffering from the fatigue of a long journey and
night-travelling.   He remained in the mesmeric
state half an hour only, yet, on awaking, he exclaimed, "Je me sens comme nouveau ne."   All sense of
lassitude was gone.

From the vital we proceed to a consideration of
the motive powers.

Any one who has attended to the cases of mesmeric sleepwaking, which I have detailed elsewhere,
must have remarked that the natural inclination of
mesmeric patients is to remain perfectly still.   It is
only by the mesmeriser's persuasion that they are
induced to move ; and their first attempts at walking
are, like those of an infant learning to use its limbs,
weak and tottering.   After a time, however, by a sort
of education from the mesmeriser, the sleepwaker
is enabled to move freely and firmly.   The mind
has evidently resumed its empire over the nerves
of motion, yet still is disconnected from the nerves
of sense ; — a circumstance which strongly illustrates
Sir Charles Bell's valuable discovery respecting
the absolute distinctness of these two sets of nerves;
and which confirms the thesis, that motion is
ever propagated downward, from the brain to the
exterior of the body ; while sensation is conveyed
upward, from the extremities of the nerves to the
brain.   I think, also, that the power which sleepwakers possess over their motory system proves
satisfactorily that, according to the ingenious sug-

gestion of Sir C. Bell, there exists a distinct muscular sense; for we see it here insulated, and acting at a time when all the other senses have ceased their functions. Nor does the muscular sense merely *act* under mesmeric sleepwaking, but it acts (from the time that the patient is habituated to his new existence) in a very perfect manner. I have seen displayed by sleepwakers an activity which they did not possess awake.

Availing myself of the attraction which draws the patient after the mesmeriser, I have made Mademoiselle M—— follow me, when mounting upon chairs or wardrobes. Another of my mesmerisées, when I got upon a chair, leapt lightly up after me; and, there being scarcely room for her to stand, remained poised in a very extraordinary manner, on the extreme edge of the chair, exhibiting, in that position, an ease and graceful firmness, quite remote from the characteristics of her usually reserved demeanour.

Additional strength as well as agility of body seems to accompany the more advanced periods of the mesmeric state. E. A——, in sleepwaking, could throw, in wrestling, a person he could not master when awake; and once he burst a locked door open with an ease that was extraordinary. The same phenomena, which occur with respect to the motor muscles in general, take place also as regards the muscles of the larynx. On first mesmerisation, the voice of the patient is weak, and it is evidently an exertion for him to speak at all; but subsequently he resumes his powers of speech, and, if musical, can sing as well as in the waking state. I should, however, observe, that the tone to which he pitches his voice in conversation depends much upon that in which the mesmeriser addresses him: should the former whisper, he whispers too, or the contrary. This is the general rule; but Anna M—— offered an exception to this. Her voice in sleepwaking was generally very feeble, and she used to complain of its extinction.

Though hitherto treating of the patient separately from the mesmeriser, it appears to me that this branch of my subject would be incomplete, did I not add to such phenomena of motion as are proper to the sleepwaker others which are caused by the mesmeriser's influence over his patient.

I have already remarked that the sleepwaker exhibits an occasional community of motion with the mesmeriser, and adopts something of his general deportment. Some additional instances of these phenomena may be interesting.

Anna M——, in her mesmeric state, mounted a staircase with me, in order to visit an invalid lady of our party. I observed that she ascended two stairs at a time, after a manner that had become habitual to me. At another time, when I had a cold, she, though free from such an affection in her natural state, coughed, in her sleepwaking, whenever I did. I extract the following from a register of our mesmeric proceedings, once kept by one of our family.

*January* 2. 1838. " The mesmeriser rubs his

hands, and Anna M—— immediately does the same. When the mesmeriser blows his nose, the patient puts her hand up to hers, as if doing the same thing. When the former began to dance, the latter did so too, imitating exactly the movements of her mesmeriser, and snapping her fingers as he did. The mesmeriser, having turned from the patient, who was sitting at a table, took up a pencil, and, without her seeing him, made some strokes upon paper. The patient took up another pencil which lay at hand, and made similar strokes."

With respect to the times when these phenomena were exhibited, I remarked that they occurred especially when any more violent muscular motion took place on my part, such as in the act of blowing my nose, of dancing, &c. ; — also, when my attention was particularly directed to the production of the phenomena. When such was the case, I have turned my back, and made gestures in perfect silence, (so as to prevent any suspicion of the patient being influenced by sight or sound,) and they have been perfectly imitated by her. Once, I agreed with a friend of mine to put my hand into my coat pocket at a certain signal from him. The sleepwaker, though standing where she could know nothing of the circumstance, thrust back her hand immediately into a similar position with my own. The same friend, who was determined to try everything to put the sleepwaker at fault, used to imitate my rubbing my hands, coughing, or blowing my nose, in order to induce Anna M—— to do the same; and thus to prove that she was guided by sound,. not by sympathy of motion. The patient, however, was never influenced in the slightest degree by these artifices. Whether such correspondences of motion result from purely mechanical causes may be much doubted; — but there is another class of motory phenomena in which the body of the mesmeriser must clearly be considered as the depository of some attractive force, entirely independent of his will; for they result simply from the position in which the patient may be placed relatively to the mesmeriser.*

Having frequently remarked that the head of a mesmerised person follows the hand of the mesmeriser, as iron does the loadstone, I, in order to submit the phenomenon to the surest proof, bandaged the eyes of E. A—— previously to mesmerising him, and filled with cotton every possible interstice that might be left between the cheek and the handkerchief, in such a manner as to convince every person, who witnessed the experiment, that for the patient to

* This mechanical attraction appears to have been first noticed by M. De Lausanne, who observed it in a patient of his own — a female, whose sensitiveness was remarkably developed. In a work on mesmerism, published at Paris, this author, who still, I believe, is living, gives a particular account of some interesting experiments relative to the subject. When M. De Lausanne was in another room, or even outside the house, the sleepwaker, above alluded to, turned her head in the exact direction of her mesmeriser. This was established by repeated trials, before witnesses. It should be remembered that the development of similar phenomena, in so striking a degree, presupposes always a rare and advanced stage of sleepwaking.

perceive by sight the motions of my hand was entirely out of the question. As soon, however, as E. A——— had entered into sleepwaking, his head followed my hand, at the distance of at least two feet, with unerring certainty. Did I place it on the left of his head, his head instantly inclined to that side; was it in front, he bent forward to meet it ; was it behind him, he leant back towards it with what would seem a painful and unnatural effort.

Again, I held my hand above his leg ; immediately the limb was attracted upwards. I continued to withdraw my hand; and the leg was so much elevated as to form an acute angle with the body of the patient. Finally, I ran quickly away to the other end of the room ; and the sleepwaker followed me quickly and unerringly.

At another time, being in one of the folds of a screen, and Anna M——— in another, I put out my hand (but not beyond the projection of the screen), when instantly the patient's hand came round to meet mine.

One evening, when I had mesmerised the same sleepwaker, and I was sitting near her, it occurred to me to inquire, " If my hand possesses so singular a property of attraction, what power of a similar kind may reside in my foot? " With this thought, I raised my foot, not in any unusual manner, but as if I were laying one leg over the other. The sole of my foot was thus presented to the patient. In a very short time she began to turn round upon her chair, so as nearly to fall into the fire. Struck with this

oddity, I made the sleepwaker rise, and again presented my foot, when she commenced a series of regular revolutions, from right to left, stopping always,
when she had half completed the circle, with a bend
and a dip, like that of the magnetic needle. The
experiment repeated always produced the same
results.

Again, when standing opposite to my patient, who
was in an upright position, I began to turn round.
She also seemed forced to revolve, but in a direction
contrary to mine. If I turned from left to right, she
turned from right to left, and *vice versâ*.

I tried the same experiments on other sleepwakers.
They also turned round when I did, and inversely to
myself; but my foot, when presented to them, instead
of making them revolve (a phenomenon apparently
confined to Anna M——), seemed simply to repel
them. I tried this with the sister of Theodore. Every
time that I held out my hand she was attracted towards me; whenever I held out my foot, she was
repelled; and, by the alternate exercise of these two
influences, she was kept oscillating to and fro like the
pendulum of a clock.

That the mesmeriser can create motion in the
sleepwaker is not the only proof of the power that he
possesses over the organisation of the latter. He is
also capable of paralysing those motory agencies
which he so much sways. But experiments of this
kind should be undertaken with the greatest caution.
Puységur relates that, having tried, from curiosity, to
paralyse the respiratory organs of a sleepwaker, he

threw the patient into a state resembling death, which lasted for several hours. Dr. Sigmond has also reported the case of a lady whom, without intending it, he threw into an alarming state of insensibility, — a circumstance which (if I may venture to say so) is explained by the Doctor's own statement of his mesmeric principles. His words are — " It is upon *the respiration* that my efforts are directed; and the principle is precisely that which is called ' stealing the breath away.' "

The account is in " The Lancet" for Dec. 1837, and I have since had no opportunity of learning any of Dr. Sigmond's subsequent experiments. I do not, however, doubt but that experience has brought him to the same conclusion as myself; — namely, that those motions and intentions of the mesmeriser which are most directed towards *equalising* the influence he wields are the best; and that we should above all avoid concentrating the action upon any vital part. Being fully persuaded of this, I have seldom ventured upon paralysing the nerves of motion, save in the less important parts of the body, — as in the hand or arm, for instance. What led me to try such experiments at all was this : — Mr. K——, of Antwerp, my first instructor in mesmerism, had exhibited to me a curious phenomenon which occurred whenever he closed the hand of one of his sleepwakers over a piece of coin (silver or copper). The hand remained rigidly shut, and the patient seemed utterly incapable of opening it, until after the dispersion of the mesmeric sleep. From a train of analogous reasoning, I was much inclined to believe that the coin had no more to do with the phenomenon in question than the magnet and the baquet with the effects which Mesmer produced upon his early patients. I accordingly closed the hand of Mademoiselle M—— without the coin, and again with it. As I had anticipated, the result in both cases was the same. Moreover, I discovered that, by breathing on the patient's hand, I could always, and under any circumstances, restore it to its functions — in fact, that myself, and not the metal, was the depository of the paralysing power, what-ever it might be.

I relate this in the hope that it may warn others, as it did me, against attributing, in mesmerism, effects to wrong causes. In all that regards mesmeric sleepwaking there is especial danger of falling into such an error, from the curious blending that it exhibits of human and of mechanical action ; and never can we adjust the balance properly, until the mesmeriser's influence over his patient be fully taken into account.

A curious circumstance, attending the mesmeriser's power over the patient's organisation, is this :— When the two are in frequent mesmeric relationship, the phenomenon is carried forward into the natural state. This I found to be the case after I had often mesmer-ised Anna M—— and E. A——    I could, at any time, fix the hand or arm of either of these persons in any position I pleased, and in all the rigidness of catalepsy. The means I used to produce this effect were as follows : — I first placed the limb as I wished

it to remain, and then brought the fingers of both
hands into contact with it, at opposite poles, as it
were, and pointing to each other, so that, supposing
they emanated a stream of galvanic or other influence, it would pass exactly through the part to be
affected. The desired result did not ensue immediately, but in two or three minutes; and its intensity
was proportioned to the length of time I acted on the
limb. The patients described a sensation of gradual
paralysis, and a deadness, as when a leg or arm, in
common parlance, goes to sleep. At length a spasmodic contraction of the muscles took place : the
hand looked white and bloodless, and the fingers un-
naturally rigid. To the touch they seemed as chords
that are pulled tight ; and, if the numbness continued, they lost all feeling. No force, short of what
would have produced mechanical injury, could undo
them. On one occasion, I placed the arm of E. A.
in a position so twisted and constrained, that it was
not possible for him to oppose much voluntary force
to any one's efforts to pull it down ; yet, when once
paralysed, there was not a person present who could
stir it. It offered the resistance of marble. Moreover, this patient was not in the least imaginative,
and ridiculed the idea that fancy could produce the
phenomenon in question. When I wished to restore
motion to a limb thus paralysed I breathed on it,
and made some magnetic passes down it, accompanied sometimes by slight friction. The return of life
and feeling was gradual as their loss, and was accompanied by those tinglings called pins and needles.

A slight numbness was generally felt for some time after the experiment. Once I persuaded E. A——— to permit me to try whether, in his waking state, I could prevent him from opening his eyes. The experiment was perfectly successful. By pressing down the upper eyelid with a finger of one hand, and placing a finger of the other upon the lower lid, I could so influence the levator muscle as (after withdrawing my hands) to keep either eye closed while the other was wide open. The patient could not, in general, shut his right eye without also shutting the left.

I have now taken a review of the condition of the bodily organs, and of the motive powers, under mesmeric sleepwaking, being at the same time fully aware how much is left to be performed by abler hands than mine. I am no physician; and a very minute inquiry into the physical state of mesmeric patients is not my province. I, however, seriously recommend to all who study medicine as a profession, especially to the younger aspirants in this branch of science, (since *they*, at least, are not grounded and rooted in antique prejudices,) to pay attention to mesmerism, after the plan that I have laid down; *i. e.* to ascertain, during its operation, the exact and peculiar state of every organ and function of the human body.

By mesmerism we best dissect man, whether mentally or physically; and, if ever the vital influences are to be understood, it is not by anatomising the dead, or by torturing the living, but by observation of our fellow-beings when in the state we call mesmeric.

I now proceed to offer some remarks upon the mental condition of patients under mesmeric sleepwaking, considering the soul in its moral, intellectual, and sensitive capacities.

First, as regards our moral being: —

That the state of mesmeric sleepwaking is a rise in man's nature, no one, who has been conversant with it, can doubt.

Separated from the usual action of the senses, the mind appears to gain juster notions, to have quite a new sense of spiritual things, and to be lifted nearer to the fountain of all good and of all truth. The great indication of this elevated state of feeling is a horror of falsehood, which I have found common to all sleepwakers. Sincerity is their especial characteristics; they cannot feign or flatter : they seem to be taken out of common life, with all its heartless forms and plausible conventions.

I proceed, by one or two instances, to show how forcibly sleepwakers are impelled to speak the thing that is, and to clear their consciences of that dissimulation which clings so much to man in his natural state.

During the Antwerp carnival, a lady, who took a sincere interest in Anna M.'s welfare, advised her not to go to the masked ball, which is usually given at that season. The night after the ball, Anna came to be mesmerised, and, though complaining of fatigue, would not own that she had acted in opposition to the advice that she had received. When, however, in sleepwaking, she acknowledged, of her own accord, that she had been at the masked ball, and said that she felt she had done wrong in practising conceal- ment, though her motive had been to avoid giving pain to her kind monitress.

A similar instance of candid confession occurred in E. A. I had given him a bottle of lotion for his eyes, which were weak at the time: he took it home with him; and a day or two afterwards, in reply to my inquiries as to the benefit received, answered in some prevaricating way, so as to make me suppose that he had used the lotion to advantage. Sub- sequently, however, being in mesmeric sleepwaking, he said, quite voluntarily, "There is something that I wish to tell you. In going home the other night I broke that bottle which you gave me. I feared you would be angry if you knew this, and I dared not own it when awake; but now I feel that I did not act rightly."

In the mesmeric state, the character of this sleep- waker presented generally a strong contrast to its wak- ing exhibition. Good talents and a good disposition had, in him, been warped by an unfortunate educa- tion; and, young as he was, he had imbibed at Paris certain infidel opinions, of the worst kind, which he scarcely studied to conceal. I asked him once, in his waking state, what he thought became of us after death; and his answer was, "Dès qu'on est mort, on n'est plus rien du tout."

This extreme ignorance on most subjects was ac- companied by a vain belief that he knew a good deal; and if one stated to him the commonest facts of phi-

losophy, (the distance of the sun from the earth, for
example,) he suspected a design of playing upon his
credulity, and entrenched himself in absolute unbelief. In sleepwaking all this was changed. His
ideas of the mind were correct, and singularly opposed to the material views he took of all questions
when in the waking state. He once chided me for
calling the soul " une *chose* ; " and said, " Ce n'est
pas une *chose*, — c'est une pensee. " " Can the soul
ever die ? " I asked. " Certainly not. It is the soul,
which is the only true existence, and which gives
existence to all we apprehend." " Whence came the
soul ? " " From God, who, by his thoughts, created
the universe." His words were, " L'âme provient de
Dieu, qui a crée l'univers par sa pensée. " " Is there
a future punishment for evil-doers ? " Undoubtedly,
a great one." " In what will it consist ? " " In seeing
themselves as they are, and God as he is."

On another occasion, I mesmerised E. A., when a
lady of great talents and feeling, and an author, well
known to English literature, were present. The latter
was suffering under a severe domestic affliction. He
had recently lost a beloved daughter; and the tone of
mind, induced by that bereavement, naturally inclined
him to question the sleepwaker on subjects relative
to a future state. In order that Mr. —— might
speak with the greater freedom, I placed him "en
rapport" as it is called, with E. A., and took but little
part in the conversation that ensued. The conversation itself I cannot accurately detail ; but the general impression that it left upon my mind can never

be effaced. The sleepwaker rose into eloquence which seemed unearthly. It was simple — it was beautiful — it was like an inspiration. He spoke of the never-dying nature of the soul; of its ransomed beatitude; of its progress through various eras of existence, during which he asserted (for here I remember his very words), " Elle conserve la memoire du passé, et des amitiés faites sur la terre; et elle a l'envie de revoir ceux qu'elle a cheris autrefois. Tout le bien de l'âme s'en va avec elle, et dure après la mort; et les justes qui se sont pleures ici bas seront reunis devant Dieu." Every one present was affected — some even to tears. It was, indeed, beautiful to see the young prophet — whose countenance had retaken an expression of candour and of childish innocence — speaking so calmly the words of holiness and of comfort, and the older listener humbly stooping to drink of the waters of refreshment from so lowly a source.

The same sleepwaker, thoroughly unsentimental in his natural state, seemed always, when mesmerised, to take a pleasure in losing himself in imaginations of another world. Beautiful are the things he has said to me respecting the soul's recognition of those it loved on earth, and of the privilege of departed friends to watch over the objects of their solicitude while toiling through the pilgrimage of life ; but were I particularly to record these *speculations*, as they would be called, I should probably be deemed a visionary, or branded as an enthusiast. It is enough to say that, under mesmeric sleepwaking, all the hard incredulity which characterised E. A. when awake was gone. His wilfulness was become submission; his pride, humility; and, in precise proportion as he seemed to know more, he appeared to esteem himself less. Often would he regret the errors of his waking hours, and speak of his natural state as of an existence apart. Often would he exclaim, in sleepwaking, " How I wish I could always see things as I do now ! " There is not a person who saw him in the mesmeric state but remarked the change for the better that his physiognomy underwent. His affections, also, were enlarged. Egotistical in general, and displaying but little sensibility, he, in the mesmeric state, showed all the warmth of a kind-hearted nature. Shortly before leaving me I mesmerised him. Immediately on passing into sleepwaking his countenance assumed an expression of the deepest sorrow, and he seemed scarcely able to speak. When asked the cause of his sadness, he said, " I am going away : how deeply I feel it ! " Restored to his waking state, he laughed, and talked, and seemed as unconcerned as usual.

That in the mesmeric state the mind's sensibility is exalted and refined, I had also a proof, when one day I was reading to Mademoiselle M—— (then in sleepwaking) Wordsworth's touching poem of Lucy Gray. Utterly insensible to poetry in her waking state, she at that time felt all its charms : her countenance varied with emotion ; she watched the progress of the little story with the deepest interest ; and, when I arrived at the stanza —

I

" The storm came on before its time,
    She wander'd up and down;
And many a hill did Lucy climb,
    But never reached the town,"

she burst into a passion of tears, and cried out,
" Oh ! cease, cease ! I can bear it no longer."
Yet, that this increased sensibility was regulated
by considerable self-control appeared manifestly on
a certain occasion, when some one, placed "en rapport"
with her in sleepwaking, made an unfortunate allusion to the death of a person to whom Mademoiselle
M. had been strongly attached. She wept much,
but, seeing me uneasy, said, " Do not mind, sir, I
will control myself ; " which she accordingly did, and,
by a moral effort, she resumed her tranquillity.

Another feature of the moral state in mesmerism
is a tendency to replace the conventional forms of
the world by a frank exhibition of general good will
and cordial regard to all men. Justinus Kerner (in
his " Leherin von Prevorst, " published at Stutgart
and Tubingen, 1832,) relates that the sleepwaker, to
whom he has given the above high-sounding appellation, would never, in the mesmeric state, address
any one in the third person plural, which is the cere-
monial manner of speaking in Germany; but always
in the second person singular, which the Germans
consider as expressive of intimacy and affection. I
have observed something of the same kind in those
sleepwakers who have come under my notice. E. A.
was once thrown into the mesmeric state in the presence of a gentleman whom, as his superior in age and

station in society, he regarded with the greatest respect. Having written, as requested, some music for this gentleman, he inscribed it thus : — " A mon cher V. O." It is curious to compare this familiar dedication with those which were written in his waking state, where every title was given its due weight, and nothing was presented save " avec les hommages les plus respectueuses."

A state of mind so simple, so religious, so tender, yet so pure, is in itself a refutation of the charge of immorality, which they who lack the charity that hopes and believes the best have attempted to bring against mesmerism. If to this be added the absolute deadness of corporeal sensation which I have shown to accompany mesmeric sleepwaking, there will be a manifest absurdity in the supposition that it is (as some one called it) a mere " voluptuous juggle," — an affair of the senses and of sexual feeling.

The attraction towards the mesmeriser, testified by the patient, which has, perhaps, contributed to give birth to this monstrous idea, is of a nature totally distinct from the promptings of passion. If compared to any love, it must be likened to self-love ; for it seems to result from that identification of the vital and nervous system of the two parties of which I have already given many examples, and which admits of still further illustration. For instance, a gentleman, whom I was mesmerising for the first time, asked me, when I had laid my hand on his forehead, whether he felt his own hand or mine placed there, declaring himself wholly unable to decide. Anna

M——, when I have been suffering from a sensation of cold, has not only complained of being chilly, but has said, " Will you warm yourself *for me* ?" Once when (unknown to her) I had a blister on my side, I observed that, during sleepwaking, she constantly carried her hand to her own side. I asked her why she did this, and she said, " There is something there that pains me, as if the skin were torn off." At another time, when I was thirsty, a sleepwaker complained also of thirst. I offered to give her some water, but she said, " No; you must drink *for me.*" My thirst being quenched, the sleepwaker's uneasiness was removed. The patient, also, during mesmerism, adopts, to a certain extent, the mesmeriser's likings and dislikings. E. A——, in his natural state, was very fond of the smell of heliotrope, which I found too sickly for my taste; but, in sleepwaking, when, through me, he was made to smell it, he found the odour unpleasing.

When we consider these and other proofs displayed by sleepwakers of sensitive and motive sympathy with their mesmeriser; when we reflect that they are actually heedless of injuries inflicted on themselves, but tremblingly alive to all that he is made to suffer; we may well imagine that he stands to them in a very peculiar and vital relation : nor can it seem wonderful that, when severed from him, they should acknowledge a schism in their being, and seem out of all unity with themselves. Besides, several circumstances prove how remote from sexual feeling is the one in question. In the first place, it is exhibited equally by every mesmerised person, without respect of age, sex, or character. The cold and the stubborn are subjugated by it as effectually as the warm and the yielding : the pride of talent exempts not from this despotism, neither does the simple innocence of childhood. Never have I seen its force more strongly exemplified than in the case of a sister of Anna M——, a child of nine or ten years of age. Having once left her on a sofa in the mesmeric state, while I went to take some tea at a table which stood near, I heard, after I had been away about five minutes, low stifled sobs proceeding from my little patient. I hastened to her, and found her crying. Being asked by me the cause of this distress, she replied, " Because you staid so long away. It makes me suffer so much."

Again, the attraction manifested by patients towards their mesmeriser is not in the least modified by circumstances, or by the relation of the parties to each other ; and it is openly exhibited by those whom I have shown to be in a state of high moral feeling.

Nothing can be more evident than that it is an instinct, not a passion : the springs of life are touched, and the powerful impulse of self-preservation is set in play. So, also, the repulsion from all others than the mesmeriser is but a measure of the attractive force which draws the patient there, where he exists even more than in himself. Illustrative of this physical necessity, I may bring forward a circumstance which was related to me by an eye-witness, whose word I have never had reason to doubt. A young man, mesmerised for the first time, became unwell under the extraordinary stimulus of mesmerism, which sometimes, if unskilfully applied, does affect persons unpleasantly. His mesmeriser was, in fact, inexperienced in his art; and, becoming alarmed at the result of his operations, attempted to quit the room. The consequences of this injudicious conduct were fearful. The patient became furious, and actually tore his own clothes and the hangings of the apartment. Had his mesmeriser succeeded in leaving him, he might have suffered seriously; but, by good fortune, a more experienced mesmeriser, who was present, forced back the inexpert practitioner, and instructed him how to allay the tempest of mental and bodily agitation which he had so unwittingly provoked. This may serve as a general type of the kind of feeling exhibited by patients towards their mesmerisers; and with this the language of sleepwakers themselves is in exact accordance. A gentleman, who saw E. A—— mesmerised by me, said to him, " Vous aimer bien d'être avec Monsieur." To which the sleepwaker replied, " Oui, mais pas toujours comme a present. D'autrefois, je peux le quitter, et je ne sens rien; maintenant je ne pouvais pas le quitter, quand même je le voudrais." In the same manner, Anna M——, who was exceedingly fond of my wife, could not quit me to go to her, when in the mesmeric state. Being good-humouredly rallied upon this, she said, " You know, dear lady, how much I love you, and that the feeling I have for my mesmeriser is sincere respect. That which keeps me near him now is quite different from any affection: it is not that I would not wish to come to you, but absolutely I cannot, unless my mesmeriser comes too." When seated between us both she declared herself quite happy. Nevertheless, it was necessary even for my wife to approach her, when mesmerised, with caution. Touching her once suddenly, without having been placed properly "en rapport" with her, she produced in the patient a violent fit of shivering.

But it may still be objected, that the mesmeriser, if so disposed, may make a wrong use of the influence which he possesses over the patient. Possibly : for, as I have once before remarked, all power is capable of abuse. A bad physician may employ his knowledge of drugs to the worst purposes; but are we therefore to have no physicians? And is mesmerism, because dangerous in evil hands, not to be consigned to any hands at all? There are some who wrest the Scriptures to their own destruction and that of others; but shall the Scriptures not be preached? A pursuit, too, like mesmerism, which, so far from conducting to gain or glory, subjects its votaries to a species of martyrdom, holds out but few temptations to the base minded. Besides, the fault rests with mesmeric patients themselves, if they give up their powers of self-control to evil, or even doubtful, guidance, or, indeed, to any foreign guidance whatever, unless before proper witnesses.

Exercised judiciously, there is not a doubt that mesmerism might be made an important instrument

of moral good to man, since it not only tends naturally to elevate the mind above sensual desires and
material objects for the time being, but even after
its immediate influence has passed away. A mesmeriser can always so strengthen the virtuous tendencies developed by his patient in sleepwaking, as to
prolong them, as it were, into the waking state. Often
has E. A —— entreated me, when I have given him
good advice in the mesmeric sleep, to impress it in
such a manner upon his mind as that it might influence him, even though imperceptibly, when awake.
Moreover, it is a curious fact that, if the mesmeriser
tells his patient to abstain from any thing at a certain
time — even from that of which he may be most
fond — the latter acts upon the injunction, in his
natural state, without being aware of the springs
that impel his conduct. To what beneficial uses this
peculiar influence might be rendered subservient I
need not suggest; nor how effectually it might wean
persons from bad habits, by rendering them positively
averse to the sins that most easily beset them.

That a proportionable increase of the intellectual
powers should accompany so marked a developement
of the moral will not seem surprising, if we reflect
that every faculty of our being plays, as it were, into
the hands of the other; and that a rise in moral
feeling almost presupposes a rise in intellectual dignity. If it be granted that "want of decency is want
of sense," it may also be conceded that a want of
moral perceptions generally is the produce and
accompaniment of a degraded intellect. The in-

creased sincerity which persons manifest in the mesmeric state would alone pronounce it the parent of a quickened reason. They perceive all the irrationality of falsehood. Other proofs, however, of the sleepwakers advance in intellect are by no means wanting.

He can discuss subjects which, in the waking state, are far beyond the scope of his capacity, and solve questions which, at other times, are to him as an unknown tongue. I observed this especially in the case of E. A——. His habitual confusion of thought was changed, during sleepwaking, into justness of apprehension; to indulge in abstract speculations seemed almost a feature of his condition; and he displayed, in this developement of his nature, a peculiar acuteness of remark which looked like intuitive sagacity. He apprehended with ease, and learnt with quickness. A hint was sufficient to put him in possession of a subject.

But it is especially by the improved condition of the memory, in the mesmeric state, that the general strengthening of the intellectual powers may be estimated. For is not, in truth, memory the life and breath of the mind, without whose quickening inspiration all the elements of thinking were but as so much dry dust? Is it not our identity which links every event wherein we have been concerned, every fragment of knowledge and of moral feeling which has come to us either from the external world or the universe of our own minds, into one harmonious and intelligible whole? It is true that memory, without the reasoning faculties, were but a vain gift; but it is equally true that, without memory, all the faculties in which we glory were entirely worthless. Their simplest exercise presupposes its abiding presence. It is the measure of our attention, as attention itself is the measure of the mind's strength or weakness.

That this important power is largely developed under mesmeric sleepwaking is a fact to which particular attention should be given; since it explains, in a natural way, certain phenomena which too many of the friends of mesmerism wonder over as miraculous, and which, consequently, too many of its adversaries reject as false. How much injury this presentation of startling circumstances, unaccompanied by any rational solution, has caused to mesmerism may be estimated by a comparison of the reception given to a fact, when rationally set forth by a philosopher, with that which it receives when wildly stated by an enthusiast. Many of the stories told by the judicious, and frankly accepted by the world in general, are every whit as wonderful as the tales of mesmeric *clair-voyance* which so shock the incredulous;—but then they are otherwise presented, and charm us by being seen through the clear glass of nature, instead of coming to us coloured by the Claude Lorraine medium of the imagination. A proof of this assertion may be found in the high estimation which Dr. Abercrombie's work on the Intellectual Powers so justly enjoys. The work itself is full of interesting and remarkable anecdotes, many of them especially relating to auto-sleepwaking —yet who dreams of doubting these? And why?

They are naturally and philosophically treated. For instance : — the author relates that a girl, who was constitutionally a sleepwalker, ignorant, and of the lowest rank, " has been known to conjugate correctly Latin verbs, *which she had probably heard in the school room of the family ;* and she was once heard to speak several sentences very correctly in French ; *at the same time stating that she heard them from a foreign gentleman, whom she had met accidentally in a shop.*" Now, conceive the story to have been told without the sentences which I have marked by italics, and the mere workings of memory are forthwith magnified into a miraculous gift of tongues, which we regard with quite other feelings than the present cautious and instructive statement. We cannot, then, be too guarded in our views of the faculties displayed by sleepwakers. The least hint of a thing, previously heard and forgotten while awake, will serve them, in the mesmeric state, as a clue to knowledge that shall seem supernatural ; but a little investigation will often clear up the apparent miracle ; and the principle, once ascertained, may be extended even to cases which baffle our most accurate research.

The memory of Anna M—— was much developed during mesmeric sleepwaking. All the reminiscences of her childish years seemed to recur to her mind ; and thence she was enabled to show an acquaintance with the past histories of the inhabitants of her native town, which, in primitive times, would have been by no means to her advantage. Under James the First she would infallibly have been burnt for a witch.

I 6

Often, in her mesmeric state, she recalled to the minds of persons, who had perhaps forgotten that they had ever seen her, particular circumstances, with which they imagined none but themselves could be acquainted. On one occasion especially, I remember a person entering the room while she was in mesmeric sleepwaking and had her eyes perfectly closed. His name was not mentioned; but, having been placed "en rapport" with him, she recognised him, *by the voice alone*, as an individual whom she had known in her very early childhood, but had not met with since. She named him, and brought back to his recollection many minute particulars relating to his family and to himself. On awaking, she was much surprised to see another person added to our party, and could by no effort of memory (even after he had spoken) make him appear other than a perfect stranger to her. It was not until he had declared himself, and, on his part, recalled to her sundry circumstances, that she could remember ever to have seen him. Yet, even then, the greater part of the particulars, with which in mesmerism she seemed so fully acquainted, could by no means be restored to her remembrance.

Another time, the same sleepwaker, being asked to write some Flemish poetry from memory, noted down with a pencil two lines, which, on awaking (though we told her the first words of the distich), she could not repeat. Being shown the lines, she at length recognised, by their general sense, something which she had heard as a child. In mesmeric sleepwaking, Anna M—— displayed a greater aptitude to learn than in her waking state. I had only recited to her a little poem of Goldsmith's, beginning, " Oh, memory ! thou fond deceiver," twice over, when she was already able to repeat it correctly.

E. A—— manifested a similar increase of mental retentiveness, more remarkable, perhaps, inasmuch as his memory (more from neglect, I imagine, than from natural incapacity) was singularly defective. We have tried him in mesmeric sleepwaking with the hardest phrases and technical expressions which the English tongue (of which he knew very little) could furnish ; and, on one occasion, a gentleman, who had travelled in the East, supplied him with a long and strange Turkish word, which he caught directly with the greatest ease.

This sleepwaker played beautifully on the flute, and was accustomed often to improvise upon that instrument with all the musical genius that he possessed ; but the charming strain, once uttered, was lost for ever. He could not repeat one of his extempore compositions. One day, I had been forced to reprove him for some fault, when, taking his flute, he poured forth some melancholy notes which seemed to express regret and penitence. I begged him to repeat the touching air, but it had wholly passed from his remembrance. Subsequently, when he was in mesmeric sleepwaking, it occurred to me to ask him if he could then write down the composition that so much pleased me. Instantly he seized music-paper and a pen, and wrote down the air which I here subjoin, and which, as far as my own memory can deter-

mine, is precisely the same as that which I had heard
two days before.

POUR LA FLUTE.

Another proof of increased memory given by this sleepwaker was his comprehending and speaking English (which otherwise he was remarkably slow to learn) far better in the mesmeric than in the natural state.

A circumstance to be remarked, as showing a connection between the mesmeric and the common sleep, is this — sleepwakers will remember those nocturnal dreams which they can by no means recall when waking. I once asked Anna M—— if she could recollect any thing in sleepwaking which, out of it, she had forgotten. She replied, " Certainly, many things; and just now I clearly remember something which I wished to recollect when I was awake, but could not. It is a dream which I had on Sunday night. (She said this on Wednesday, 14th February, 1838.) I screamed out in my sleep, and alarmed my mother; but I could not call to mind what I had been dreaming about: now, however, I know what it was. I thought that a very frightful old woman entered the room : she was dressed in scarlet, and had a white fillet bound across her forehead. She came slowly in, and, coming up to me in bed, laid her hand upon my chest. I tried to scream, but could not : at last, however, I screamed, and woke my mother." Afterwards, when the sleepwaker had returned to her natural state, she was questioned about the dream ; but it had again past from her memory. Wishing to establish the circumstance by as much evidence as possible, I took an opportunity of speaking to her mother on the subject, when the latter corroborated all that Anna had affirmed, and added that her daughter was in the habit of relating her dreams to her family at breakfast; but of this particular one she had been able to give no account.

E. A. could also, in sleepwaking, recall the visions of the night, which had gone from him as the dream from the Babylonian monarch. Moreover, he had been, as a child, a natural sleepwaker, and he assured me, when in the mesmeric state, that all the minutest circumstances of his early sleepwaking recurred to him. When asked if the mind ever ceased to think in slumber, he replied, " Never for an instant : the soul is wise, and learns much during sleep. It reflects on all it has seen and heard, and profits more by this than in the daytime. We cannot, indeed, be aware of this, because we are in a different state when waking, and forget what we think of by night. Yet, he continued, have you not remarked that, if you read a thing over-night, you remember it better on awaking the next morning? This is the fruit of the soul's labour (c'est le fruit du travail de l'âme), though we do not know whence it comes." Asked again about dreams, he replied, " In general they are a kind of recreation of the soul, an exercise of its inventive faculty ; but we are only acquainted with the results. As a carpenter takes a tree and saws it into boards, and then from those boards constructs something which he had previously determined, so the soul arranges the mechanism of the brain to form a dream; — and the dream itself, which we imagine to last a long time, passes, in fact, with inconceivable rapidity. It only occupies the moment immediately previous to our waking. Other dreams, again, are, as it were, forced upon the soul by something disordered in the brain: but they all agree in this — they immediately precede our waking."

I relate the above, as curiously consonant with a circumstance in the life of Lavalette. He had in prison a horrible dream, in which he saw an infernal procession pass by his cell. Whole armies seemed to defile before him ; but the riders were all skeletons, and the horses, stripped of their skins, presented a mass of raw and bleeding flesh. A clanking of chains accompanied the terrific vision, which seemed to be of infinite duration. He was awakened by the gaoler entering his dungeon, when he found that the dream, which had been an " eternity to thought," was, in reality, the impression of a moment, caused by the dropping of the chains in unfastening his door.

The last phenomenon of memory, which may here be mentioned as regarding the sleepwaker alone, is that, when the mesmeric sleep is perfect, the subsequent oblivion is perfect too. Of the truth of this fact, it has been generally asserted that there is no other assurance than the allegations of the sleepwakers themselves. I cannot concede this point. Place a mesmeric patient under circumstances different to those in which he went to sleep, but without hinting that you expect any display of surprise on his part, and he must be shrewd indeed, as well as a consumate actor, if he can counterfeit to the life such an astonishment on waking as he must exhibit naturally, supposing him really to have forgotten the events of his sleepwaking. Mademoiselle M——, under circumstances the more remarkable, because not at all arranged for the specific purpose, was once evidently almost alarmed into illness by finding herself in a different apartment (whither, it may be remembered, she had been brought to see an invalid lady) from that in which she had been mesmerised. She started, shuddered, looked wildly around her, exclaiming, " Good God ! what is this ?  How did I come here ?  This is too much ; " and was thrown into a fit of trembling which we had much difficulty to allay.

Again, supposing a sleepwaker to be under the same roof with yourself, his mesmeriser, and exposed to your constant observation, the chances are that, if he feigns, he will sometimes be caught tripping. A deceiver cannot be always on his guard ; and what a power of memory does it presuppose in the fictitious sleepwaker to imagine that he can arrange and separate in his own mind those circumstances of which he may or may not speak ! Passing a great portion of his time in sleepwaking, will it not be difficult, nay impossible, for him never to advert to that which has engaged his attention so much and has formed, indeed, half his life ? I conceive, then, that when I assert that, during four months, E. A. was staying in my house, and constantly mesmerised by myself, with the eyes of a whole family placed in guard upon him, yet that never, in a single instance, he was known to mingle his mesmeric with his ordinary existence by word or action, I do, in fact, show that the assertions of sleepwakers, are not the sole evidence that they actually forget the occurrences of their mesmeric state. We have tried in every possible way to put this to the test; we have plotted against the sleepwaker, as if, indeed, we had wished to convict him of imposition on this point; we have laid traps for him in conversation; but the most searching ordeal only served to convince us of his sincerity. Be it observed, moreover, that, in his mesmeric state, I conversed with him frequently on subjects which he was incapable of discussing in his waking hours, and that, after the sleepwaking was past, I found him, with respect to the same subjects, as unenlightened as ever. Now I know from his character that, had he retained any little piece of information acquired in his mesmeric state, he could not have resisted the temptation of displaying it, at some time or other, for our edification and astonishment. What adds to the probability of this is that, his memory being in some respects defective, he would often retail to persons the very story, or the very bit of knowledge, which he had learned from themselves, with a perfect unconsciousness that he was only giving back a man his own. Besides, he talked much and reflected little; so that, had he wished to feign oblivion of his mesmeric state, every chance was against the success of his attempt. Surely all these proofs presumptive (on which I have dwelt the more, because arguments of this nature seem hitherto to have been withheld) amount nearly to proof positive that sleepwaking has its own memory and its own consciousness, which do not overpass the boundaries of its duration.

The fact that, under peculiar conditions, we may enjoy the exercise of all our faculties, yet retain no recollection of having done so, may be applied, as I imagine, to the solution of many curious questions in metaphysics — such as that one of Locke's, "Whether the soul thinks always?" which that great philosopher has decided in the negative, on the ground that after-consciousness is the sole testimony we can have of the mind's activity; and that, since we are by no means conscious that we think always, we ought not to assume that we do think always. But in some cases, and most especially in that of mesmeric sleepwaking, we are forced to admit, for the activity of our minds, an evidence which Locke has omitted to notice; namely, the testimony of our fellow-beings (a testimony, indeed, on which we build half our knowledge), assuring us that the state which, once ended, appears a blank to us, was marked by energy and activity of the highest order. Hence, setting aside our own want of memory of our past intelligence, as an inadequate proof that such intelligence never existed, we manifestly arrive at a presumption that what we have forgotten in one case we may have forgotten in many more; and that the mind is probably always active, though we cannot always assure ourselves of its activity. Such a probability will be peculiarly grateful to those who deem that ideas are constitutive of the mind, and that Locke was fundamentally wrong when he called " thinking the action and not the essence of the soul ;" being so far misled, by a false analogy between things material and things spiritual, as to conceive " that the perception of ideas is to the soul what motion is to the body." They who would not

> " lose,
> Though full of pain, this intellectual being,
> These thoughts that wander through eternity,"

yet who believe that to cease to think is to cease to be, must rejoice in all that tends to prove the continuity of thought, its unceasing and inexpressible activity. It is also remarkable that the state of mesmeric sleepwaking, though one of oblivion, leaves always, as I have observed, a sort of agreeable savour on the mind, which renders persons of every age, sex, and temperament, desirous to enter it again. It has appeared to me as if the soul rejoiced in it, as in an element to which it was " native and endued," and eagerly pursued it, as a foretaste of its development through eternity. We might also enter upon some curious speculations, suggested by this severing of existences ; this forgetfulness in one state of that which we have done in another. It is an answer to our inquiry, whether Lazarus, when restored to the body, brought back with him the secrets of another world ; it leads us to conjecture that they, who have

seemed dead, yet have been recalled to this morbid
existence, may have visited the world of shadows,
although they know it not.* It makes the river Lethe
seem no fable, and half realises to us the inspired
words of the poet: —

> " Our birth is but a sleep and a forgetting ;
> The soul, that rises with us, our life's star,
> Hath had elsewhere its setting,
> And cometh from afar."

In a practically religious point of view, also, the phenomena regarding memory, under mesmeric sleepwaking, are full of instruction.    The prodigious increase in this state of the recollective powers seems to
indicate that, in our normal existence, they are but
partially developed; and that hence the moralist's
suggestion, that all our lives, with all their good and
evil, will, in some future era of being, be spread
before us as an open scroll, is not to be slighted as a
fiction, but to be duly regarded as a probability, of which
we have the presumptive evidence in our own hands.
I proceed to speak of phenomena of memory in the
sleepwaker, resulting from the mesmeriser's influence.

1st. Strange and inexplicable as the fact may
seem, it is undoubtedly true that the mesmeriser has
the power of impressing some things upon the
memory of his patient, in such a manner as that the
latter shall retain them, and them alone, on awaking.

* I have a relation who lay for twenty-four hours in a state
of seeming death, and who was wept for as dead.

Thus, at my desire, E. A. used to remember English words which I had taught him during his sleepwaking. Persons who were present when he was mesmerised would amuse themselves by proposing to him, through me, words the most difficult for a foreigner to pronounce; such, for instance, as "boatswain's whistle," which a gentleman one day gave him as a puzzle, yet which he repeated correctly, and retained after his sleepwaking. Anna M. when mesmerised, if asked to remember anything, used to tap her forehead on either side, just where she had two singular prominencies, and say, "I will do my best to fix it in my brain." After this she always recollected what she had been desired to retain. I did not find, however, this faculty extend to circumstances connected with the unusual mode of sensation apparently opened to sleepwakers. Even if so requested, neither E. A. nor Anna M. could give any account, on awaking, of their perceptions during mesmerism, nor how objects were presented to their minds. The subjects adapted to their after-memory seemed to be chiefly of a verbal nature, such as a few lines of poetry, or the particular day fixed for a subsequent mesmerisation.

This possibility of making sleepwakers remember what they have learned in the mesmeric state (and learned, too, with superior ease and quickness) might, I imagine, be turned to important uses. I have read, though where I have forgotten, of a lady who acquired a new language through mesmeric recollection; and my experience inclines me to believe that such might be the case.

2dly. The mesmeriser, by the expression of his de-sire, can make the patient remember certain things at a particular time, subsequent to sleepwaking, and not till the very moment fixed beforehand for the recurrence of the suspended recollection.

Thus E. A., in mesmerism, was told to remember, at nine the following morning, and not till then, that he should come to fetch one of our party to hear a beautiful mass that was to be performed in a church at Antwerp. The person with whom he lodged, and in whose room he slept, informed me subsequently that, when the clock struck nine, E. A. was in bed, when suddenly he started up, saying, "It is very odd; but I seem to remember that I am to take Mrs. —— to church." He then dressed quickly, and came to fulfil his engagement. I asked him in what manner the idea had recurred to his mind, if it was like the recollection of something heard in a dream? He replied that it did not resemble this in the least, but was a very curious and indescribable sensation, as if a thought flitted past him from he knew not where, and fled again so rapidly that to trace its origin was impossible. In sleepwaking he could not give me a much better account of the phenomenon. It was, he said, a something pre-arranged in the mechan-ism of the brain, and which, like the spring of a watch, was set in motion at a given moment. There was nothing very satisfactory in this explanation. As to the phenomenon itself, I and the members of my family have witnessed it again and again. We have tried it in minute circumstances, difficult of remembrance, such as (on the day succeeding the sleep-waking) to sit in a particular chair, or begin with a particular dish at dinner; to cut a potato, when first taken on the plate, into four pieces, &c. ; and the result was always satisfactory. I have also tried if the circumstances, desired to be remembered, could be recalled to the sleepwaker before the proper hour ; but there was evidently no means of doing so. The very first time I told Anna M. to remember anything at a fixed hour, I desired her to call to mind, at three the following day, that she must ask one of our party to take a walk. After her awaking, some one inadvertently said, " Do you remember any thing about coming to-morrow to walk with us ?" She said, after reflection, " No, indeed, I remember nothing about that; and besides I have an engagement for to-morrow." Soon after three, however, the next day, Anna appeared out of breath, and with a face of wonder, exclaiming, " Do tell me ; was I to come here or not ?" She then related that she had been out to dinner, and that, in the midst of that meal, something seemed to tell her that she had promised to come to us ; and that, such was the force of the impression, she could not help leaving her party and hastening to ascertain the truth from ourselves.

Having now described in what manner the memory is affected in the mesmeric sleep, both as regards its spontaneous exercise and its direction through the mesmeriser, I proceed to state other remarkable peculiarities in the metaphysical condition of mesmeric patients, without adverting to which my view

K

of mesmeric sleepwaking, as a condition of man, would be incomplete.

The peculiarities alluded to regard the action of the whole mind of the sleepwaker, and not of any particular faculty. They may all be ranged under two heads: —

1st, With respect to the patient himself. These are peculiarities of consciousness.

2ndly, Of will, resulting from the patient's relation to the mesmeriser.

The peculiarities of mesmeric consciousness will be treated of in the succeeding book, as being explanatory of many circumstances connected with our subject.

The peculiarities of mesmeric will may be expressed briefly thus: — That lower degree of will which we call volition, and which chiefly regulates the motive functions, is active, but will (in its freedom and absolute sense) is passive, in the sleepwaker, through the influence of the mesmeriser.

The motory adroitness of persons in the mesmeric state proves the first part of my assertion. The second branch of it is susceptible of equal demonstration, as the following reflections must render evident.

Common sleep, it will be granted, is the indication of an abjured will — voluntarily abjured, it is true, in some cases ; but, in others, suffering a forced abdication, as when the sentinel, who knows that, if found slumbering on his post, his very life must pay the penalty, is vanquished by the slumber he would shun. In sleep itself, also, we are no longer our own masters, but at the mercy of our dreams.

" Reason then retires
Into her private cell, when Nature rests.
Oft, in her absence, mimic Fancy wakes
To imitate her; but, misjoining shapes,
Wild work produces oft."

In this state we seem to commit crimes which, waking, we regard with abhorrence; and often the impotence of our will is shown by an utter inability to throw off the yoke that oppresses us. We would wake — but cannot: we strive to cry — but we find no voice : we endeavour to move our limbs — but they remain idle.

If common sleep, then, be so manifestly the index of a suspended will, the mesmeric slumber is not less clearly an indication of a will that is over-ruled and held in abeyance by a force that is external to itself. The proof is simple. The mesmeric sleepwaker cannot awake without the mesmeriser's aid. The sub-jugation of his will is, therefore, even more complete than under the condition of common sleep; for, after certain efforts we often succeed in throwing off the latter when it becomes disagreeable to us, or the casual touch of any one who witnesses our perturbed slumber may recal us to ourselves; but the mesmeric patient can neither throw off his bonds, nor be de-livered from them by any other than the mesmer-iser. I have seen such efforts vainly made to arouse a sleepwaker otherwise than by the mesmeriser, that I have been led to conceive that he might be torn limb from limb (as certain animals may be in certain states of torpor) without awaking. As with the mes-

K 2

meriser his sleep began, so, it appears, by the mesmeriser alone it can end; and thus the peculiar slumber, called mesmeric, is not only the proof of a suspended will, but of a will subdued, and actually held subject, by another.

But it is not only the duration of the mesmeric sleep which proves this predominance of a foreign will : the conduct of the sleepwaker, while it lasts, is an additional warranty that the rudder of his being having changed masters, the whole vessel is also under another's direction. At the command of the mesmeriser he walks when he would naturally repose, and brings his faculties to bear upon any point, at a moment when otherwise he seems powerless to exercise them.

Summing up the evidence, we find the argument run thus : —

Natural sleep is the indication of a suspended will. A sleep that begins and ends by another must be allowed to show a will that is held in abeyance by external agency.

A sleep, during which a person is made to act by another, demonstrates a will that is swayed by that other in precise proportion as the obedience is prompt and absolute.

The state of things resulting from this sway on the one side and submission on the other is most remarkable. We have the phenomenon before us of an existence at once dual and single; for, when the sleepwaker's capacities are acting under the immediate direction of the mesmeriser, the latter may be considered as

MESMERIC SLEEPWAKING.      197

making up together with him the complement of one
full being, whereof the mesmeriser supplies the willing
and the conscious portion, and the patient the intellectual part. The one impels, the other obeys impulsion : the one designs, the other executes : the
one sets in motion a machine, (and what a machine !
the mind of man, with all its complicated marvels ! )
the other is the machine itself, instinct with life, as was
the fiery car of Kehama, innately active, yet guided
by the volition of another !

This is no poetic fiction. The facts are before us.
The volition of a sleepwaker is swayed in an arbitrary
fashion, fully to conceive which, we have only to consider the difference between the mesmeric subjugation
of the will, and any other that occurs in our ordinary
state of existence. I may tell my servant to do such
or such a thing. When he obeys, it may be said, in
a certain sense, tnat his will is subject to mine ; yet,
all the time, he does actually retain the functions of
his own will in perfection. In truth, he wills as much as
I do, and on the same principles : good to be sought,
uneasiness to be shunned, is the efficient cause of
volition in either case. I command ; his mind, with
rapidity, strikes a balance with itself, and finds it
better to obey. If, however, the balance should incline to the other side, my servant has it in his power
to disobey, and to manifest that his will is in free and
unimpaired operation. Moreover, he may obey me
from fear, and not by any means from desire, or
identity of his will with mine. I sway the outward
and not the inward man. But, in mesmerism,

K 3

it is not only the *manifestation*, but the very *function* of the will which is dominated. It cannot, if it would, rise to freedom: it is under a more than moral restraint. Its motive force, desire, is evidently changed, since, as I have shown, the sleepwaker adopts, *pro tempore*, the likings and dislikings of the mesmeriser. Hence, when the patient is engaged in an act of obedience, there is an identi·ficaticn between the mesmeriser's will and his own; a phenomenon which can never obtain in the case of simple obedience. The inner springs are touched, and not the mere mechanism swayed.

Such a view of the state of mesmeric sleepwaking may seem unpleasing, and incompatible with its being a rise in man's condition; but, let it be remembered, we do not rob the soul of any one faculty, we only change its mode of action, and that mode of action has its own undeviating law. In order to sleep, we must abdicate our will; yet, in order to enjoy intellectual activity, it is plain that we must *substitute* a presiding will; for it is the absence of this which makes slumber a state of disorganised and inconsequent thoughts. Thus the terms are absolute. If we desire to be freed from the thrall of the senses, yet to retain the mental faculties, we must abjure our own will, yet find an intellectual substitute.

Again, we may have observed that the superior activity of any one power of the mind greatly depends on the non-action of certain others. With some individuals fancy revels at the expense of reason; with others, the cultivation of the exact sciences deadens the imagination ; and this inequality of power results naturally from the imperfection of our nature. Were we strong throughout, we should be as gods. But, in mesmeric sleepwaking, the question is not whether we shall cancel one of our faculties in order to increase another ; but whether we shall, in a degree, change their directive force, in order to obtain a general and immense expansion of them all. Large advantages demand large sacrifices ; and the price to be paid for the increase of mental powers and of sensitive capabilities, which attends upon the state of mesmeric sleepwaking, is a certain forfeiture of the will. Does this seem a hard condition ? How many who have suffered from self-direction would be glad to place the government of their being in other hands ! How many a one, who " has felt the weight of too much liberty," would gladly cease to be

> " Lord of himself — that heritage of woe —
> That doubtful empire, which the human breast
> But holds to rob the heart within of rest."

This, at least, is my view of the limitations of the will in mesmeric sleepwaking; it may be accepted or rejected, but the facts, on which it is founded, remain the same.

The power which the mesmeriser has over his patient is, indeed, as great as it is undeniable, and involves, in my apprehension, an immense responsibility on the part of the former. Should he direct the patient's attention to frivolous or evil things, he

K 4

might do harm, for which he would be justly answerable. But, fortunately, as I have before remarked,
there is, in mesmeric sleepwaking, a natural elevation
of the mind above what is base and sensual; and the
powers that appertain to reason, though the subjects
on which they shall exercise themselves may be much
at the mesmeriser's disposal, are neither to be clouded
nor controlled. Nor is this to be wondered at. The
mesmeriser can hardly be expected to command in
another that which he can by no means dispose of in
himself. In him, in every one of us, Reason, though
too often we would quench her voice, will speak as
loudly as conscience, which, indeed, is but a more
rapid reason; and he who takes the evil path beholds,
and reluctantly approves, the better way. This shows
a godlike capability in reason, apart from volition:
it seems less to belong to ourselves than to be heaven
speaking in us; and, in proportion as we obey its
dictates, we are exalted in the scale of creation. Its
peculiar luminousness, then, under the mesmeric conditions, is at once a safeguard to the sleepwaker, and
a clear record of the most valuable birthright of man.
Many persons, as I am aware, have disliked the idea
of being mesmerised, from the fear of making wrong
revelations, when deprived of their usual consciousness. But this is a vain apprehension. It is true
that mesmeric patients act from impulse, but then their
impulses are good. They will say severe but whole-
some truths to persons' faces; but their instinctive
sense of right forbids them to speak ill of any one
behind his back. Anna M——, with all her know-

ledge of circumstances relating to others, never sea-
soned her talk with scandalous disclosures, never
uttered a word that could injure a human being ; and
once, being pressed to relate something which might
have compromised another person, not only kept
silence on that head, but reproved the curiosity of
the querist.

Before proceeding to consider the mind under
mesmerism, in its relation to external things, it
seems necessary to notice a class of its intelli-
gences which may be called instinctive, and which,
as seeming to occupy the middle ground between
mind and matter, may properly claim our consideration in this part of our inquiry. Instinct being
the very reverse of artificial acquirement, a high
degree of the one is, under ordinary circumstances,
incompatible with the existence of the other. In
proportion as we are educated we recede from our
intuitive perceptions. But, as in many instances, so
in this, mesmeric sleepwaking reconciles two seeming
opposites, which co-exist not in any other known
state of being. Retaining the advantages acquired
by art, the mesmeric patient is restored to those
which, in a more primitive condition, we might in-
herit from nature.

The instinctive knowledge of sleepwakers appears
to be chiefly

 1. Of remedies.
 2. Of all that relates to the mesmeric state.
 3. Of the elapse of time.

*First,* The instinct of remedies, evinced by sleepwakers, has been frequently remarked and illustrated. As I have never studied mesmerism in a medical point of view, I have naturally had but few opportunities of verifying allegations to this effect. The following instances, however, of the mesmeric powers of prescribing, fell under my notice, and will, perhaps, be the less suspected by the incredulous, as having presented themselves unsought to one who is not of the medical profession.

The first time that I succeeded in mesmerising E. A——, a gentleman present told me to ask him, while in sleepwaking, what remedy a lady of their mutual acquaintance should apply for the cure of a chronic sore throat with which she had been for some time afflicted. The question being put to the sleepwaker, he immediately replied, " Mademoiselle De —— should, on going to bed every night, envelope her throat in a linseed poultice, which should be kept moist the whole night through." He then added some very particular directions about the making and applying of the poultice, &c. When the reader reflects that this piece of medical advice came from a lively boy of fifteen, who was never ill in his life, and who hated every thing appertaining to disorders and their remedies, he will, I think, experience somewhat of the surprise with which I listened to the young sleepwaker. Conceiving that he might not be original in the remedy which he had proposed, I asked the gentleman, who had told me to consult him, whether Mademoiselle De —— had been recommended any thing of the kind. He assured me

that such had not been the case, but that she had
tried many other things with no effect.    When E. A.
awoke and we told him of his medical exploits, he
was the first to laugh at his own oddity: however, it
was an oddity at which Mademoiselle De ——— had
reason to rejoice; for the remedy prescribed for her
in sleepwaking removed the painful ailment from
which she had so long suffered.

The above circumstance gave me more confidence
in the remedial instincts of sleepwakers, and induced
me, on one occasion, to ask a young lady whom I
had mesmerised what measures she should recommend me to take for subduing a pain in my side, to
which I had long been subject, in consequence of a
blow received there in a fall from horseback.    The
sleepwaker, without hesitation, told me that I ought
to wear a perpetual blister on the part for some time,
and gave the minutest directions about the size, application, &c. of the irritant.    Yet I did not have
recourse to it until I had consulted a medical man,
who assured me that I could do no better than follow
Miss ———'s advice, and that, had the lady in question
taken out a diploma, she could not have counselled
me more wisely.    Yet the fair doctress was a shy and
delicate girl of sixteen, who, when awake, was shocked
at herself for having touched on medical matters in
her sleepwaking.    The application which she recommended me was of great service, and subdued the
chronic pain from which I had suffered incessantly
for many months.

Under the head of Case X., mentioned in the

preceding book, I have related that I mesmerised a
gentleman who was not in good health; and I promised to detail some additional particulars respecting
him. I now redeem that pledge. Mr. W. had long
been in what is called an ailing state. Without being precisely ill, he felt generally incommoded. He
had no idea whatever of the cause of his indisposition,
and he had consulted no one respecting it; having, as
he said, a particular dislike to medicines of all kinds.
While he was sleeping mesmerically, it occurred to
me to ask him whether he was more cognisant of his
malady than when in his natural state. He replied
directly, " Yes. I see now exactly what is the matter
with me; and I am happy to say that there is no
organic complaint. My disorder is a weakness of
the lower intestine, which proceeds from sedentary
habits, and which can only be cured by alterative
measures." He then prescribed for himself a course
of the waters and baths of Aix-la-Chapelle, and a
certain regimen (which I carefully noted); and finally
pronounced with decision that, in strictly following
these rules, he should be radically cured before the
end of the year. When the patient was awake, I told
him all that he had been saying in sleepwaking, at
which he was much astonished; and assured me that
the idea of taking baths, or drinking the waters of
Aix, had never entered his head while in his natural
state. I urged him to pursue the curative plan which
he had traced out for himself. This he promised to
do; and last autumn (I had mesmerised him in the
spring of the same year) I received a letter from

him, announcing that, by following my advice, or rather his own, he was nearly restored to health.

*Secondly,* The instinct of sleepwakers, relating to all that concerns the mesmeric state, is one of the first that seems to awaken within them. As soon as they have passed into sleepwaking, they will indicate the best means whereby the mesmeriser may deepen their slumber, soothe them, or remove any slight uneasiness they may feel.

Anna M——, on one occasion, while I was mesmerising some water for her, took the glass from my hand, and said, " You do not employ the best method. I will show you what you must do." She then placed one hand on the top, the other at the bottom, of the tumbler, held it so for a short time, and then again applied her hands to the sides of the glass, in a very methodical manner. Another time, being asked, in sleepwaking, if she could say how a person present ought to be mesmerised for a pain from which he was suffering, she proceeded to show how the operation should be performed, and executed the mesmeric passes as if she had studied them all her life. Curious to see how far this kind of instinct would extend, I, one evening, requested her, while in the mesmeric state herself, to mesmerise a younger sister, who was present. She did not hesitate to do this ; and, placing her sister in a proper position, proceeded to mesmerise her in the most skilful manner. The most singular thing is that she used gestures which were perfectly new to me, but which I at once perceived to be very effective. Frequently she

breathed upon her own hand, and then laid it on her
sister's forehead. In a short time the little girl
began to be affected with sleep, when my sleepwaker
told me that I must aid her to complete the work.
She took one of my hands, and instructed me to
mesmerise with the other, while she herself continued
to mesmerise with one hand. The effect of this
double operation upon our young patient seemed to
be great. Her head was alternately attracted towards
each of her mesmerisers, and she was soon in a profound sleep. Nothing could be more curious than to
see the two sisters sitting opposite each other, both
with their eyes shut, and yet, by the expression of
their countenances, appearing to look at each other.
I now went away to another part of the room, when
Anna M—— got up, and walked to just half-way
between her sister and myself, but she seemed
arrested there by the attraction of contending forces,
and so she remained, turning from me to her sister,
and *vice versa*, as if she knew not to which she should
go, till I put an end to this curious scene by returning to my sleepwaker, and begging her to awake her
sister. This, however, she declared that she could not
do; but added, " The moment that you awake me, my
sister will wake also." The event justified her assertion. Anna M—— seemed to possess, when in the
mesmeric state, a curious power of divining which
persons of the company, who might happen to be present during her mesmerisation, would be capable of
mesmeric sleepwaking. In such particulars as these
I have never known her err. It has also happened

Maclure lith. 5 Wellington St Strand

that, while she was mesmerised, some other person
in the room has mesmerised another patient. At
these times she would indicate, with exact precision,
the degrees of sleep through which the other patient
was passing; and often, when every one else has supposed the slumber to be complete, she has declared it
was not so, and was always right in her judgments.
I should observe that, in her natural state, she had
none of this knowledge; and once, when I asked her
to mesmerise some one, declared her inability to do
any thing of the kind.

*Thirdly.* The appreciation of time displayed by persons in the mesmeric state is remarkable. I have never
known perfect sleepwakers overpass the exact moment
at which they may have been told to remind the mesmeriser to awaken them; and yet, directly after returning to their natural state, they will, if questioned,
make the widest guesses of the time. I extract the
following from our mesmeric register : — " Anna, on
being asked the hour, says it is ten o'clock : the
cathedral bell strikes ten five seconds after. She did
not know the time when the mesmeriser began his
operations, yet can now tell correctly that she has
been sleeping an hour and five minutes. — On being
asked when she would awake, she named half-past
ten ; and, before we had heard the first chime of the
cathedral, she interrupted the conversation suddenly,
saying, with a start, Ah ! you ought to awake me
now." I used to remark that, when this sleepwaker
was *asked* the time, she did not reply so correctly as
above ; but when, as was often the case, she spon-

taneously called out the hour, she never erred.   The
least degree of effort on her part seemed to spoil the
instinct.

I now proceed to consider the mind under mesmerism, in relation to its susceptibility to external things.

When it is remembered that the bodily senses, in
the mesmeric sleep, have been shown to be as dead,
an inquiry into the sensations of sleepwakers may
seem inconsistent in the highest degree ; but we may
perhaps discover,  on reflection, that the apparent
contradiction is only to be laid to the charge of our
own erroneous notions.   Did persons clearly perceive
the simple fact, that sensation is not seated in the
senses, but in the mind, they would be less astonished
at hearing of a means of sensation apart from the
usual action of the senses ; but there are few — very
few—who discern this important truth (which, indeed,
lies at the base of all metaphysical knowledge) with
such perfection as to be wholly free from a certain
confusion of thought respecting it.   Thus, the preparatives of sensation have been studied as sensation
itself; but, as Sir James Mackintosh has admirably
observed,  "All the changes in our organs, which
can be likened to other material phenomena, are
nothing more than *antecedents and pre-requisites of
perception,* bearing not the faintest likeness to it:
as much *outward,* in relation to the thinking principle, as if they occurred in any other part of
matter; and of which the entire comprehension, if it
were attained, would not bring us a step nearer to
the nature of thought."

Nothing can be truer than this ; but our minds and bodies are, in this life, so strongly identified, that it is a hard matter for us sufficiently to discern the huge gulf between mind and matter, and to perceive that we may throw in matter for ever, without filling up the abyss of separation. God alone has solved that dilemma ; and, by attaching the external senses to the soul, has built a bridge for us, arching the chasm across. He has thus brought us into conscious connection with matter and certain of its qualities, primary and secondary ; but, if we could be made aware of the same things in another way, is it not plain that the great end would be equally answered ? If the soul be brought into relationship with matter and its properties, the means are but of small importance, either as regards their nature or their number. So that the visible universe, which is but an expression of God's ideas, be in some measure read and comprehended by us; so that there be a language established between man and his Creator ; the particular types and configurations of that language are of no more consequence than the forms of our written or printed alphabets. This is not the place to dwell further upon these reflections (to be resumed hereafter) ; but the mere smatterer in metaphysics, who has learned that odour, colour, savour, sound, are absolutely only known to us as sensations, and can be said to have no proper existence but in ourselves, should by no means be alarmed at my viewing sensation in its results, rather than in its pre-requisites ; while the truest metaphysician will be the best prepared to ex-

amine, without prejudice, a mode of perception
differing from the normal. There are, however,
weaker minds, for whose sake I must here expose
a fallacy which has prevented many well-meaning
persons from even venturing to think upon mesmerism. It has been argued, with a show of plausibility,
" God gave us our senses for such and such uses:
is it not, then, an impiety to say that we can do as
well without them? Is not their beautiful contrivance, their unrivalled mechanism, thus thrown into
contempt? Shall God be said to create needlessly,
and to endow without wisdom?" Reasoning like
this is an *à priori* condemnation of mesmerism; —
but is it just? A child might refute it. It proceeds
all along upon false assumptions. Does the advocate of mesmerism assert that we can do as well without our senses in the normal state? Certainly not
Does he any more deny that our present senses are
exquisitely adapted to that state in which, being our
natural, the greater proportion of our time must
necessarily be passed? Far from it : he acknowledges
this and far more. He reaches a cause of gratitude
to God which is beyond the scope of the narrow-
minded bigot. For what nobler purpose, he may
ask, could a change in our mode of perception be
permitted, than to show the necessitarian that the
arrangements of the Almighty depend upon the
almighty will alone ; that what is might easily
have been otherwise ; and that infinite wisdom can
vary its resources infinitely? Believing, thus, that
to witness a *change* in our mode of perception may

be allowed for the instruction of the sceptic, he may also conjecture that the privilege of beholding an *improvement* therein may be accorded for the encouragement of the believer. He can well imagine that the beneficent Being, who ever in the present gives indications of the future, has placed within us a palpable proof of our high immortal destiny; and will permit the soul, even in this restricted sphere, to view the temporary expansion of its yet folded wings.

This may be deemed too vague; but to the colder reasoner I should suggest, that, in truth, mesmerism does in no way *contradict* our former experience: it merely enlarges it. We do not learn that our senses are not the organs of sensation to us in our usual state; but we learn an additional truth — namely, that, in another and anomalous condition, we find some unwonted medium of sensation. Nor by this are nature's established principles necessarily violated, or even infringed. They who believe that the perceptive peculiarities of mesmerism proceed (as Coleridge seems to hint they may do) from "a metastasis of specific functions of the nervous energy" may still agree to that definition of sensation, which a great philosopher has pronounced the only one that can be given, namely, "A state or affection of the mind, arising immediately and solely from a state or affection of the body." *

It is important, also, to remark, that the inform-

---

* Brown's Philosophy of the Human Mind.

ation respecting external objects, obtained through
the new action of the senses in mesmerism, though
sometimes exceeding, does not at all vary from, that
which we receive by the usual channels. Thus we
have a double witness to the truth and reality of the
things around us : we feel, with Descartes, that God
is not the author of a lie; and, far from being led to
doubt or to under-value our present organs of sensation, are more than ever disposed to confide in
them : scepticism itself allowing the potency of evidence which springs from two separate yet concurrent
sources.

But, whether the preceding remarks be admitted
to be just or not, the truth must be stated. Mesmeric sleepwaking has its own mode of perception;
the peculiarities of which I now proceed, to the best
of my ability, to represent.

In the first place, the mode of sensation in mesmerised persons is (like their mode of motion) of a
twofold nature : the one resulting, as it seems, from
a species of nervous communication with the mesmeriser, the other, from the peculiar state of their
own nervous system under an exciting and developing stimulus. To discriminate between these two
modes of sensation is important, but by no means
easy. The intimate union between the patient and
his mesmeriser, subsisting throughout the mesmeric
trance, complicates the phenomena, and renders them
hard to disentangle. Moreover there are decep-.
tive appearances, which are calculated to perplex
an inquiry as to the exact medium through which

sensation is conveyed to the mesmerised person. For instance : — we will suppose the mesmeriser to place something in the patient's hand ; — after a time the latter can tell the form of the object, *apparently* by the usual sense of touch. Now this is in diametrical opposition to the fact, that pins thrust into the patient's hand excite no sensation there. How shall we reconcile the seeming contradiction ? One moment we see the nerves of sensation useless and dead ; the next, we see them apparently acting as usual. All this is very paradoxical ; but, if we cling to what we have *proved*, and take some evident principle, thence educed, to guide us through the darkling maze, we shall perhaps find that the opposite poles belong to one common orb of reasoning, and are, indeed, the axis upon which the whole revolves.

*First.* We have established, by a series of experiments, that the nerves of sensation, in mesmeric patients, are of themselves quite estranged from their usual modes of action.

*Secondly.* We may deduce from this, that, if the patient experiences sensation, it must be, whatever appearances might lead one to suppose, by some other than the usual mode of nervous action.

*Thirdly.* Facts come in aid of this deduction.

Let us observe a mesmerised person holding any thing in his hand, after a command to examine it. How unconscious, how unmeaning, is the action with which he turns the object about ! How palpable is his hesitation, as if the mere attempt to use the accustomed channel of sensation were embarrassing! Can we indeed affirm, should he correctly name that which he holds, that it is by *touch* he has formed his judgment? Having once proved that his sense of touch is null, surely it is wiser to conjecture that his tactual perceptions now come to him in some other way. Agreeably to this conjecture, I may remark that my sleepwakers have often failed to recognise by touch objects which they have at other times seemed to distinguish rather by a species of vision. Once Anna M—— was, for a time, made to believe that a piece of wood that had been given her to hold was a piece of snow. Another time, an orange was placed in her hand. She told what it was, but observed, " You may think I know this by feeling : it is not so. In the mesmeric state I have only one method of knowing things ; and whether I see them, smell them, or feel them, it is all the same."*

I explain, then, the circumstance that mesmeric patients sometimes handle objects, by mere habits of motion ; — the circumstance that they occasionally recognise objects so handled, by the new perceptions, which, as I shall hereafter show, are opened to them in their new condition.

* This is curiously consonant with the account which the sleepwaker of Prevorst gave of her own sensations.

" She was eating some soup while her eyes were shut, and she said, I can find whatever I will upon my plate with the spoon — I know well where it is ; but I cannot tell whether I see or feel it ; — and so with regard to all other objects : I know not whether I see or feel them.' " — *Die Leherin von Prevorst*, vol. i. page 152.

Should this be granted, our path to a comprehension of the mystery will be at least smoothed and prepared; and there will be no obstacle to our subscribing to a proposition, which goes far to explain and reconcile the antithetical phenomena of mesmerism, namely, that the conditions of *ordinary* sensation are only restored to mesmeric sleepwakers *through* their mesmeriser.

In demonstration of this, I need not re urge the deadness of sleepwakers to ordinary sensation. I have only to bring forward some singular proofs, hitherto held back, that impressions of the external world not only reach the mesmeric patient, when transmitted through the mesmeriser, but, in such cases, are actually transmitted through the usual nervous channels.

Did any one strike or hurt me in any part of the body, when Anna M—— was in sleepwaking, she immediately carried her hand to a *corresponding* part of her own person. Thus, she would rub her own shoulder when mine was smarting with a blow, manifesting that the actual nerves of that part were, *pro tempore*, restored to their functions. Once an incredulous person came near me unawares, and trod upon my foot, which was quite hidden under a chair. The sleepwaker instantly darted down her hand, and rubbed her own foot, with an expression of pain. Again, if my hair was pulled from behind, Anna directly raised her hand to the back of her head. A pin thrust into my hand elicited an equal demonstration of sympathy.

I have already remarked that, when the mesmeriser eats, or drinks, or smells anything, his patients go
through the same motions, as if the impact of the
substances were on their own nerves. But this, it
may be said, might be referred to the simultaneity of
motion which I have shown to exist occasionally
between the sleepwaker and the mesmeriser. I have,
however, a very strong proof that the former has
really an impression on the nerves of taste, corresponding with that of the latter. Three of my sleepwakers (on whom alone I tried the experiment) could
in no way distinguish substances when placed in their
own mouths, nor discriminate between a piece of apple
and a piece of cheese; but, the moment that I was
eating, they, seeming to eat also, could tell me what
I had in my mouth. Once I tried this, before many
witnesses, on the sister of Theodore, with some pieces
of fig which I had carefully concealed, and the experiment answered perfectly.

Again, Anna M—— heard my watch ticking
when I held it to my own ear, though not when she
held it to her own. In the former case, she assured
me that she heard the sound exactly as if the watch
were close to her own ear.

We are then, I think, justified in concluding that,
in certain cases, the actual capacity of the nerves to
convey impressions is restored to mesmeric sleepwakers through the mesmeriser, but in no case
spontaneously. Indeed, the very thraldom of their
will seems to forbid the latter supposition: and it is
not extravagant to suppose that the mesmeriser, who,

by facts, has been shown capable of paralysing the nerves of his patient, should also have the power of restoring to them the conditions of sensation.

It may not be amiss, however, to answer here an objection that has been raised by some who have not sufficiently considered the nature of mesmeric sleepwaking. It has been asked, why, if there be such a sympathy between the patient and the mesmeriser, the latter does not also share the sleepwaker's sensations? For one plain reason; the one is active, the other passive; there is union, but not reciprocity. The mesmeriser is not in mesmeric sleepwaking, the patient is; and only he who is in the mesmeric state can appreciate sensations in the mesmeric manner. The mesmeriser must be considered as the bestower of a stimulus which exalts another to a particular pitch of feeling; and the patient as the recipient of that stimulus, which if he were to re-bestow, he must fall back into his old habitual state. Thus there can be no re-action on the mesmeriser. A proper view of the relationship between the parties will at once clear up the difficulty.

I proceed to show what are the principal phenomena of sensation which more properly belong to the patient himself.

First. In some cases, the class of perceptions, dependent upon a particular sense, are restricted in number only to be increased in degree.

Anna M——, in sleepwaking, ordinarily heard nothing but my voice and the voices of those who were in a certain contact with me; but her appre-

L

ciation of such musical sounds as she was made to hear was much greater than in the waking state. I was in the habit of singing to her when mesmerised; and, though at other times she cared little for music, she then evinced a pleasure which indicated an extreme susceptibility to musical impressions. She listened with the deepest attention : as the measure changed, her countenance became animated or serious; and she marked such parts of the performance as were most pleasing to her by a motion of her head or by raising her finger. It was remarkable that my voice, when I was singing, acted, in relation to the sleepwaker, as a conductor to other sounds. Then, and then only, she appeared to be sensible of the most trifling noises in the apartment, and was impatient of them to the highest degree ; holding up her forefinger, as if to enjoin silence, and uttering prolonged *hushes.* On one occasion, a person joined me in singing, when Anna testified the strongest disgust, making wry faces, and gestures of repugnance.

We then tried the experiment of making the person sing alone; but the sleepwaker no longer showed dislike, and, on being questioned, said that she heard nothing. Occasionally this patient manifested a slight agitation or oppression in the mesmeric state ; but these uneasy feelings invariably passed away under the influence of music. When brought into " rapport " with other performers than her mesmeriser, she still exhibited an unusual degree of musical sensibility ; marking the best passages of a composition with so much taste, and beating time so accurately, as once to make a musician suppose that she was intimately acquainted with his art. She always, however, heard a voice more distinctly than the piano. The conduction of the sound was thus accomplished:— I laid one hand upon the performer's shoulder, while she herself touched him with one hand and myself with the other. We have often tried what effect music would have upon the patient when awake ; and it was evident that her musical feeling had passed away with her mesmeric state.

Perhaps I ought here to record a phenomenon, relative to the sense of hearing, which spontaneously presented itself. Anna M——, in sleepwaking, had my watch in her hand, in order to tell the hour; but, disregarding this, she seemed to be amusing herself with the watch, alternately carrying it from her ear to the pit of her stomach, when suddenly she exclaimed, "This is wonderful! I cannot hear the watch tick when at my ear, but, placed below my chest, I hear it." Other experiments showed that she could also hear the ticking of the watch when it was held close to her forehead, or to *my own ear.* I should observe that, in her waking state, she knew nothing whatever of the alleged transposition of the senses in mesmerism, and that I had not much taste for developments in the " cerebrum abdominale."

E. A., who, in sleepwaking, could hear nothing when persons shouted close to his ear, did yet evidently retain, in certain cases, a very delicate perception of sound. This was proved by his being able still to play the flute with his ordinary precision and even more than his usual brilliancy. Often, too, he composed original airs during his sleepwaking with a great fertility of invention; and, though it was not easy for him, accomplished musician as he was, to display more musical talent in his new state than he evinced commonly, yet, on some occasions, he demonstrated that his discrimination of sounds had attained, under mesmerism, a higher development.

*First.* He played on the piano, correctly, an air, which, awake, he played but imperfectly, not having the same power of execution on the piano as on the flute — his peculiar instrument.

*Secondly.* Playing an air on the piano, in sleepwaking, he adapted to a particular note a new chord, which I at once perceived to be a great improvement on the original. Not willing, however, to trust to my own judgment, I consulted a musician on the point; when he declared the chord played in sleepwaking to be the best, and indeed the very one which was required by the laws of harmony. Curious to observe whether E. A. would discover this of himself, in the waking state, I never mentioned the subject before him; and I found that, unless sleeping mesmerically, he continued to play the air with the erroneous chord.

The second peculiarity of perception in the state of mesmeric sleepwaking, which it occurs to me to point out, is —

A sensibility to influences which in no way affect persons in the waking condition.

A prism, which I held in my hand, and turned

with the point towards the patient, produced, in two several cases, startings and slight shudderings.

Again, I concealed a magnet in my hand, with such precautions that the patient could have no idea of any thing unusual, unless through some development of feeling. The patient (Anna M——) started each time that I extended towards her the hand which held the magnet, and exhibited convulsive movements in the hands. When questioned respecting her sensations, she replied, " I feel as if I were beginning to be mesmerised, but not pleasantly — I feel all cold. "

Desirous to try if the different poles of the magnet would produce different effects, I held the south pole to the forehead of the sleepwaker, when she made sundry oscillations with her head, which at length slowly advanced towards the magnet. On presenting the north pole, her head began to retreat gently, and with an unsteady motion, like that of the magnetic needle. The same experiments, tried upon two other sleepwakers, gave the same results. Sometimes, when I have presented the north pole of the magnet, the patient's head has retreated by jerks, as if driven backward by successive shocks. The south pole, on the contrary, has always seemed to exercise a very powerful attractive force over sleepwakers; yet, as far as I can judge, they dislike to be submitted to mineral magnetism, which, as they affirm, produces in them disagreeable sensations. I have known one individual, who was sensible to the action of a magnet lying on a table at some distance and covered

L 3

by a piece of paper, and who begged it might be removed. I once asked a sleepwaker whether she could perceive any analogy between animal and mineral magnetism, when she replied, " There is an analogy — but the latter is of a coarser nature. " Minerals, in general, seem also disagreeable to sleep-wakers. Once, I gave a pair of scissors to Anna M—— (then in the mesmeric state), in order to cut some thread belonging to a work she was engaged in. Having cut the thread, she shuddered and threw the scissors from her. When asked if she did this because of the old proverb, that " scissors cut love, " she smiled, and said, " In my present state, I have no such superstitions; but I threw the scissors away, because, when mesmerised, I cannot bear to touch metal. " *

The effects which precious stones produce upon sleepwakers are also curious. In three cases, where I had the opportunity of making experiments of the kind, I found a certain correspondence of sensation, which inclines me to believe that a more extended observation might lead to some certain and interesting results.

The diamond, when presented to the forehead of a sleepwaker, seemed invariably to excite agreeable feelings : the opal had a soothing effect ; the emerald

* Caspar Hauser, whose nervous sensibility had been exalted by an unnatural seclusion from the stimulus of external things, was extremely alive to metallic influences, and could tell, through obstacles, where a needle was placed; pointing to the exact spot, and saying, " It draws me here."

gave a slightly unpleasing sensation; and the sapphire one that was positively painful. What is singular is that, if the last-named stone was applied to Anna M——'s forehead, she complained of its roughness, though it was polished perfectly smooth, — while, on the contrary, on contact with the diamond, which was cut in facets, she experienced a sensation of smoothness. One sleepwaker loved the diamond so much as to lean forward after it, when I held it in my hand, and to rub her forehead against it. In general, however, I did not touch the patients with the gems, but held them, concealed in my hand, at a few inches distance from the forehead ; and I changed their order sufficiently often to prove that the sleepwaker's judgment of them was not accidental.[*]

Another proof of the increased sensibility of sleepwakers is the ease and certainty with which they dis-

---

[*] The following is extracted from our mesmeric diary : — " Being shown several precious stones, Anna put her forehead towards them, and started, more or less, at each. Some, she says, produce agreeable, others disagreeable feelings. Being asked if it is because they are cold, she answers, " Oh no ! far from it. Some are very hot, and others scratch me as if they would tear the skin off my face. After several experiments, she declared that the opal gave a soft feeling, but was neither particularly agreeable nor disagreeable. The sapphire was hot, and very disagreeable. The Brazilian diamond had a pleasing effect, as if it were the hand of her mesmeriser. The emerald she made faces and pouted at, saying it was not agreeable. When the stones were held to her forehead, concealed in the mesmeriser's hand, she knew them all again, and gave the same account of her feelings."

L 4

tinguish water that has been breathed on or touched by the mesmeriser. I have sometimes thus *mesmerised*, as it is called, a glass of water, half an hour before it was presented to the sleepwaker, amongst three others, exactly similar and filled to exactly the same point. The result of one of these trials shall be given, as taken down at the time by an eye-witness of the experiment : — " Four glasses of water being presented on a tray to Anna M——, one of which was mesmerised, she passes her hands over them all, feeling the rims, and fixes on the mesmer-ised one. The order of the glasses being changed behind the mesmeriser's back, who stood so as to screen the tray from the patient, she feels them all again, and fixes on the mesmerised water a second time. Changed again, again she finds it, and drinks half of the contents of the tumbler, saying it will do her good. "

This sleepwaker seemed to have two methods of dis-tinguishing the mesmerised water — the one, as above described, by running her fingers over the glasses; the other by carrying each glass to her lips, and slightly tasting the water it contained. In the latter case, she invariably marked both her knowledge and her preference of the mesmerised water, by taking a deep draught immediately on tasting it. When asked how it seemed to taste to her, she replied, " So agreeably, that it is impossible to describe it."

On one occasion, a person who suspected that the mesmerised water was discovered through some slight difference in the taste or look, from its having been handled, tried to deceive the sleepwaker, by presenting, amongst the other glasses, one that contained the weakest possible solution of toast and water; but the instincts of the sleepwaker were in no way disturbed. She chose the mesmerised water unerringly as before.

Whatever I had touched she distinguished with equal certainty, and would only accept such articles of diet as were transmitted through my hand. Once, when I was giving her some cherries, a friend, who was incredulous, tried in every way to place the fruit in her hand, in such a manner as to make her suppose that I was continuing to present it, but she threw away instantly every cherry that my friend had touched, making a face expressive of dislike. Is this, by the way, more extraordinary than the acuteness of sense by which the dog recognises its master's property, or do we only deem it so because less within the scope of our daily experience?

But it was not only as regarded objects that belonged to myself that Anna M—— exhibited extraordinary powers of discrimination. More than once, the pocket-handkerchiefs of five persons have been tossed to her indiscriminately, (sometimes they have been all cambric and all nearly alike,) yet, after they had been jumbled together on her lap, she has quickly and correctly restored each to its several owner.

Occasionally, this finer perception, manifested by Anna M—— in sleepwaking, was, like certain other of the mesmeric faculties, continued forward, for a brief moment, into the natural state. I and another

L 5

person have presented the sleepwaker with two
flowers, or two strips of paper, as nearly alike as
possible, discriminating them by some little mark
which could only be known to ourselves; — we have
seen her hold these carelessly in her hands, and
frequently change their position; yet can we testify
that, if bid to remember, and asked, immediately on
awaking, who gave her the articles in question, she
could give them back to their respective donors
without hesitation, though without knowing how or
by what impulse.

E. A. manifested similar faculties. On being
asked how he distinguished the mesmerised water, he
said, " J'y vois une espèce de lumière."

All this demonstrates the truth of Laplace's assertion, that, " of all the instruments by which the
finer emanations of the natural world can be detected, the most delicate are the nerves; especially
when their sensibility has been developed through
particular causes *," and I am thence led to suggest
that, instead of quarrelling about the exalted sensation
of sleepwakers, we should endeavour to turn it to
account. Even should we never succeed in fathom-
ing its principle, there is an obvious probability that
it may be rendered practically useful. If worthless
as an object of analysis, it may at least be valuable as
a test. But it is the nature of man to insist upon a
thing being just as he would make it, or to have
none of it. We reject the benefit that we cannot
comprehend; and, like children crying for the moon,

* Théorie Analytique du Calcul des Probabilités.

are rendered unhappy by seeing an object, however beautiful, which cannot be brought down to our level and given us as a plaything.

Yet, on the Continent, men of science have, in some instances, been wise enough to profit by the extraordinary perceptions of the mesmeric state. It is related, in " Die Leherin von Prevorst," that the sleepwaker, so called, was able to detect in certain stones a certain ingredient which always affected her in a particular manner ; and that a mineralogist made much use of this discrimination ; subsequent experiment completely establishing the capacity of his informer.*

Thirdly. The organs of vision being under quite other than the normal conditions, a faculty analogous to sight is developed in mesmeric sleepwaking.

In approaching the phenomena that result from such an overstepping of the ordinary limits of our nature, I deeply feel the difficulties that bar my way. That persons can hear and answer questions while in a kind of slumber may be conceived; but that they should see with their eyes shut comes not within the compass of our faith. It is contrary to nature — it is utterly incredible.

Even they who behold the fact are, when it is first presented to them, incapable of reasoning for very wonder. They perceive that the mesmeric pa-

* The same sleepwaker distinguished between twelve different kinds of grapes, and their several degrees of excellence, astringency, &c. — *Leherin von Prevorst*, vol. i. p. 81., published at Stutgart and Tübingen.

tient's mode of vision is different from any within the range of their experience; but what that mode may be appears to them a mystery unfathomable.

Such were my own feelings on first witnessing an exhibition of what is called *clair-voyance* in mesmerism; and I now endeavour to recal them, in order that I may sympathise more with the astonishment or incredulity of my readers over the statements which I am about to make. Should they suspend their belief as to certain phenomena till they have actually beheld the alleged wonders with their own eyes, I can scarcely blame them for a doubt which is, perhaps, but proper caution. Rather must I address them in the words of Treviranus, when speaking to Coleridge of mesmeric marvels, — " I have seen what I am certain I would not have believed on your telling, and in all reason, therefore, I can neither expect nor wish that you should believe on mine." Even they who are most firmly convinced that mesmerism has its own mode of vision must be content to accept the phenomenon, without having that surety of it which personal experience alone can give of any thing. The mesmeriser witnesses the wonder, but does not feel it in himself: the sleepwaker, who is the subject of it, seems incapable of analysing his new sensations while they last, still more of remembering them when they are over. The state of mesmerism is to him as death.* He cannot, when he awakes,

* I once asked a young lady, who was in sleepwaking, why, on awaking, she forgot all that she did with such apparent intelligence. Her answer was, " Dés qu'on est magnetisé on est comme mort."

reveal the mysteries of that great deep. His mesmeric feelings are to him as though they had never been; and less favoured, in this respect, even than they who have beheld him in his unusual condition, he is forced to take his own actions upon trust, and to exercise his own faith, while he draws so largely upon the realising faculty in others.

It is manifest, then, that we cannot believe in the *clair-voyance* of sleepwakers, in the same manner that we believe and know that we ourselves see with our eyes. It is a fact which transcends our present understanding.

To what end, then, it may be asked, should I state phenomena which will be believed by few, and perfectly comprehended by none? Because many things that are mysteries are, nevertheless, profitable subjects of contemplation. Whatever is beyond our actual state of being is confessedly out of the pale of empirical knowledge; yet shall we, on that account, banish the higher developments of nature from our thoughts, or even from our own scientific examination? Were all our ideas confined to that which we certainly know, the domain of our intellect would be limited indeed. Besides, by careful study, we may always *extend*, though we cannot *complete*, our apprehension of things above us; and, by discovering their analogy to things already known, bring them at least nearer to our experience. Clearly then, where there is so much room for progress, it is our duty to advance, remembering that the point where we should abandon enterprise has not yet been decided. In

the present instance, the objects that I propose to
myself are these : — To show in our present being
the elements of a future existence; to prove that our
actual senses are limitations of our percipient power;
and that, in proportion as the mind is detached from
them, it acts more largely, overcoming obstacles
which, by means of our usual organs, it is impossible
to surmount.   It may be, also, that the facts on
which I found man's capacity of development, being
supported by good testimony, may convince some
who have hitherto been incredulous, and may supply
others, who already believe, with additional light to
guide them further on their way.

I have already stated that I have seen sleepwakers
descry objects, when their eyes were, to all appearance, perfectly closed ; but my experiments on this
head were not so rigorous, but that it might still be
objected that mesmeric patients, like certain politi-
cians,

" See through all things with their half-shut eyes."

I now proceed to show that, in many cases, such
a supposition is untenable ; and that the mesmeric
sleepwaker may have a mode of vision to which the
usual conditions of sight are altogether wanting.

The first time that I mesmerised Anna M——, a
workbox, which she had never seen before, was held
before her.   She stooped her forehead towards it, in
a manner that struck me, and immediately named
what it was.   The box having been opened, the
sleepwaker again bent her forehead till it was nearly

parallel with its surface; then rapidly named the various objects it contained; and, taking them up one by one in her hand, seemed desirous of examining them more particularly. But, to my surprise, she waved the articles about before her, as if trying in what point of view she could best descry them, holding them to various parts of her face and forehead, and exclaiming, as if perplexed, " Where, then, *are* my eyes?" At length she seemed best satisfied, when holding objects before her forehead, at the distance of a few inches, declaring that she saw them most distinctly there. In order to put her assertions to the proof, I held my watch before the forehead of the sleepwaker, without descending it to the level of her eyes. She took it from me, and, not lowering it in the least, held it so turned as that it formed an acute angle with her forehead, immediately above the eyebrow. It is to be remarked that she thus presented the watch to her forehead, first on the right side, then on the left, as if to submit it to the scrutiny of a double organ. After this she named the exact hour and minute. The hands having been altered, she found the time with equal correctness.

A poppy being held before her forehead, she said, " I see a red flower; but I do not exactly know its name."

Remembering that an experienced mesmeriser had told me that sleepwakers, in general, perform most readily anything which gives them pleasure in their waking state, and observing, in conformity with this statement, that Anna M——, who was an expert needle-woman, took particular interest in the work-box and its appendages, I proposed to her to proceed with a piece of work which was at hand. She immediately took the work, and, holding it always on a level with her forehead, went on methodically with the hem of the piece of muslin I had given her. When necessary, she turned down a new fold ; and, in every respect, performed her task as well as she could have done awake. The work, submitted to female judgment, was pronounced to be a capital piece of sempstress-craft, the stitches being even and not one of them dropped.

After this, we often gave Anna M—— work to do in the mesmeric state, when parties of ten or twelve persons have been present, to witness her extraordinary development of vision. She continued to hold every thing to her forehead ; and with her hands raised to that level, in a position which, under ordinary circumstances, would be difficult and painful, has embroidered delicate flowers upon muslin, and even threaded her needle, without apparent effort.

As it is my sincere desire to give a correct picture of mesmeric sleepwaking, I would on no account represent this power of vision as greater than it really was, or omit the inconsistencies which attended its exercise. That it was by no means even or constant cannot but be acknowledged, though I am by no means prepared to develope the cause of its caprices. Thus Anna, though giving incontestable proofs of vision by the forehead, could not be brought

to distinguish printed or written letters in the mesmeric state, except on one occasion, when she read
her own name, which I had written in a large hand,
and held at once before her forehead. It seemed
to me that her new visual faculty was always in its
best condition when spontaneously exerted, and that
any effort on her part, any over-anxiety to fulfil our
requisitions, marred it altogether. I have often
asked her to name an object, which I have allowed
her to examine as she would; but she has not named
it, though apparently striving earnestly to do so.
Again, she has indicated other things spontaneously
when it was quite impossible for her to have discerned them in the ordinary manner. It was when
she was sitting quietly, and apparently forgetful that
she was an object of observation, that she displayed
the most remarkable phenomena of vision. One
instance however is better than a thousand assertions.
She was sitting with her head so much bent down as
to bring the upper part of her forehead parallel to the
wall of the apartment. In this position, with her
eyes closed, it was impossible for her to have seen, in
any usual way, objects that were immediately fronting
her. So placed, I observed her smile, and asked
her why she did so? "I am smiling," she said,
"because I am pleased to see Mrs. —— opposite to
me." "You see her, then, well?" I inquired. Yes;
she has a cup of tea in her hand." Upon this, the
lady in question adroitly changed the cup of tea for
a book; upon which Anna immediately remarked,
"But now she has taken a book." The lady then

opened the book, and held it by the two sides, spread out exactly on a level with the forehead of the sleep- waker, who said directly, "Oh ! she holds the book quite open by its two ends."

This experiment, neither suggested nor in any way conducted by myself, was interesting to me in no trifling degree, and was convincing to all who witnessed it.

Another singular circumstance was, that no one could put on an ugly mask that lay about the room, and to which Anna, in her mesmeric state, had a great aversion, without her testifying, by faces ex- pressive of dislike, that she was aware of the circum- stance. We have tried this, when the sleepwaker was occupied by other things, and with every pre- caution, of making no noise, &c.; yet the result was always the same.

When placed before a looking-glass, she could indicate, more correctly than at any other time, the gestures of persons standing near her, and seen by reflection. I have pulled out her comb, and she has arranged her hair again perfectly before a mirror, holding her forehead parallel to its surface. Being asked if she saw herself with her eyes open or shut, she replied, " Open, to be sure ;" and, when I reasoned with her on this point, she replied, " I see as if my eyes were open; and so they must appear to me open." It is singular that another sleepwaker gave me exactly the same answers under the same circum- stances. I shall refer again to the subject, which is of metaphysical importance.

A gentleman who was once present during a mesmerisation of Anna M——, being placed "en rapport"
with her, laid his hand upon her forehead, when she
exclaimed, " Why do you cover my eyes ? " He then
touched her eyes, and asked, " What part of your face
am I touching now ? " The sleepwaker seemed perplexed, and at length answered, " It is a part of my
cheek, is it not ? "

When asked to point out where different persons
were placed in the apartment, during her sleepwaking, she never failed to do so, however their
respective positions might be changed ; leaning her
forehead forward all the time, and presenting it to
each individual. At the instant of recognising each
person she always gave one or two convulsive starts,
as if her forehead came in contact with some invisible
thing.

The account that she gave of her visual perceptions
was sufficiently confused. These are her own words,
relative to this subject, taken down, on one occasion,
by a friend : — " It is all clear through my forehead.
Sometimes I see *so* clear ! But then, again, there is a
sort of light cloud that comes over the clearness, and
then I can hardly see anything. I do not see as with
two eyes, but *here* (passing her hand across her forehead), with my brain.'

Already, in various accounts of experiments, I have
mentioned E. A——, a boy aged fifteen, whom I
had opportunities of frequently mesmerising. This
patient, of all whom I have ever seen, manifested in
sleepwaking the most extraordinary development of

visual power. As I have before said, he had been, in childhood, a natural sleepwaker; and I now add, on the testimony of his father, that he would sometimes rise in the night, take out his flute, (an instrument which he was studying professionally,) place music before him, and play from notes, continuing to turn over the leaves of the music-book correctly, although his eyes were closely shut. On one occasion, while his father was watching him in a paroxysm of this kind, the only light in the apartment, a lamp, went suddenly out; but the sleepwaker continued to play as before, and was heard to turn over the leaves of his music until he had come to the end of the piece, which, moreover, he could not execute without book. Thus, it will be perceived, there was in this sleepwaker a sort of natural ground for the development of extraordinary vision; and it was this consideration which led me, after having heard the anecdotes above related, to request the boy to submit himself to the effects of mesmerism. He consented to a trial, the result of which will show how careful persons ought to be in deciding, from one or two unsuccessful experiments, that a patient is insusceptible of mesmerisation. I had mesmerised E. A——— twice (each time an hour), and he had shown no symptom of being affected by any influence. I should not have made a third trial, but for the following circumstance :—I had thrown Anna M——— into mesmeric sleepwaking before a large party of persons, amongst whom was E. A———. I think I have every reason to affirm that the sleepwaker did

not know anything whatever respecting E. A——,
not even his name, nor that I had attempted to
mesmerise him ; yet, being accidently near him, she
said to me, (as if guided by that extraordinary
instinct, respecting all that relates to the mesmeric
state, which I have before noticed as characteristic
of true sleepwaking,) "You should mesmerise this
young man : he is a natural sleepwaker; and will
become very *clair-voyant.*" I answered that I had
already tried to mesmerise the boy, and that I had
failed in my object. "You should try again," replied the sleepwaker, "and you will succeed." But
before relating how truly this prophecy was accomplished, and its further results, I must observe, in
order to remove all suspicion of Anna M——'s sincerity, that in recommending me to mesmerise E.
A—— she was speaking strongly against her own
interest, and even her own general feelings ; for, with
a sort of sentiment that may be called mesmeric
jealousy, she never could endure me to mesmerise
any one but herself. Then, again, I was endeavouring to educate her powers of mesmeric vision, in the
hope that she might gain the prize offered by the
academy of medicine, at Paris, to a sleepwaker who
could read without the aid of the eyes ; and, knowing
that attention to any one else must, in a degree, dis-
tract me from this object, she had every reason to
fear a rival in my mesmeric graces. Moreover, in
order to avoid giving her the least uneasiness, I
carefully concealed from her my relations with E.
A—— ; and as, when awake, she forgot wholly the

events of her sleepwaking, she never knew (as far as I can judge) either that I had mesmerised E. A—— at any time, or that there was any question of my doing so.

Having said thus much in justice to Anna M——, (a precaution not unnecessary in the present hostility to mesmerism,) I proceed to state that, the third time I mesmerised E. A——, he, at the end of an hour, passed so far into sleepwaking as to be able to answer questions without awaking — questions of which he retained no recollection in his natural state. On a fourth mesmerisation, he manifested all the characteristic symptoms of mesmeric sleepwaking; was able to move about with tolerable ease; and began to display those extraordinary phenomena of vision which I have prepared my reader to expect. These may be divided into two classes — namely, such as presented themselves spontaneously, and such as were developed during a course of strict experiment.

With regard to both it may be observed that there was exactly that progress in their development which attends the education of a new faculty. At first, the patient could only descry the larger objects around him, or such as most interested him, or to which he was the most habituated. Thus, though able, in the early stages of his sleepwaking, to discriminate between the persons present in an apartment, and though testifying, in all that related to music, great powers of sight, (for, from the first, he could, while mesmerised, write out music with precision,) yet, for a long period, he found considerable difficulty in reading from a book — always complained of the smallness of the type, and could rarely be prevailed upon to look at more than two or three words at a time. Subsequently, his eyes being always firmly shut, (as far as the strictest observation could determine,) he was able to read any number of words in the minutest type with perfect ease, and to discern small and large objects, near or distant, with exactly the same facility of vision which is possessed by a waking person. In proof of this I may mention that I and the members of my family have seen him, when in the mesmeric state, thread a small needle, and sew a button on his coat, and, again, distinguish minute letters on a seal which a gentleman showed him, and which I could not make out myself. At another time, in mesmerism, he played on the flute a piece of music which he had never seen before, from a book that was set up before him at the distance of some feet; and once being mesmerised out of doors, and led to a spot where he had never been, he described all its features, and indicated the form and position of the distant mountains. This power of perception, analogous to sight, seemed principally to reside in the forehead. Whatever objects he took up to examine he immediately carried there; and once, in the presence of Dr. Foissac, at Paris, he, being given a set of eye-glasses which he had never seen when awake, of eight different colours, shut up in a tortoise-shell case, unfolded them, and applying one, at hazard, to his forehead, without descending it to the level of his

eyes, exclaimed, " Every thing appears blue to me !"
at the same time, boy-like, imitating the gestures of
a Parisian dandy, and observing that he should like
to show off his pretty *lorgnette* in the street.    The
glass which he had accidentally chosen was, in fact,
blue.    Subsequently, he at various times has named
the principal tints of the eight glasses correctly, when
presented to his forehead in any order.    The same
result took place when his eyes were bandaged.    It
was, however, remarkable that a powerful magnifying
glass being placed before his forehead was not perceived by him to enlarge objects, though he read in
a book, through the glass, with perfect ease.

Though the power of vision was greatest in the
forehead, yet at times, and especially when he
was excited, and not in any way called upon
to exhibit, (for such requisitions often seemed to
fetter his faculties,) he seemed to see on every side
of him, as if his head were one organ of visual perception.    This is no exaggeration, as the following
instance will show : — He was once sitting on a sofa,
in the mesmeric state, when a gentleman with whom
he was well acquainted came behind the sofa, and
made all kinds of antics.    On this, the sleepwaker
exclaimed, " Oh, Mr. D——— ! do not suppose I cannot see you : you are now doing so and so (describing
all Mr. D———'s gestures).    You have now taken a paper-
cutter into your hand, and now a knife.    Indeed, you
had better go away, and not make yourself so ridiculous."    Another time, he was sitting at a table,
writing music, with his back to the door, when a

servant entered the apartment, " Oh, Mademoiselle L——! is that you? he said. How quietly you stand there with your arms folded." He was quite correct in all he said. Directly after this I took up a bottle from a table behind the patient, and held it up to the back of his head, asking him if he knew what I held. He instantly replied, " A bottle, to be sure. "

Having mentioned the principal phenomena of vision which E. A. spontaneously presented, I proceed to state those which he manifested under strict experiment.

Well convinced that doubts would exist, not only in the minds of others, but in my own, respecting the patient's manner of seeing, so long as there was the remotest possibility of his discerning objects through any aperture, however small, between the eyelids, I studied to put this out of the question by precautions so severe as to forbid imposition, and so varied as to supply that accumulated and concurring evidence which circumstances that contradict experience imperatively require.

1st. I laid the patient on a sofa, in bright daylight, with his face turned towards the window, and made him lean his head back until I could see completely under his eyes. When he was so placed, I could have detected the slightest gleam of the eye through the smallest opening of the lids. I have then given him a book, from which he has read with ease, (holding it nearly parallel to his forehead,) while all the time I fixed my eyes earnestly on his,

M

and yet could perceive not the slightest tendency in them to unclose.

2dly. I laid the palms of my hands, the fingers pointing upwards, on the eyes of various persons, in such a manner as that the projecting parts of each hand should exactly fit into the concavities about the eyes. These persons assured me that, with their eyes so covered, they could see nothing whatever. I have given them cards or books in their hands, but by no efforts on their parts could they distinguish these objects. I have repeated the same experiment again and again upon E. A., in a state of sleepwaking, and never found that the palms of my hands in any way impeded his vision. He could see cards, or read in books, under the above circumstances, with perfect ease. I never felt any motion beneath my hand, as if the patient were trying to open his eyes; nor did he evince the slightest inclination to draw his head back from the pressure of my hands.

3dly. Standing behind the patient, I have laid my closed fingers over his eyes horizontally, or I have forcibly pressed down his lids with one finger of each hand. This, which, when tried on others, effectually impeded their sight, made no alteration in the visual perceptions of the sleepwaker. In order to avoid repetition, I here state, generally, that the efficacy of the means I employed to stop the eyes of my patient was always first proved upon the eyes of indifferent persons. I should remark, moreover, that the patient had no objection to permit any one, who was previously placed "en rapport" with him, to try upon him all the above experiments. I therefore usually engaged other persons to do this, in preference to myself, as the most satisfactory way of convincing them that there was no complicity between me and my sleepwaker, and of bringing the matter home to their own business and bosoms. This manner of proceeding has shaken many from their incredulity.

4thly. Having filled a couple of china eye-glasses with wadding, I, or some other person, held them firmly to the patient's closed eyes when in sleepwaking. This also made no difference in his visual perceptions. When the same eye-glasses have been applied without the wadding, notwithstanding their perfect opacity, the patient has declared that he could see the light very plainly through them ; and that they were so transparent that he could not conceive why we imagined they should prevent him from seeing.

5thly. I have tried various methods of bandaging the patient's eyes. I have tied a broad and thick silk handkerchief over them, and then I have held down with my fingers, or the palms of my hands, the whole of the bottom part of the bandage. This method seems to me as perfect as any. It did not at all impede the sleepwaker's vision. In addition to this, (the same result always ensuing,) I have laid strips of wadding over the eyes before applying the handkerchief, and I have firmly secured every possible interstice between it and the cheek with cotton. In the presence of Dr. Foissac, strips of diachylum were added to all the above apparatus, in order to

M 2

fasten down the edges of the handkerchief to the cheek; but the sleepwaker saw as well as ever. On several occasions I bandaged his eyes, adding the cotton and the wadding *before* beginning to mesmerise him, when he has assured me that he could not distinguish day from night. Then, having passed into sleepwaking, he has immediately given proofs of perfect vision—quite as perfect, indeed, as that enjoyed by persons whose eyes are open and unbound. Again, on awaking, (the bandage never having been stirred during the whole period of his sleepwaking,) he has found himself in perfect darkness. The transition was marked. One moment, drawn by the strong attraction of my presence, he was following me about the room, through intricacies of chairs and tables, with perfect ease; the next, he was standing helpless, not caring to be near me, and, if called upon, unable to move, except with the groping hesitation of a blindfold person. I remarked that he did not wake so easily with the bandage on as when he had no bandage. The action of the transverse passes that I used to that effect seemed modified by the interposing substance. The striking proofs of vision that the patient gave, when properly bandaged, were, that he read in books, and distinguished cards, their colours, suit, &c., often playing with me at various games upon them. I remarked that in sleepwaking he was quite adroit at the game of cassino, which I had almost vainly tried to teach him in the waking state. It will be allowed that for a person, even bandaged in a slovenly manner, to perceive, at a glance, the combinations on the board, would be no easy matter; yet this he did with rapidity, completely bandaged as he was.

6thly. I threw over the patient's head two thick and large towels, which covered him in front down to the hips. Through these he has read, holding the book at an angle with his forehead, and has distinguished cards with perfect accuracy. This kind of experiment was occasionally varied. Sometimes the sleepwaker has been bandaged, and, in addition to this, a towel has been thrown over his head; but the result was equally satisfactory. This power, however, seemed to have its limits. The addition of a third towel greatly impeded the patient's vision, yet even thus he has distinguished cards. On one occasion, a visiter, instead of covering up the patient's eyes, enveloped the object to be seen in the folds of a napkin. The experimenter, in order, if possible, to mislead myself, the sleepwaker, and all who were present, gave us to understand that he had placed one card only in the napkin, (he had performed the operation with his back turned,) but the patient was not to be deceived. At first, indeed, he seemed puzzled; but even this transient perplexity elicited a curious proof that he saw not only through the triple folds of the napkin, but through the back of one of the cards. He said, " There seems to me to be a *five*, but the points are not of the same colours — " Oh!" he exclaimed, after a pause, " how could I be so stupid! there are *two* cards. One is the ace

M 3

of hearts, the other the four of clubs." He was perfectly right. The four of clubs had its face uppermost, the ace was laid under it, and, in order to form a five, the sleepwaker must have seen the ace underneath the other card.

7thly. One day I was curious to observe whether my patient could discriminate colours or forms, on apparently flat surfaces, by the touch, as the blind sometimes have been known to do. To this end, I held an open book over a little round French paper box, which had no distinction of top or bottom, except that there was a picture on what was meant for the upper side. I gave the sleepwaker the box into his hand, (holding the book always so as to screen the object from him,) and asked if he knew whereabouts the picture was. He immediately turned it right side uppermost, and with his finger pointed out the principal features of the little painting — a shepherdess sitting under a tree, &c. "Then," I said, "you do perceive by your fingers." "No," he replied, "but I see *through* the book." Desirous of testing this more perfectly, I held open a large book exactly before his face, and, taking up, at hazard, a card from a pack that lay near me, I kept the figured side concealed against my hand, until I could turn it towards the sleepwaker, immediately behind the book. Stooping his forehead forward, till it nearly touched the back of the volume, he told correctly what the card was. This experiment I frequently repeated, with equal success ; sometimes varying it,

by bandaging the patient's eyes before I held the book before them. The additional obstacle seemed to make no difference. Sometimes I have placed a card, with due precaution, in the midst of a book which I kept open only by the interposition of a finger, holding the face of the card pressed against the leaves of the book, and thus entirely concealing it. I have then held the book upright before the patient, who has bent his forehead forward, as usual, till it was parallel to the cover of the book, and has then told the card correctly. What is singular is that, if I withdrew my finger and quite closed the book, the experiment failed. The sleepwaker said that the reason of this was that the vibrations of the medium, by which he pretended to perceive objects, were too much intercepted by the perfectly closed book. I should observe, once for all, that, when I record the sayings of my sleepwakers, it is not because I give faith to them, but for reasons to which I shall advert hereafter. Indeed, nothing can be more contra-dictory or unsatisfying than the account which mes-merised persons render of their own mode of sensa-tion. For instance, E. A. told me once that he saw *through* the book that was held between him and an object; but at another time he propounded a long and confused theory, by which he made it appear that his sight did not penetrate the obstacle, but re-ceived an impression from certain rays, that did not come to his eye in direct lines, but were bent round the edges of the book.

8thly. Another class of precautions which I have

taken, to convince myself that the *mechanism* of the eye had nothing to do with the mesmeric mode of vision, consisted in spreading out a pack of cards on a table, face downward, (frequently changing the pack, lest any of the cards might, in any way, become known to the sleepwaker,) and in sliding a card, at random, onto the palm of my hand, so that its faced side was never for a moment exposed until I had lifted it above the patient's head, when I presented it at once to his forehead or occiput. In the same way I have taken up a written paper, previously prepared, or a book, which I have not opened until it was above the level of the patient's eyes; and I have held these objects to various parts of his head. The following are the results of observations made under the above circumstances : —

The sleepwaker's power of perception seemed to become weaker as it withdrew from the region of ordinary sight. I have known him read half a page from a book which was held immediately above the eyebrow ; and he found but little difficulty in telling any number of cards which were held in the same position. He did not at all like to have objects held behind him ; saying that perception by the occiput was very fatiguing, and cost him an effort which did him harm. It certainly is a fact that, after he had told things in this way, he used to complain of uneasiness on awaking. Still he gave indubitable proofs that he could occasionally exercise, by means of the back or other parts of his head, a faculty analogous to sight. The following are the most striking instances of this:—

I wrote in my pocket-book (holding it above the patient's head) the words, " Voulez-vous aller à Milan?" I then presented the writing immediately to his occiput. He immediately called out, " Oh! je vois la le nom d'une grande ville!" and then, word by word, he seemed to make out the whole sentence, which he at length repeated correctly.

On another occasion, I took up a book at random, and opened it just behind the patient's ear, holding the book parallel to the side of his head. I had opened it at the beginning of a chapter. His first exclamation was, " There's a great deal of white on that page." I said nothing, but continued to hold the book as before, when he began to repeat some numbers that were on the page, — " 1425, 1426." Then, without pausing, he read, distinctly and correctly, the title of the chapter, which was as follows : — " *Scénes Historiques. — La Main Droite du Sire de Giac* (1425, 1426.) "* He continued with one or two words of the chapter itself, but suddenly broke off, in a way that was habitual with him, pushed away the book and declared himself fatigued.

In telling cards by means of various parts of his head, he liked to observe certain conditions which were executed either by himself or at his direction. His favourite mode of proceeding was to lift his own hand above his head, and to take the card from me,

* Taken from a work called " Dodécation, ou le Livre des douze." Bruxelles, 1836.

which he held at a certain distance from the part
where I told him to exercise his perception, observing that no one ever put a thing they wanted to see
*close* to their eyes.   When I held the card myself,
and approached it nearer than he liked, he always
gave indubitable proofs of being aware of the circumstance, begging me to place the object farther
off.   Sometimes, when he found a difficulty in ascertaining the card, he would beg me to breathe on it;
and, when I had done so, he would tell it directly.
At other times, he would hold the card horizontally
above his head; and then, without stirring the centre
of the card from its place, would dip down first one
end and then the other, like the two extremities of a
see-saw.   This he called " Le moyen electrique. "
He was generally successful in telling cards in this
manner, sometimes on the first operation, sometimes
after two or three repetitions of it.

Occasionally I have placed the patient's own hand
on a table with the palm uppermost, and have glided
onto it a card drawn at random from the pack, with
the face downward.   Then I have told him to what
part of his head he should lift the card; but it has
often happened that, while in the act of conveying
the card (the back always uppermost), he has said,
" I need not trouble myself to take the card any farther, for I know already what it is."   On these
occasions he was sure to be right; and, notwithstanding his assertions to the contrary, I was by this, and
other minute circumstances, led to imagine that his
whole nervous system shared something of the per-

cipient faculty which more peculiarly manifested itself in the forehead. His own account of this was, that he received perceptions through the motions of a fluid, which was conducted by the brain generally, but was transmitted with the greatest facility through the orbits of the eyes and the region immediately above them. He declared that the ball of the eye had no share in the production of his mesmeric vision, but that the impression was made direct upon the brain; moreover, that the sensations caused in him by the motions of the fluid resembled those excited by light. When cards, &c., were held behind his head, he rarely distinguished the whole object at once; but (as in the case of a card, for instance) he would first name the suit and colour, then count the pips one by one, and, finally, name the whole card. When questioned (not by myself, but by a person who took interest in my experiments) why he did not tell the object at once, he replied, "The light that seems to strike upon it falls partially, and first I see one part of the card, then another."

9thly. I have been making still larger and larger demands upon the belief of my reader, and now I am about, perhaps, to revolt him altogether; but, having once undertaken to give a faithful relation of that which I have witnessed, can I draw back from the accomplishment of my task? No; while conscious that I only speak the strictest truth, I am prepared even to have that truth doubted. Time will " bring in its revenges."

Remembering that E. A——, on his father's testimony, had, in natural sleepwaking, seemed to perceive objects in total darkness, I was curious to ascertain whether, in mesmeric sleepwaking, he would manifest a similar phenomenon of sensation. I therefore, having mesmerised him, took him with me into a dark press or closet, of which I employed a friend to hold to the door, in such a manner as that no ray of light should penetrate through crevice or keyhole. Then, like the hero of " The Curse of Kehama,"

" I open'd my eyes and I closed them,
And the blackness and blank were the same."

My utmost efforts to see my hand only produced those sparks and flashes which waver before the eye in complete obscurity.   Having thus ascertained the perfect darkness of the closet, I drew a card at hazard, from a pack with which I had provided myself, and presented it to the sleepwaker.   He said it was so and so.   I repeated this to my friend, whom I then told to open the door.   The admission of light established the correctness of the sleepwaker : it was the card he had named.   The experiment repeated four times gave the same satisfactory result.   This peculiar development of vision was, like the other faculties of the sleepwaker, capable of improvement through exercise.   At first, he seemed unable to read in the dark ; then, like a person learning the alphabet, he came to distinguish large single letters, which I had printed for him on card ; and at length he could make out whole sentences of even small print. While thus engaged in deciphering letters, or in ascertaining cards, the patient always held one of my hands, and sometimes laid it on his brow, affirming that it increased his *clair-voyance.* He would also beg me to breathe upon the objects which he desired to see. He used to declare that, the more complete the darkness was, the better he could exercise his new mode of perception; asserting that, when in the dark, he did not come to the knowledge of objects in the same manner as when he was in the light: " Quand je suis dans l'obscurité," he said, " il y a une lumière qui sort de mon cerveau, et qui tappe justement sur l'objet ; tandis que, dans la lumière, l'impression monte depuis l'objet jusqu'à mon cerveau." Often, when I could not see a ray of light, he used to complain that the closet was not dark enough, and, in order to thicken the obscurity, he would wrap up his head in a dressing-gown which hung in the closet. At other times he would thrust his head into the remotest corner of the press. His perception of colours, when exercised in obscurity, sustained but little alteration. He has named correctly the different tints of a set of coloured glasses. It was, however, worthy of remark that he was apt to mistake between the harmonic colours, green and red, not only when he was in the dark, but when his eyes were bandaged.

Many persons can bear testimony to the accuracy of the above experiments ; and I refer to the Appendix for proofs that I sought for witnesses and invited scrutiny, feeling that such things as I had to narrate could scarcely be credited on the word of a single person.

Desirous of neither exaggerating nor detracting from the powers of mesmeric vision, I subjoin a few general remarks upon what I observed of its development in the case of E. A——, begging the reader at the same time to remark the coincidence between the feelings of this sleepwaker and of Mademoiselle M——. Like the latter, he declared that he was not always " i' the vein " to exercise his faculty, and that it was a power which came and went. He would sometimes say to me, " Do not think that I can always see a thing when I desire it; on the contrary, the more I wish the less I can do." In conformity with this, I observed that he always did things best when summoned to them without preparation, without any idea of difficulty being raised in his mind. It was also evident that he did not like being experimented upon, and hence I was obliged to use a little management with him, in order to make it appear that he was acting without constraint. Very often it happened that he would seem stupid, and unable to perceive a single thing, when, all at once, the power would, as it were, descend upon him; and he would suddenly read a whole page from a book, or tell five or six cards running. In the latter case, he was not accustomed to name the cards separately, as he looked at them, but afterwards, from memory. The first time he did this, I thought he did not know the cards. I showed him one, and he said, without telling what it was, " Now show me another." This also was put aside, and so on to five; when suddenly, to my no small astonishment, he repeated from memory the whole

five, correctly in their order.  This was one of his
self-originated caprices.  Another was, that he would
occasionally refuse to tell a card which he, however,
manifestly knew.  That he did so I discovered in the
following manner : — He had refused to tell a card,
saying, " I know it, however."  " Indeed," I replied,
" I doubt that."  " Well," he said, " ask me what
the card is, as soon as I awake.  Impress on my mind
that I am to remember it then, and I promise you
then to tell it you."  He kept his word exactly.  It
is not to be supposed, however, but that, by exercising
my influence over the will of the sleepwaker, I could
at any time compel him to execute whatever was in
the compass of his ability; but I preferred allowing
his mesmeric character to develope itself spontaneously.  By pursuing a similar course with all my
sleepwakers, I was enabled to ascertain that the mesmeric state, though broadly marked, has, in addition
to its striking and invariable characteristics, slight
shades of difference which individualise it in the
individual cases.  These generally consist in some
little caprice, or, rather, odd development of sensitiveness.  For instance, E. A—— could never endure
spades amongst the cards I showed him, and Theodore's sister shuddered and showed marked dislike if
wadding were presented to her, even in a box.

I have also remarked that sleepwakers, when once
disturbed by the presence of objects that influence
them disagreeably, become indisposed to a further
exertion of their faculties.  Any thing like a doubt of
their sincerity will also distress them exceedingly,

and obstruct the exercise of their powers. Moreover, they display extraordinary penetration in discovering which of the persons around them entertain feelings of incredulity or suspicion ; and, should they have to encounter a large amount of unbelief and hostility to mesmerism, they will become reserved and irritable, and will fail in everything they undertake. Thus I have known E. A——, after having told a card held behind a book, seem quite incapable of repeating the effort, from the moment that a certain person, who was sceptical about mesmerism, entered the apartment. I found, too, that the sensibility of the sleepwaker might be exhausted by a multiplicity of experiments, or their too rapid repetition. Sometimes, after having named many objects correctly, he would begin to make mistakes, and evidently to guess instead of to perceive. At other moments, he would push impatiently away from him the cards, books, &c. that were presented to him, and exclaim, " Maintenant je ne puis plus." Again, when allowed to remain quiet for awhile, he would recover his *clair-voyance*, in the same manner that the nervous energy of persons in the normal state, when impaired through over-excitement, returns to its pristine functions after an interval of repose. That my own state of mind, or body, or both, influenced the sleepwaker, it was impossible to doubt. Nor can this seem surprising, when it is considered that I was the depository and the dispenser of the agency which threw the patient into the condition of *clair-voyance*. I observed that on days when my thoughts were pre-occupied, or my health a little out of order, E. A——, in the mesmeric state, was dull, spiritless, and disinclined to exertion. The variations of the atmosphere seemed also to affect him. In dry clear weather his mesmeric faculties were best developed; on damp misty days they were less alert; and when thunder-storms were passing they were singularly disordered. Mesmeric passes, renewed at intervals during the period of sleepwaking, were, under all circumstances, favourable to the improvement or the maintaining of the patient's *clair-voyance.*

It is absolutely essential that the experimentalist in mesmerism should be acquainted with particulars such as the above, and should inform himself of all the conditions under which mesmeric sleepwaking is either promoted or impeded. This has been too much forgotten. Many failures, which have stamped mesmerism as an imposition, may be attributed, I am convinced, to the action of disturbing causes, or the absence of those circumstances which are requisite to ensure success. That this has not been acknowledged on all hands is, perhaps, as much the fault of mesmerisers themselves as of their opponents. The former, proud of the faculties of their patients, do not like to admit that these faculties are variable and liable to a number of restrictions. They therefore fail to forewarn those whom they invite to witness their proceedings that the whole exhibition may chance to be a failure, and that the *clair-voyant* of to-day may be nothing remarkable to-morrow. What is the consequence of this mistaken disingenuousness? Even they, who, if duly advertised of the true state of things, would be the first to acquiesce in the necessities of the case, are revolted by finding a discrepancy between the performance and the promise — the fact and their expectation of the fact. Mesmerisers, then, cannot be too careful in stating all the drawbacks to their success; and, at the same time, every person should, in all fairness concede to mesmeric experiments the same privilege which is accorded to all others, namely, a precognition of those causes which may render them difficult or impossible to be repeated. At present, it may be safely asserted that never was any subject capable of physical experiment submitted to such unjust requisitions as that of mesmerism. It has been expected to give the same results at all times and under all circumstances. The truth, however, is that mesmeric sleepwaking does not only present different degrees in different persons, but in the same. The patient may at one time be mesmerised, but not to *clair-voyance*; at another he may display the most admirable faculties of the mesmeric state.

Having ascertained, in this and the previous case, that persons under the mesmeric conditions could exercise a faculty analogous to sight, without the intervention of the ordinary apparatus of vision, I was desirous still further to inquire how far the optic nerve played a part in this development of the sentient powers. In order to solve this question, it appeared to me essential to mesmerise a person in whom the optic nerves were inefficient or destroyed. Should such a person be found to see in the mesmeric state, it would thenceforth be evident that man might possibly, in certain states, exhibit a perception of objects of sight which could have nothing in common with the system of ordinary vision.

Soon after the idea had arisen in my mind, accident threw in my way a lad of nineteen years of age, a Swiss peasant, who for three years had nearly lost the faculty of sight.   His eyes betrayed but little appearance of disorder ;   and the gradual decay of vision which he had experienced was attributed to a paralysis of the optic nerve, resulting from a scrofulous tendency in the constitution of the patient. The boy, whom I shall call by his christian name of Johann, was intelligent, mild-tempered, extremely sincere, and extremely unimaginative.  He had never heard of mesmerism till I spoke of it before him, and I then only so far enlightened him on the subject as to tell him that it was something which might, perhaps, benefit his sight.   At first he betrayed some little reluctance to submit himself to experiment, asking me if I were going to perform some very painful operation upon him ; but, when he found that the whole affair consisted in sitting quiet, and letting me hold his hands, he no longer felt any apprehension.

Before beginning to mesmerise, I ascertained, with as much precision as possible, the patient's degree of blindness.   I found that he yet could see enough to perceive any large obstacle that stood in his way.   If a person came directly before him, he was aware of the circumstance, but he could not at all distinguish whether the individual were man or

woman.   I even put this to the proof.   A lady of our
society stood before him, and he addressed her as
" mein herr" (sir).   In bright sunshine, he could
see a white object, or the colour scarlet, when in a
considerable mass, but made mistakes as to the other
colours.   Between small objects he could not at all
discriminate.   I held before him successively a book,
a box, and a bunch of keys, and he could not distinguish between them.   In each case he saw something, he said, like a shadow, but he could not tell
what.   He could not read one letter of the largest
print by means of eyesight ; but he was very adroit in
reading by touch, in books prepared expressly for the
blind, running his fingers over the raised characters
with great rapidity, and thus acquiring a perception
of them.   Whatever trifling degree of vision he possessed could only be exercised on very near objects :
those which were at a distance from him he perceived
not at all.   I ascertained that he could not see a
cottage at the end of our garden, not more than a
hundred yards off from where we were standing.

These points being satisfactorily proved, I placed
my patient in the proper position, and began to mesmerise.   Five minutes had scarcely elapsed when I
found that I produced a manifest effect upon the boy.
He began to shiver at regular intervals, as if affected
by a succession of slight electric shocks.   By degrees
this tremour subsided, the patient's eyes gradually
closed, and in about a quarter of an hour he replied
to an inquiry on my part, " Ich schlaffe, aber nicht
ganz tief." (I sleep, but not soundly.)   Upon this

I endeavoured to deepen the patient's slumber by
the mesmeric passes, when suddenly he exclaimed,
his eyes being closed all the time, " I see ! —I see your
hand ! I see your head." In order to put this to the
proof, I held my head in various positions, which he
followed with his finger : again, he told me accurately
whether my hand was shut or open. " But," he
said, on being further questioned, " I do not see distinctly. I see, as it were, sunbeams (sonnen strahlen)
which dazzle me." " Do you think," I asked, " that
mesmerism will do you good ? " " Ja freilich" (yes
certainly) he replied: " repeated often enough it
would cure me of my blindness."

Afraid of fatiguing my patient, I did not trouble
him with experiments ; and, his one o'clock dinner
being ready for him, I dispersed his magnetic sleep.
After he had dined I took him into the garden. As
we were passing before some beehives he suddenly
stopped, and seemed to look earnestly at them : —
" What is it you see ? " I asked. " A row of beehives," he replied directly, and continued, " Oh !
this is wonderful ! I have not seen such things for
three years." Of course I was extremely surprised ;
for, though I had imagined that a long course of mesmerisation might benefit the boy, I was entirely un-
prepared for so rapid an improvement in his vision.
My chief object had been to develop the faculty of
sight in sleepwaking ; and I can assure my readers
that this increase of visual power in the natural state
was to me a kind of miracle, as astonishing as it was
unsought. My poor patient was in a state of abso-

lute enchantment. He grinned from ear to ear, and called out, " Das ist prächtig ! " (That is charming.) Two ladies now passed before us, when he said, " Da sind zwei frauenzimmer ! " (There go two ladies.) " How dressed ? " I asked. " Their clothes are of a dark colour," he replied. This was true. I took my patient to a summer-house that commanded an extensive prospect. I fear almost to state it, but, nevertheless, it is perfectly true, that he saw and pointed out the situation of a village in the valley below us. I now brought Johann back to the house, when, in the presence of several members of my family, he recognised, at first sight, several small objects, (a flower-pot, I remember, amongst other things,) and not only saw a little girl, one of our farmer's children, sitting on the steps of a door, but also mentioned that she had a round cap on her head. In the house I showed Johann a book, which it will be remembered he could not distinguish before mesmerisation, and he named the object. But, though making great efforts, he could not read one letter in the book. Having ascertained this, I once more threw Johann into the mesmeric state, with a view to discovering how far a second mesmerisation would strengthen his natural eyesight. As soon as I had awaked him, at the interval of half an hour, I presented him with the same book, (one of Marryat's novels,) when he accurately told me the larger letters of the title-page, which were as follow : — " OUTWARD BOUND." Johann, belonging to an institution of the blind, situated at some distance from our residence, I had, unhappily, only the opportunity of mesmerising him three times subsequently to the above successful trial. The establishment, also, of which he was a member, changed masters; and, its new director having prejudices on the score of mesmerism, there were difficulties purposely thrown in the way of my following up that which I had so auspiciously begun.

The following is the general result of my after experiments : —

On first passing into the mesmeric state, Johann always spoke of a kind of internal light, which he compared to sunbeams, diffusing itself over the region of the forehead.

Whenever I pointed the tips of my fingers towards his closed eyes, at the distance of about two inches, with a quick darting motion, he had the sensation of a flashing light, and sparks of fire passing, as it were, before him.

Being led up, accidentally, to a large mirror, when in sleepwaking, he called out that he saw " eine grosse klarheit." (A great clearness.) Nevertheless, the mirror was in the shade. After this, I conducted him to a glass door that led into the garden, through which the light of day was shining brightly, but he made no remark; and, on being questioned, declared that he was not sensible of any peculiar light. Again taken up to the mirror, he again said that he saw before him much light and clearness. By whatever route I led him up to the looking-glass, he was always aware when he came before it, though his eyes were perfectly closed.

Occasionally I presented the points of my fingers to the mirror, in the same manner as to his forehead, in order to ascertain whether he would perceive any thing like reflected sparks, but the experiment did not succeed.

Music seemed to have a pleasing effect upon him when in the mesmeric state; and the sound of my voice always palpably increased the depth of his slumber.

On first awaking from mesmeric sleepwaking, the patient's powers of vision were always stronger than at any other time; but, in addition to this temporary benefit, there was a gradual bettering of his eyesight, which, though less striking, was more valuable from its permanence. Even the external appearance of his eyes was improved, in the course of mesmerism, to a degree which attracted the notice, and excited the wonder, of the master of the institution to which Johann belonged.

On one occasion, being rather indisposed, I found that I could not influence Johann so forcibly as usual; so that, after long mesmerisation, I had only brought him as far as an imperfect sleep, in which he retained his consciousness. Having met with an account of Dr. Elliotson's experiments, by which it is proved that the mesmeric agency is capable of increase by means of other individuals co-operating with the mesmeriser; having also experienced the truth of this when mesmerising the little sister of Mademoiselle M——, conjointly with herself; I requested a friend, who was present, to aid me, by motions of the hand, in deepening the patient's slumber. Each of us held a hand of Johann, and each of us manipulated with the hand that remained at liberty. The effect was very remarkable. In a short time the patient passed into complete sleepwaking; but that there was a remission of the mesmeric influence, whenever my friend ceased to be in contact with me, was proved by this: Johann's head did not then follow my hand so readily; and, at such moments, when questioned, he said that he did not sleep so profoundly. The patient being still unconscious, I, being always in contact with him, drank half a glass of sherry, when he exclaimed, spontaneously, " Das ist wohl stark. Das steigt mir im kopf." (That is very strong. It mounts into my head.)

The last time that I mesmerised Johann was in the evening, by candle, or, rather, lamp light. On this occasion he manifested an extraordinary increase in mesmeric *clair-voyance*, giving proofs that he had sensations, analogous to sight, of a far stronger nature than those which his visual organs could afford him in the waking state. With ease he indicated the relative positions of the party present, consisting of three persons besides myself; and, though the several individuals often and silently exchanged places, he continued to show that he was acquainted with the exact situation of each. Occasionally he would remark, and always with perfect correctness, that a lady was smiling and pointing her finger at him. Three dahlias, which were respectively of a bright

N

scarlet, deep crimson, and yellow colour, were held
before his closed eyelids. He discriminated between
them with singular accuracy, saying, " Das ist feuer-
roth, das ist dunkel-roth, und das ist gelb. " He
also distinguished a large leaf, which was held before
him, to be green.

The lady above alluded to handed me a nosegay,
directing me, in English, what to do with it. Agreeably to her request, I gave the nosegay, consisting of
red geranium, white stock, and other flowers, to
Johann, telling him that he must select some of the
red flowers to give to the lady. He instantly and
accurately separated the geranium from the other
flowers : "and now, I said, you must add some of
the white to your bouquet. " This he also did with
equal readiness.

Again, he told the letters, B, M, and O, which I
wrote in a large printing hand on pieces of card and
held before his closed lids. When led before the
mirror, which was then in deep shade, being at the
farther end of an apartment forty feet in length and
lighted only by a single lamp, he, as usual, expressed his perception of " etwas hell und heiter "
(something clear and bright) ; but, when brought
close up to the lamp, he made no observation of the
kind. Again I took him to a glass over the chimney-piece, on which the light of the lamp fell strongly,
when he cried out, " viel licht, viel licht! " (Much
light, much light !) While the patient was still in
the mesmeric state, tea was brought in. I ate some
dry toast, while holding Johann's hand. He imi-

tated the movements of mastication; and, on being asked what he tasted, replied, " Bread of some kind." Upon this one of the party present, without speaking, gave me quickly a piece of sugar, signing me to substitute it for the toast. This I did, and the sound which I made in eating was not perceptibly changed, yet Johann instantly and spontaneously exclaimed, " I taste something sweet."

I here close the proofs which I have to offer, that the perfect mesmeric sleepwaker does indeed possess a faculty of perception apart from the mere external mechanism of the senses. The means which I have been led to take, in order to convince myself of this fact, have, besides conducing to the end which I had in view, been productive of other results, which some persons may deem more important to the welfare of humanity than any discovery of new modes of sensation or of extraordinary developments of vision. I have established beyond a doubt that the action of mesmerism is highly remedial in affections of that precious organ whereby we enjoy

" Or sight of vernal bloom, or summer's rose,
Or flocks, or herds, or human face divine. "

Not only as regards Johann have I proved this, but also in the case of E. A——, who, when I first began to mesmerise him, suffered so much from a weakness in his eyes that he could neither see to read nor write by candlelight. At first, indeed, before I became fully aware of the efficacy of mesmerism as a remedy, E. A——, by my advice, occasionally used a lotion of rose-water for his eyes;

but in sleepwaking he declared that mesmerism would cure him without the employment of any additional means. The lotion, therefore, was discontinued, almost in the beginning of the mesmeric treatment; and, under the beneficial influence of mesmerism alone, the patient not only recovered the healthy action of his sight, but, before he left me, attained to the enjoyment of a remarkable strength of vision.

Again, I once mesmerised another boy, nearly blind, belonging to the same institution of which Johann was a member. After about a quarter of an hour's mesmerisation, he fell into a profound sleep, which lasted for nearly four hours. Though the patient gave no other token of being mesmerically affected, his sleep was evidently the result of mesmeric action, for neither calling to him nor shaking him could dispel it; and he walked without awaking from a chair to a sofa, where he remained till he awoke. This sleep was followed by an improvement in his powers of vision ; and, when I saw him at the institution some time afterwards, he told me that the benefit had been lasting, for that he then saw better than before I had mesmerised him, adding that he much wished I could be permitted to mesmerise him again.

These are facts which can scarcely be regarded with indifference, even by those who will not see in mesmerism a promise that transcends this life and the purposes of our present organisation.

# BOOK III.

## SECT. I.

CONFORMITY OF MESMERISM WITH OUR GENERAL
EXPERIENCE.

HAVING now described the state of mesmeric sleepwaking and its accompanying manifestations, I am
desirous of elucidating it by such reasonings as may
prepare the minds of men to receive this peculiar
phasis of our nature as a confirmed fact, not as an
insulated phenomenon; — as a link in the eternal
chain of things, not as an interruption to the universal order. With this end in view, it is necessary
to inquire under what conditions men believe that
they comprehend any thing whatever; and, this being
ascertained, it is clear that we must endeavour to place
under those conditions the object we wish to explain,
and to adapt our arguments thereupon to the disposition of the listener's mind. Attaching something
mysterious to the idea of causes in general, and to
all that relates to the explanation of unusual circumstances, we are too apt to forget that causes are themselves but facts, and that to explain is only to make
the unknown clear by a reference to that which we

N 3

already know. But, if we rightly consider the springs
whence arises our seeming comprehension of any
subject, we shall find that they all reduce themselves
to three very natural and intelligible sources, namely,
traditional faith — personal experience — and an idea
of adequate power.

For, I ask, when is it that we rest contented with
our knowledge of a fact?

Undeniably, then, 1st, When we have accepted it,
as a received truth, from infancy.

2dly. When we have felt it in our own persons.

3dly. When we imagine that we can refer it to an
adequate cause.

For, as to the first condition of our belief, let us
consider what a multitude of things we trust in
merely because we have heard them from our nurse's
lips, and though, these things should be mere errors,
we still find that to tear them from the hold they
have upon our minds is always difficult — often impossible. In such a case, we do not reason upon our
notions; — we do not strive to render them more
clear; — we are indifferent about mounting to their
source; — we rest in them with simple confidence.
The very religion of many persons reposes upon such
a basis, and this, relating as it does to our highest
interests, may serve as an instance of the power of
early association and of the traditional credence upon
which I am insisting.

Secondly; it is no less evident that we accept, on
the testimony of our personal experience, whatever
we find assimilated thereunto. No one dreams of

doubting that other men may sleep, eat, walk, or write. The force of this principle may be gathered from the strength of its antagonist feeling. We have an actual repugnance to credit that which we have not felt ourselves. To give an example of this. There is a singular state of mind, which is known to some, and which has been adverted to by various poets (as, for instance, Coleridge) —

> " Which makes the present, while the flash doth last,
> Seem but the semblance of an unknown past,"

when, according to Tennyson's more particular description of this mental phenomenon, —

> " We ebb into a former life, or seem
> To lapse far back in a confused dream,
> To states of mystical similitude.
> If one but speaks, or hems, or stirs his chair,
> Ever the wonder waxeth more and more ;
> So that we say — ' All this hath been before ' —
> All this hath been, I know not when, or where."

Now let any one, who has experienced this perplexing reference to events which he seems to have rehearsed in some prior state of existence, speak of the feeling to another who is wholly ignorant of it. The latter will very likely exclaim, as a friend of mine to whom I described this mood of mind actually did exclaim, — " Had I ever felt any thing like what you mention, I should think myself fit for Bedlam."

On the other hand, he, who knows the sensation and can sympathise with it, will listen to a similar statement without surprise, and will probably say — " I can well believe this, for I have felt it myself."

A more common instance of this incredulity respecting all that lies beyond our own experience is our reluctance to believe that pains, which we have never known, can really be as acute as they are represented to us. " He jests at scars, who never felt a wound," and he, who has been always free from the toothach, will almost laugh at a friend under a paroxysm of that torment; not because he is hardhearted, but because he cannot conceive the matter deeply enough to be serious.

There is an actual inherent propensity in human nature to make personal knowledge the measure of truth, so strong as to incline us to believe any thing that has reference even to our prior experience. I have read somewhere of an old woman whose son, returning from foreign parts, related to her the wonders he had seen. He spoke of flying fish, and of a burning mountain; but the cautious mother cried out, "No, no, Jack! I know what travellers' tales are! *That* I'll never believe!" At last, after many similar truths rejected, the sailor, in despair, hatched a lie, and said, " In Jamaica I saw sugar growing on trees, and rivers of rum." " Ay, ay, child," exclaimed the beldam, " *Now* you speak sense. *That* I'll well believe, for I *know* that rum and sugar come from Jamaica! " Now, in mesmerism, it is just thus. There is much rejected that accords not with the pre-

vious experience of persons, yet perhaps every one
has his rum and sugar reservation — some greater
marvel than those which he refuses to credit.  After
all, it is *not* the marvellousness of mesmerism that
shocks; it is only its discrepancy with each individual's experience or figment of the brain.

In the third place, we need reflect but little in
order to perceive that we rest contented with our
knowledge of events so long as the idea of an
adequate cause for them is present to our minds.
Yet this tendency of our nature to refer events not
only to a cause, but to an adequate cause, has not
been sufficiently distinguished from the simple propensity (also inherent in our intellectual structure),
which inclines us, when we see an effect with which
we are familiar, to suppose that its usual antecedent
either is, or has been, also, present ; for we are not
only inclined to take things in connection, and to view
the relationship between them, but likewise to account
for the phenomena we behold by tracing them to
their primary and efficient impulses.  When we fail
to ascribe an event to an adequate agency, we are
troubled and discontented, and a restless idea of the
supernatural (here truest) haunts and agitates us : on
the other hand, whenever we deem that we can assign
a sufficient cause to an appearance, however startling
and unusual, we are satisfied, and seek to know no
further.  For this reason, they, who are in the habit
of overlooking all the intermediate processes of nature,
and who resolve every thing at once into the power
of God, enjoy a pleasure distinct from the merely

N 5

pious one of attributing every thing to one beneficent
Creator. They rest in a sufficient cause; they are
tormented by no importunate questionings of spirit;
their answer to every enigma of the universe is ever
ready. They who are not of this happy, but (it may be
suspected) indolent frame of mind, and who, at the
same time, have not knowledge sufficient to ascend,
surely and gradually, the steps of nature's temple,
are actually forced into an irregular flight upwards,
and, rather than ascribe perplexing appearances to an
inadequate cause, will attribute them to an unknown
and mysterious one (the inadequacy of which cannot
be put to the proof at least), to some cloud-born
agency, which flatters their imagination with a sense
of power. Hence, where gross ignorance prevails,
superstition, with all its notions of demoniac influences,
takes its rise; hence, in darker ages, the learned have
been deemed wizards, and the man of science has
been supposed to have signed a contract with Satan.
Hence, even in our own day, some, whose minds are
in arrear with time, will attribute the whole of mesmeric agencies to the machinations of the evil one,
being unaware how much they are influenced in their
judgment by the propensity we all have to cut a knot
when we cannot untie it. But, after all, the fondness
of some persons for diabolic agency is not a little
surprising and unaccountable. For my own part I
had always rather rest in God than in the devil.
Observe we, too, that many a person will deem himself vastly enlightened by a fine word, and that hence
knowledge is too often turned into a few high-sounding

phrases.  Wherefore is this ?  A pompous term may
stand as the representative of an efficient force; and,
in the case where the event to be accounted for appears very wondrous, the alleged cause always pleases
the ignorant better, the more incomprehensibly it is
expressed.  But, where phenomena come, or seem to
come, nearer the level of our experience, then we are
apt to explain them by some familiar force, which,
however, may have as little to do with them as any
cause with which we are entirely unacquainted.  This,
however, is hard to be perceived, for, whatever we
know and have proved to be capable of effecting something, we are inclined to exalt into a power and to
force into offices alien to its sphere of operation.  For
instance, every man has felt that the frequent repetition of any action increases the tendency to it.  This
phenomenon of our nature we personify under the
name of Habit, and recognise the potency of its influence, binding us, as the Lilliputians did Gulliver,
not by the individual strength, but by the number, of
its subtle bands.  Again, when one part of our bodies
is deranged, another often partakes in the affection,
or, as we express it, suffers by sympathy.  Sympathy,
then, must be accounted as a power, explanatory of
many mysteries of our being; so must association —
so must imitation, and a host of other shadowy im-
personations, which are but names for certain mental
or bodily states.  To words like these, when hard
pushed for an explanation of any thing, we can resort
with much complacency ; not that, after having done
so, we understand the matter one whit better than

before, but we have referred it to a principle with which we are in part acquainted, and which, deeming it the efficient cause of some things, we suppose may also produce the one in question.

These strivings after adequate causation I consider to be the upward and inevitable tendency of mind to its sources of reason, teaching us by internal admonitions that all phenomena whatever emanate from thought; but, as I shall have occasion hereafter to make some remarks on power, and whence our notion of it is derived, I will, at present, advance no deeper into this subject, having, I trust, said enough to show that we rest satisfied with our knowledge of events in proportion as we think that we can refer them to an adequate cause; and that consequently the surest means to obtain general credence for any fact is to exhibit it in connection with a cause which shall at once be perceived adequate to its production.

Applying the above reflections to mesmerism and its phenomena, we cannot but perceive that of the three great sources of human conviction which we have been considering, — namely, traditional faith, personal experience, and adequate agency, — there is not one which has been called into play in a manner favourable to our subject.

Whatever may be the case on the Continent, it will be admitted that in England the present generation have been educated in a profound contempt and disbelief of mesmerism. Thus from traditional faith it receives no support, but the contrary; for the lessons of our childhood have raised a barrier between

it and us, the strength of which no one can have felt
more forcibly than myself.

From personal experience mesmerism is not less
removed; and it must be confessed that writers on
the subject, in general, have not contributed to bring
it nearer to the home-felt conviction of mankind. Too
often they have endeavoured to illustrate it by a reference to marvels as far distant from personal experience as itself. In its behalf, ancient chronicles have
been ransacked; classical authorities have been marshalled in formidable array; the tomb has given up
its vampires, and Egyptian mysteries have been dragged to light, — for what purpose? Mesmerism is
not at all rendered more credible by being bol-
stered up by histories which themselves are doubtful; nor is it explained (*ignotum per ignotius*) by
means of phenomena that surpass human comprehension.

In the third place, it appears to me that mesmerism
has not been viewed in connection with an adequate
cause, since, as I have elsewhere observed, men of
science would refer it, with all its intellectual phenomena, to a simply physical agency; while men of
imagination would force even its physical affections
into the domain of pure spiritualism.

It should then be the object of a writer on mesmerism to supply his subject with those advantages
of which concurring circumstances have tended to
deprive it. Others, I feel confident, will effect far
more than I can in this cause; yet I trust I shall
not be deemed presumptuous in bringing forward

some considerations which have offered themselves
to my mind, tending to approximate at least the new
condition of man to those grounds of credibility on
which I have shown our quiet possession of truths
in general to repose.

Of the three claims to confidence, wherein mesmerism is deficient, the hardest to supply is undoubtedly
that of early association. Yet even here it may be
urged that we have all, from childhood, supposed
natural sleepwaking to be an undoubted phenomenon of man's constitution. There are many, perhaps, who remember to have heard tell of some
sleepwalker, whose feats have been the wonder of
their native village, who has been known to rise from
his bed, and to display, in slumber, even more than
his ordinary activity, crossing, it may be, some perilous ford, or balancing himself on the edge of a precipice, where the waking eye would sicken and the
conscious brain would reel. Who does not believe in
the existence of such a state? Doctors have descanted
upon it with the precision of medical lore; metaphysicians have examined it as a curious feature of
humanity; and the light and gay, regarding it as a mere
matter of amusement, have flocked to see its mimicry
in dramatic representation, enhanced by all the charms
of music and the fascinations of genius. Nay, in the
higher departments of scenic art, it has illustrated one
of the sublimest creations of the poetic mind: — I allude,
of course, to the tragedy of Macbeth, in which Shakspeare, who painted every shade and variety of our nature, has not left untouched this its most peculiar and

interesting development.    In brief, it is one of the
truths which cannot be torn from the texture of our
minds, or erased from the book and volume of the
brain.

This being the case, is there any good reason why
we should exclude from our belief so similar a state
of man as is mesmeric sleepwaking; — or, rather, is
not our conduct, on this occasion, the most striking
proof of the really inefficient grounds on which the
greater part of mankind believe or disbelieve anything?    How greatly we are inconsistent with ourselves !  We hear, without one feeling of doubt, tales of
sleepwaking, quite as wonderful as the following account of Lady Macbeth's sleepwaking :—" I have seen
her rise from her bed, throw her night-gown upon her,
unlock her closet, take forth paper, fold it, write upon
it, read it, afterwards seal it, and again return to bed ;
— yet, all this while, in a most fast sleep :"—but,
should the very same things be stated respecting a
person under mesmerism, the name seems suddenly
to make all the difference : we no longer believe; we
cry out that it is all an imposture ; — our whole being
is in a state of hostility and agitation.   Is this wise?
Is this in accord with human dignity of thought?    I
would more particularly ask medical men whether
they have not been accustomed to regard as true not
only certain phenomena of natural sleepwaking, to
the full as wonderful as any which mesmerism presents, but also of catalepsy — a state under which the
most extraordinary developments of nervous sensibility may take place; and I would then urge them to

examine well the real causes of their denying in one
case what they accede to in another.   Also I would
inquire, with what patience they who have attended,
in a medical capacity, many an unfortunate wretch
dying of hydrophobia, can listen to the prabble of a
man who contends that human nature is liable to no
such affliction, merely because he has heard from
childhood the fact spoken of as doubtful ?   If no good
answer can be given to these questions,  surely a reasonable being may suspect that there is some great
error at the root of all these discrepancies and con-
tradictions in feeling and belief; and it may be contended that, if we accept traditional faith as the
ground of our confidence in certain particulars, we
must retain it still as corroborative of the same particulars, whatever change in their nomenclature may
occur.

In truth, when we demand acceptance for many
of the phenomena of mesmerism, we only require a
sacrifice of men's prejudices, and not a dereliction
from their ancient principles of belief.

The mention of natural sleepwaking leads us
almost insensibly into realms that border closely upon
those of personal experience.   There are few of us
probably who have not, in early years, had some
touch of this malady, so incident to childhood, or, at
least, so far beheld its effects in a brother, sister, or
young relative, as to be convinced that persons may,
being all the time asleep, perform many of the actions
of their waking hours, and answer short questions,
either rationally or with a certain ingenuity.   Many

also have perhaps had the opportunity of hearing
stories relative to sleepwaking from the lips of the
parties actually concerned in them.   At least, I can
affirm that, in the course of my own experience, I
have met with several persons, who have either been
sleepwakers themselves, or have been eye-witnesses
of sleepwaking in others.   On such individual testimony, I present to my reader the following nar-
rations, which show at once the connection and the
difference  between  natural  and  mesmeric  sleepwaking.

Mr. Matthieu, president of the Academy of Painting  at Louvain, told me that, not many years since,
having an essay to write for what the French call a
*concours*, he was in an anxious state of mind.   His
thoughts did not flow readily, and, after many efforts
at composition, he could  not  produce anything to
please himself.   What was his astonishment, one
morning, to find upon his table, in his own handwriting, the beginning of an excellent dissertation
upon the subject which had lately occupied him !
He could not  doubt  that this was the work of his
sleeping hours ; — but, curious to ascertain how far
his mind would proceed with its train of thought
under the same conditions which had originated it, he
left the manuscript untouched, until he could inspect
it again after the interval of another night.   The result of the experiment showed that he had resumed
the pen in sleep, and had continued his composition
in a perfectly connected manner.

Another gentleman, Mr. Z—— of Neufchâtel, related to me the following instances of his own sleep-waking. When he was about seventeen, his father being dead, the management of an extensive mercantile establishment principally devolved upon himself. One day he had transacted a great deal of business, and still there were certain invoices to be written before the departure of some merchandise early on the following morning. Worn out with fatigue, he went to bed, having given orders to be called in time to supply the deficient inventories. At day-break, according to his directions, he was awakened by a servant, and went immediately to his mother (who generally assisted him in his labours), informing her that he was ready to begin the invoices. "What can you mean?" she exclaimed; "do not you know that you wrote them last night? and here they are!" — At the same time she placed in his hand the papers properly executed, and in his own handwriting. On inquiry it appeared that Mr. Z——, half an hour after he had retired to his room on the preceding evening, had returned, dressed as usual, to his mother's apartment, where he sat down before a writing-table and completed the documents in question; Madam Z——, nothing doubting of his being awake at the time, though, on subsequent reflection, she remembered to have observed something fixed and strange about his eyes (they were open during his sleepwaking), — an appearance which she had attributed to her son's fatigue and want of sleep.

On another occasion — not more than a few months ago — Mr. Z—— formed, over-night, an engagement to cross the mountains with a party of friends, who were to call for him early in the morning. This arrangement gave rise to the following curious circumstances, which subsequently were made evident by collating the different particulars of the case. It happened that, on the same night, when Mr. Z——— was fast asleep in bed, a party of young men, strangers to him, stopped under his window accidentally, and, being not very sober, laughed and talked loudly for some time. The noise influencing the sleepwaker, gave him the idea that his friends were come to call him. Under this impression he arose, threw open his window, and addressed the party in familiar terms. The young men, being in an irritable and touchy state, replied to him in no agreeable manner, until a quarrel arose between the sleeping man and the group below, who were in no better possession of their senses than himself. The tumult was only appeased by the watchman dispersing the rioters, and counselling the sleepwaker (not suspected to be such) to return to bed. The next day Mr. Z——— retained no recollection whatever of the events of the night, and made, as agreed upon, the expedition with his friends. On his return, however, to Neufchâtel, two of the gay party, who had been under his window, waited on him, and, to his utter astonishment, begged him to explain why he had called to them, and, in a manner, insulted them, on such a night. "How could I insult you, when I never saw you before?" was Mr. Z———'s exclamation ; and great was the perplexity of both the parties, until Mr. Z———, remembering his addiction to sleepwaking, suggested an explanation, which every circumstance tended to confirm.

Mr. M———, also of Neufchâtel, has a brother, who is subject to sleepwaking. The following is an amusing specimen of his exploits in that way. Mr. M——— was on a journey with him, and one night, when they occupied the same apartment, the sleep-waker awoke the former by loud and earnest talking. There was a light in the room, and Mr. M——— (as he told me) could see his brother standing upright on the edge of his bed, occasionally leaning forward and peering over (though his eyes were shut) with the expression of a man who earnestly gazes after some-thing. " What is the matter with you, John ? " asked Mr. M———. " Oh," replied John, " take care of your steps, for Heaven's sake ! Do not you see that we are just on the brink of a horrible precipice? "—" Not at all," rejoined Mr. M——— ; you are under a delu-sion, and I pray you to calm yourself ! "—" What ! " said the sleepwaker, " you say there is no preci-pice? I 'll throw a great stone down to prove it to you. — There ! " he cried, snatching off his night-cap, and flinging it on the floor, — " there it goes ! — and what an awful depth the precipice must be, for the stone makes no noise ! "

Such facts as the above will perhaps be credited — just because they are not headed " Mesmerism." They are at least sufficient to show that many of the features of mesmeric sleepwaking have prece-dents which are to be met with in acknowledged phenomena common to man. These, however, rest

rather on personal evidence than on personal experience. The sleepwaker, when once awake, does
not remember the occurrences or the sensations of
his sleepwaking, and the spectator who bears testimony to his actions is a stranger to their inward
sources.

I would, however, give mesmerism a stronger hold
upon human conviction than can be supplied by any
external proof. I would introduce it into the very
bosom and private feelings of each individual, by
showing that parts of it, so far from being removed
from our daily life, are altogether familiar to us,
though under another guise and bearing other
names.

I would prove this — 1st. As regards the production of the mesmeric state. 2dly. As regards the
state itself.

First, then, let us inquire — Are the means by
which the mesmeric slumber is produced in any way
approximated to our experience?

The subsidiary agents, which have been found most
efficacious in aiding one person to influence another
in the manner called mesmeric, are the eye and the
hand.

Now, I would ask, who is there that has not felt the
influence of the human eye — that window of the
soul? What is there which equally with itself arrests
attention, fascinates, fixes, calms, and subdues? Anger
is quelled before its steady gaze; even madness becomes docile beneath its dominion. The late Dr.
Willis used to say that by the eye chiefly he controlled

his most refractory patients. Acting at once physically and morally, it seems to affect the nervous system in a real and peculiar manner; for brutes, which cannot be supposed to be misled by imagination, are extremely sensible to the influence of the human eye. Nay, even to certain of the brute creation a remarkable power of eye seems to belong. To say nothing of some doubtful stories respecting the fascination which serpents exercise over weaker creatures, I have seen a bird fluttering before a cat, as if held and bound by her watchful gaze, and so almost constrained to offer itself a voluntary sacrifice to its enemy.

When we consider these things we can scarcely marvel that powers even magical have been attributed, by ignorance and superstition, to the wonderful organ of which we are speaking; nor that diseases, and even death, should have been supposed, in some cases, to have originated in the stroke of an evil eye.

Now, be it remembered, that the doctrine of mesmerism is that the mesmeriser's force should become predominant over that of his patient; and mesmeric phenomena do really show, as I have formerly demonstrated, that the mesmeriser's will sways the patient's volition in a very peculiar manner. The relation between the mesmeriser and the patient is that of activity to passivity. The one is to guide, direct, and influence; the other is to receive, and to be influenced. The first process, then, of mesmerism may be called a strife between the two forces of mesmeriser and patient. There is a will to be subdued, and a body

to be brought into subjection; yet this is to be accomplished in all the tranquillity of a religious silence.
What instrument is adequate to such an effect?
Surely we should at once name the eye as the great
agent in this strife. By a sort of natural instinct, we
*do* use it, where we wish to prevail. The meanest,
striving with another, will call the first blenching of
his adversary's gaze, — victory. Men of energetic
characters and calculated for dominion have always
been remarkable for the potency of their glances,
and have been celebrated as not only " Lords of the
lion port," but also of " the eagle eye." Few could
endure the searching gaze of Napoleon. The rulers
of the domain of intellect have been as famous as the
kings of the earth for a keen regard. The glance of
Dr. Johnson from under shaggy brows availed no less
than the eloquence of the sage's tongue. I remember
that Dr. Parr, who was a sort of Johnson in his way,
was very proud of his own visual prowess, and I
once heard him boast that his eye had saved the
nation.

Pitt, he declared, was about to bring forward some
disastrous measure, when he, having previous knowledge of the statesman's evil intentions, repaired to
the House of Commons. " I fixed, sir," said he,
" my eye upon Pitt. I gave him one of my looks
(every body knows my looks), and he could not utter
a word, sir. He rose, stammered, and sat down
again, — and there I kept him off his legs till Fox
had taken possession of the ear of the house; and so
the measure, when brought forward, was nugatory.

Yes, yes, I think I *may* say that my eye saved the
nation."

In fact, we perceive that Dr. Parr mesmerised Pitt.
But, jesting apart, I need only refer to my reader's own
feelings for a testimony to the all but miraculous
powers of the human eye.  It seems to be life and
will made visible ; and, when life is fled, we may exclaim with the poet —

> " Oh, o'er the eye Death most exerts his might,
> And hurls the spirit from her throne of light."

If the eye be powerful to influence and to subdue,
the human hand is not less remarkable for its soothing properties.  We naturally apply our hands to any
part of our bodies in which we feel pain, and how
often are we sensible of a mitigation in our sufferings !
The mother soothes her child to sleep by the application of her hand, and calms it into patience, when it
has received an injury, by her gentle touches.  A
lady, on whose word I can rely, told me that, during
a heavy malady, she could gain no sleep except
when holding her husband's hand, and that the moment he withdrew it, however gently, she awoke.
Similar phenomena, if oftener attended to, might
doubtless press upon our remembrance and justify
mesmerism by the analogies of common life.  Touching
for the toothach is a sort of charm that is often re-
sorted to by the vulgar with success ; and in Con-
stantinople, I have been informed, there are men
who actually make it their business to cure toothach in this manner.  Once it was supposed that a

certain insect, being crushed and held to a painful
tooth, acted as an anodyne, and learned papers were
accordingly written on the virtues of this particular
remedy.  Unfortunately, however, it was soon discovered that the finger which applied the insect was
the real depository of the soothing virtue ; for the
desired effect took place just as well without the insect as with it.

A travelled friend of mine tells me that he has
seen a sort of mesmerism exercised by the dancing
Dervises of the East upon each other.  After whirl-
ing round in their mystic and exciting dance, they
are apt to fall upon the ground in strong convulsions.
On such occasions, one of the calmer brethren comes
up to the extatic patient, and, by gentle touches of
the hand, invariably succeeds in soothing him to repose.

That the hand should be an agent in dispensing
the human or the mesmeric influence can hardly
seem wonderful to those who consider how much the
hand is identified with, and distinctive of, man.
Considering the perfection of its form, and the dis-
criminative delicacy of its touch, we can almost
pardon those who have gone so far as to affirm that
the hand alone is the efficient cause of our supremacy
over the brute creation.  The tactile nerves, with
which it is so abundantly supplied, seem peculiarly
to adapt it to the task of transmitting an influence,
which, in some mysterious way, yet manifestly,
seems to be connected with the nervous and vital
forces.  In some old systems of physiology each

part of the body had its separate spirits: there were the spirits of the knees, the spirits of the feet, &c. Could we adopt so fanciful a notion, we should say that the spirits of the hand are particularly potent. After the eye, the fingers seem to enjoy a fuller share of the sentient and motory power than any other part of the body. Seldom are they at rest — seldom are their functions suspended. The marvels of the painters', sculptors', musicians' skill depend upon their agency. Apart from their recording powers, the headwork of the poet and philosopher would be vain: to them belong all the triumphs of architecture, all the splendid results of industry; in fine, all the success of all the ingenuity of man. Their free and constant exercise seems more than any thing to betoken life: the baby's clutched hand is a type of its unfolded being. The purposeless action of the same organ is a presage of death: " After I saw him fumble with the sheets," says Hostess Quickly, " and play with flowers, and smile upon his fingers' ends, I knew there was but one way."

Shall the general feeling of mankind be in vain appealed to? If that be true which Shakspeare has affirmed — " One touch of nature makes the whole world kin," — the influence of the human hand must be universally acknowledged. It has a language of its own; it can appeal from man to man; it can bless, and it can curse: the most ancient belief connects it with authority and power. The holding up of Moses's hands gave victory to the Israelites.

" And it came to pass, when Moses held up his hand, that Israel prevailed; and, when he let down his hand, Amalek prevailed." *

Gifts of healing, not less than of power, belong to the hand by prescriptive right. If the potency of the royal touch in curing the king's evil be but a super-stition, let us remember that it took its origin from a holy source : Christ and his disciples laid their hands upon the sick, and they were healed. The miracles of our Lord were remarkably accompanied by actions of the hand †, as if they were in some manner con-nected with that external means. In restoring sight and hearing, he touched the ears and eyes of the afflicted persons. Even the imparting of the gift of the Holy Spirit followed the imposition of hands; and this external ensign of a spiritual agency is still retained in our church. Who that has undergone or witnessed the beautiful rite of confirmation but has felt its power?

The eye and the hand, then, appear to be fitting instruments in transmitting a potential and remedial agency.

If we seek for such a general instance of the in-fluence of one human being on another as may seem like that mutual loss and gain and interchange of vital force which is the principal wonder in mes-merism, we have only to look at the effects produced when young persons sleep with old. Since the days

* Exod. xvii. 11.
† See Matt. viii. 15. ; Mark vii. 33., and viii. 25.

of King David *, it has been known that the latter
are strengthened at the expense of the former. Some
painful instances of this have fallen under my own
observation : one, in which the future well-being of a
person very dear to me was compromised ; and I
was acquainted with an infirm old lady, who was so
aware of the benefit that she derived from sleeping
with young persons, that, with a sort of horrid vam-
pirism, she always obliged her maid to share her bed :
thus successively destroying the health of several
attendants.†

Even amongst animals it has been found that the
young cannot be too closely associated with the old
without suffering detriment. Young horses, standing
in a stable beside old ones, become less healthy.

* 1 Kings, ch. i.
† The celebrated German physiologist, Hufeland, has remarked the longevity of schoolmasters, and attributes it to their
living so constantly amidst the healthy emanations of young
persons.

## SECT. II.

### ON THE MESMERIC CONSCIOUSNESS.

I NOW proceed to assimilate the state of mesmeric sleepwaking itself to our intimate personal convictions, striving to demonstrate that all its phases and phenomena are only intense degrees of known and even ordinary conditions of man. I would render it not only comprehensible, but familiar — not understood merely, but also felt.

I have elsewhere remarked that there are, in the mesmeric state, peculiarities of consciousness, of which I postponed the consideration until I could adduce them as explanatory of mesmeric sleepwaking. To this portion of my subject I now turn my attention. But, before proceeding to demonstrate that the peculiar conditions of mesmeric consciousness arise from novelty of combination — not of principle, it is absolutely essential that I should fix in what sense I use the word consciousness, — a precaution the more necessary on account of the different significations which have been attached to it by various metaphysicians; some making it a distinct faculty of the soul, others a necessary accompaniment of every mental operation and in no way to be distinguished from the operation itself.

Examining, however, whence springs this diversity

o 3

of notion respecting consciousness, we find that all
the confusion has arisen from this — Consciousness
is susceptible of various developments, which have
never been properly distinguished into their several
grades; one person applying the generic term in
question to one only of its specific degrees; another
equally restricting it to a different class of its operations. Dr Thomas Brown, indeed, has some admirable remarks upon the subject, and has well distinguished between the simple consciousness of the
moment and remembered consciousness. But he
has omitted to observe that there is again a difference
between these two actions of consciousness and a
third — namely, reflective consciousness, or internal
observation. It is true that he once casually observes,
" consciousness is only another name for internal observation ; " but herein I conceive him to speak with
less than his usual accuracy; for consciousness is *not*
another name for internal observation, but internal
observation is one of the operations of consciousness,
which, as not identical with its parent, should not
be involved with it in one common definition.
Simply to feel, or simply to pass again through a
succession of former feelings with a sense of their
relation to one personal identity, is not the same as
to be self-regardant and watchful of our own sensations as they arise. Under the last circumstances
the mind is manifestly in another state and tone of
feeling. Besides, nothing can be clearer than that
the mind *may* act without internal observation, while
without simple consciousness it *cannot* act; since, as

Dr. Brown himself logically observes, "The consciousness which we have of our transient thoughts or sensations is nothing more than the thoughts or sensations themselves, which could not be thoughts or sensations if they were not felt." But Dr. Brown errs as much, perhaps, from simplicity, as others from multiplicity, of system. Nothing, indeed, can be more admirable than his view of the mind — not as a bundle of faculties, but as a unity, capable of passing into various consecutive states. But, had not this great philosopher been as anxious to reduce all mental phenomena whatever to suggestion, as Locke was to compress the world of intellect into the two faculties of sensation and reflection, I think he would have admitted distinctions which his theory caused him to overlook. Nature, however, will not thus be restricted. She is simple, but her effects are manifold ; and the very necessities of our language demand as distinctive a reference to her various operations, as if the differences were actual, and of kind, instead of degree. Göthe, in his Morphology of Plants, has beautifully shown that all their various parts are but developments of one original principle; yet what confusion in botany would arise were some naturalists to apply the generic word *plant* to the leaves, others to the blossoms, of a vegetable ! Yet an absurdity as great as this is committed when we apply the generic term consciousness indiscriminately to all the different manifestations, degrees, and varieties of that action of intelligence. The safer plan is to mark difference, where difference can be

discerned, and to give distinct names to distinct developments, whether of matter or of mind. In doing
this, however, we have to guard against unnecessary
multiplication, and we should never omit to trace
the original source amidst the mazes of its various
channels. Simplicity of principle — variety of development, — these are the points to be kept in view.
Thus, while we admit sensation, perception, attention, reflection, to be excellent terms to express certain acts of the mind, we must not forget that these
so-called faculties are but different forms and manifestations of consciousness. A little consideration
will show that they are truly so.

A sensation of which the mind is not conscious is
no sensation at all; neither can we feel without perceiving, nor perceive without, in some degree, attending to that which we perceive. When we say that
we attend to a thing, which before we barely perceived, it is not that consciousness, perception, and
attention are going on together, but that consciousness has advanced through the stage of perception
into that of attention, — attention only expressing a
higher degree of consciousness, while reflection,
again, indicates a different operation of the same.
This last is attention directed towards ourselves,
instead of external things; but the principle is un-
changed. A reflection is just as much the consciousness of the moment as is a simple sensation.

If it be asked, then, in what sense I use the term
consciousness, I answer, as a general expression for
every act and state of the mind. I do not say with

Reid that it is "*an* operation of the understanding "
(though I do grant him that it cannot be "logically
defined "— for how can we define existence? ), but I
affirm that it is *the* operation, not of the understanding only, but of the whole man.   To explain
myself still further, I conceive that consciousness
stands in the same relation to the mind that physical
force does to the body.   As the least of our motions
implies a degree of the latter, so does the most
trifling mental change infer a degree of the former.
Let it be remembered, however, that consciousness,
like physical force, has no real existence out of the
subject that manifests it.   Nevertheless, for purposes
of analysis, we must not only speak of it separately,
but distinguish its various intensity of action.   In
order to do this, yet as much as possible preserve an
unbroken unity, I conceive that the best nomenclature for the varieties of consciousness would be
one that should ever recall the primary quality, yet
distinguish, by means of epithets, its various manifestations.   Omitting its minor developments, I
propose, therefore, to classify thus the operations of
the agency in question : —

Simple consciousness, — that is to say, the mind's
action in those absent and dreamy moods, when
much thought is accompanied by no reflection, and
is succeeded by no memory of the subjects of its
meditation.

Retrospective consciousness, — the mind's action,
when it passes through a series of former thoughts

and sensations, without making them objects of scrutiny.

Introspective consciousness, — the mind's action, when self-regardant. It is distinguished from mere memory in two marked particulars. It immediately succeeds the thought, on which it casts a reflective glance, and it has ourselves for its object. It is a state in which thought, and observation of thought, succeed each other so rapidly, and with such even alternation, as to seem identical.

But the end which I have in view compels me still further to observe that the mind has a physical as well as intellectual sensibility, and takes cognisance of the motions and sensual conditions of the body to which it is attached. Its observation of these seems to constitute a consciousness apart. The proof may be drawn from our simplest actions. When walking and thinking on various subjects, we never cease for a moment to direct our limbs properly, yet never once interrupt our train of thought, unless some unusual circumstance should call upon us for an unusual attention to our manner of progressing. Now allowing, as axiomatic, that bodily motions (the vital and involuntary of course excepted) spring from the mind, and that, cause ceasing, effect also ceases, we must perceive that to every act of the body an act of the mind is requisite. Of how many acts of the mind, then, is walking composed! consequently of how many simple consciousnesses! But other consciousnesses have been shown progressing identically with these. While occupied in walking, we pre-

served a chain of intellectual thought as consistent
and as unbroken as the physical one whereby our
external actions were directed.

This is a phenomenon of our nature, in the consideration of which we may indeed cast out

> " Philosophy's deep sea-line, but to find
> Truth's ocean fathomless."

But, while we confess it to be a mystery, we should
also value it as a blessing; for, were we forced ever
to attend to those motions, which may now be called
automatic, we should scarcely have leisure for any
other mental process. If the mind that now guides
the pen wherewith these lines are traced were occupied in the mere act of writing, what would become
of the thoughts which it desires to record and to
transmit to others?

That the physical consciousness is capable, like
the intellectual, of a triple development, is also
manifest.

It has its simple action; as when each motion of
our bodies is carelessly performed, and cannot be
remembered so as to be exactly repeated.

It has its retrospective action; as when we remember and repeat, yet do not observe our motion
(in the case of playing an air from memory, for example, while we are thinking of something else).
And,

It has its introspective action; as when we attend
to our footsteps, in descending, with precaution, a
snow-covered and slippery hill.

It would be easy to show the various states produced by the mixture of these elements in various
proportions, tracing the degrees from the constant
self-scrutiny, both mental and physical, that some
persons carry on in society, when they observe their
every least word, fearful to utter aught amiss, and
their own least gesture, lest they should commit an
awkwardness, to the unmixed and simple consciousness of reverie. But this would lead me too far from
my present purpose. The two states above mentioned may be taken as the extreme points of consciousness, while its medium condition may be found
in our usual and habitual modes of conduct, when
we are sufficiently attentive to our motions, not to
knock our heads against a wall, and to our thoughts,
as to be sure we are thinking.

Regarding these states according to their value,
we find that in none of them would man ever be
capable of much intellectual exertion. In the first,
the attention is too generally diffused, in its opposite
too restricted, in the medium too alternating. Would
we find man's distinguishing stamp of mental supe-
riority, we must seek it in a fourth state — namely,
that of abstraction, in which the pure intellectuality
reigns nearly alone, and almost free from any disturbance of the introspective consciousness, which,
being of itself an act, annihilates, *pro tempore*, all
other acts. Were we perpetually to exercise the reflex act of the mind, and to pause upon our thoughts
with self-observation, our train of ideas would halt,
and fall to pieces for want of connection.

But this is not all. Any admixture of the introspective consciousness detracts from the perfection of
once-acquired and habitual motions, as much as it
spoils the freedom and bold expansion of our thoughts.
Of this we may soon convince ourselves. Though
generally insensible to the act of breathing, we may,
by attention, become aware of the process. What
follows? An immediate sense of uneasiness, and an
interruption of that regular motion which seems to
go on so well by itself. Again, that winking of the
eye, whereby the organ is healthily preserved, becomes
a torment, if we think about it. If these instances be
objected to by persons who would confound the above
motions with such as are vital and involuntary, I may
appeal to a known phenomenon, which regards the
voluntary motion of our fingers. Every musician
must have felt that, when he has learnt to play a
piece of music by heart, if he *thinks* upon the direction of his fingers, he is apt to play false. Let him
trust to the simply memorial consciousness of his physical being, and he does not err. I have even known
persons successfully consult this memory of the fingers
(as it may be called) when they have in vain questioned their intellectual memory. I remember a lady
trying to recall an air, which once she could execute
on the piano, and she could not do this, until she
had (as she expressed it) " ceased thinking about it,
and let her fingers go of themselves." Again: the
operations of memory are impeded by the introspective consciousness. As Darwin, in his Zoonomia,
observes, — " We frequently experience, when we are

doubtful about the spelling of a word, that the greater voluntary exertion we use — that is, *the more intensely we think about it,* the further we are from regaining the lost association, which readily recurs when we have become careless about it."

Introspective consciousness, then, appears equally to mar our liberty and our memory both of thought and of motion; and consequently it should seem that, in proportion as we can be exempted from its inter-ference, we must attain a higher state of intellect and of corporeal activity.

This we may surmise; but proof is not wanting to confirm it to us. The state of the philosopher who solved the problem of the universe was avowedly a state of abstraction and of self-forgetfulness; and it is equally well known that natural sleepwakers, who can never be supposed capable of self-scrutiny, will achieve feats, which would be the horror of their waking moments. They will stand, self-balanced, on the ridge of a house, where, under the usual con-ditions of consciousness, they could not preserve their equilibrium for a single moment: they will cross a roaring torrent on a single plank; — but, if suddenly awakened to a contemplation of themselves or their situation, they will lose their footing, or perhaps die of alarm. Are these examples too far removed from general experience? We will then bring the matter at once home to every one's personal feelings. What is it that accompanies and adds to the awkwardness caused by timidity? An over-watchfulness, a care that mars itself — in fine, the too predominant pre-

sence of the introspective consciousness. The shy
scarcely ever forget themselves, as it is called —
make them do so, and their deportment is at once
improved.

Consciousness having now been defined, and its
principal varieties of action noted, it will be instructive to observe how far it is peculiarly modified in
mesmeric sleepwaking, and how far it is acting in
harmony with known and established laws. On
examination, I imagine it will appear that mesmerism
is not that violation of nature's order which it has by
some been deemed to be; — but that, on the contrary,
the changes, which it induces in our being, are in strict
accordance with well-known principles, which are of
the commonest application, and which may be easily
tested in ourselves.

First, then, we have gathered, from the observations which have lately been brought forward, that,
in proportion as introspective consciouness is annulled,
the powers of thought and of motion are developed.

Now that the mesmeric state, compared by some to
that of sleep, by others to that of delirium, can be a
state of self-watchfulness, no one will, I think, maintain. It is what is vulgarly called a state of complete
unconsciousness. In corroboration of this, it may be
observed that mesmerised persons speak with a freedom, instances of which being related to them in their
waking condition cause them surprise and even vexa-
tion. Anna M——, in sleepwaking, once lectured
me on burdening myself with so many effects in
travelling, and yet, in her natural state, almost wept

to think she had been guilty of so much rudeness. It was plain that, when awake, she was able to *reflect* upon her own conduct in a manner which she could not while sleeping mesmerically.

Again : sleepwakers seem wholly unable to scrutinise their own sensations, and, if questioned concerning them, will (as I have in other places re- marked) give confused and even irrational replies. Moreover, if asked to be self-observant, when perform- ing any thing, their actions come to a stand still, and the very efforts which they make at reflection prove their inability to exercise that faculty.

Introspective consciousness, then, may be pro- nounced to be absent in the mesmeric slumber; and, consequent upon that absence, we find — exactly as in the normal condition of man — an improvement both of thought and motion in the mesmeric sleep- waker. Proofs of this will be found in former pages of this work. Not only have mesmerised persons been shown to reason with a perspicuity of which they were incapable at other times, but to display a grace and freedom of gesture wholly different from the constraint of their habitual deportment.

Thus the perfection of motion, and superior coher- ence of thought, which mesmeric sleepwakers dis- play, in connection with the absence of introspective consciousness, is but a higher grade of a known con- dition, — the extension and not the alteration of a principle.

The second law of our normal condition, with which I would show the mesmeric state to coincide, is — that, In proportion as the intellectual consciousness is heightened in its character and spirituality, the physical consciousness is deadened and depressed.

The accuracy of this position may be tested by every one under the most common circumstances of life. In states of even trifling abstraction we become insensible to slight bodily discomforts ; and, while meditating inwardly on themes that are interesting to us, we frequently maintain one position until our nerves are compressed and our circulation hindered to a degree sufficient to compel us back to our physical sensations. Again : when we are suffering from slight headach, or other temporary derangements of the system, should any subject strongly rivet our attention, we become, by degrees, less and less sensible of the inconvenience, and, at length (as it is phrased) forget our pains. The principle runs through all mankind — from infancy to age. A child falls, receives a hurt, and begins to cry. Speak to it — arouse its mental activity, and its tears cease long before the mere mechanical injury that set them flowing has been done away. Pursuing the same law to its higher instances of application, we arrive at the holy abstraction of religious men and the elevated faith of martyrs, who have smiled on bodily anguish to the last, not only triumphing over, but becomiug insensible to, agony, while their songs of praise and joy ascended to heaven from the torturing wheel or the consuming flame.

But this entire spiritualising of humanity is of rare
occurrence, and, should we desire to behold the more
striking conquest of mind over matter, it is plain that
we must seek them in some other than our normal
condition, — amidst our loftiest abstractions, as we
are at present constituted, the body drags us down to
earth. Those busy inlets of conscious sensation,
which yield us so many pleasures and so many pains,
are continually marring the higher processes of our
intellect by their exquisite capacity to receive impressions. A Newton's chain of thought may be for
ever broken by the buzzing of an unlucky fly; and
links so subtle, when once severed, how difficult are
they to be again united! The power of total abstraction were, to great minds, power indeed! In
sleep, it is true, our senses are under a strict warden-
ship, but then reason is also fettered. Yet even from
sleep we may gather how much stronger is the force
of mind, when no longer dissipated by external
things. Every one who has dreamed of sorrow and
alarm will confess with what unusual awfulness their
stern images have stood forth from amidst the surrounding blankness of slumber. It is only in sleep
that we can become, as it were, one thought of
agony. Our impressions are then violent, because
they are unmitigated. There is none of that counter-
poise of the senses, which, under true waking sorrow,
we acknowledge to be so merciful a part of our constitution; for, could we live under the pressure of
some afflictions, did not external things sometimes
take us off from ourselves? Occasionally, also, we

enjoy in sleep, happiness which appears to be more
than earthly, —

" Dreams such as lull the hermit in his shades."

Yet, strong as are the bands which slumber can cast
around us, the senses are not always so entirely fettered but that their imperfect action shall dimly
influence the nature of our visions. Moreover, a
mere touch, a casual noise, may at any time disperse
the most agreeable pictures of sleep's shadowy creation, and recall us at once to the world and to the
pressure of all its trifles, —

" Heavy as frost, and deep, almost, as life."

Where then shall we find an abstraction so profound,
yet so peculiar, as to bind the senses fast, yet leave
the intellect a wider scope, a vigour more effective?

These conditions we find alone united in the mesmeric slumber. More perfect than any that opiates
can produce, it is also so far beyond the reach of
accidental dispersion that painful surgical operations
have been performed during its continuance. Yet,
on the other hand, reason, which in the natural sleep
is either quenched, or but feebly active, burns
brighter in that which we are now considering.

That mesmeric patients speak always truth, is no
doubtful proof of this, — they perceive all the irrationality of falsehood. Be it observed, too, that the
insensibility of the body under mesmerism, bears no
resemblance whatever to the deadness caused by

narcotics or any thing that affects the mere mechanism of existence. The mind of the mesmeric
sleepwaker is not benumbed together with the body,
but it has retired into its inner fastnesses, with all its
susceptibilities and functions unimpaired. It medi-
tates, and is absorbed, and gives to the countenance
(even should the physiognomy in general be of no
elevated cast), a loftier character, and a look of inward contemplation, which has frequently struck even
the most careless observer of the mesmeric slumber.

To the state of abstraction, then, a state so well
known in its inferior degrees to all persons, mesmeric sleepwaking bears no trifling analogy, and it
exhibits the two phenomena of intellectual activity and
corporeal deadness, heightened indeed to an extraordinary pitch, yet still preserving the same coexist-
ence which they present in ordinary life.

A third law of our being, with which mesmerism is
in perfect harmony, is the following.

Consciousness, whether it relate to sensation or to
intellectual exertion, acts more forcibly the more it is
brought to bear upon a single point.

Who has not felt that the senses, by their simultaneous action, are restrictions the one upon the
other? How seldom can we be wholly absorbed in
the pleasures of the eye! How seldom, when listening to sweet sounds, can we become "all ear!"
There is almost always a something to be deducted
from our feelings by the interference of some other
sense than that we desire to exercise. Were all our
capacities of sensation concentrated upon any one

property of matter, we may judge how much stronger would be the force of our perceptions, by observing that, where one sense is actually wanting (as in the blind or deaf, for instance), the vivacity of impressions received through the other organs is greatly increased. We may remark, too, in ourselves, that not only the stillness of night, but its obscurity, which concentrates our attention upon objects of hearing, quickens the auditory powers in an extraordinary degree.

Again, as regards the exercise of our mental capacities, experience inculcates, as one of the precepts of wisdom, that, if we wish to do a thing perfectly, we must do it singly ; and, though stories have been told of great, or would-be great men, who have played games at chess and dictated letters to secretaries at the same time, it has not been said that the chess would not have been better played and the letter better composed, had the attention been directed solely either to one or the other. Napoleon had an extraordinary power of directing his mind, when needful, to a multitude of simultaneous occupations ; but, when tracing the plan of a battle, he sate alone in his tent, and watched while others slept. Besides, correctly speaking, it is not pretended that the mind *can* do more than one thing at once, it merely leaps from one object of attention to another with astonishing swiftness and versatility; while, in that very versatility it loses force, and abjures that continuity which is the element of greatness.

When then we find that there is a remarkable

singleness in the consciousness of mesmeric sleepwakers, and that whatever they do absorbs them
wholly, we shall perhaps have less reason to be
astonished at the great eminence to which their perceptive and reasoning faculties attain. Indeed, the
concentration of the mind, in mesmerism, is such as
I should imagine to be unattainable under any other
condition, — all the bodily organs being, as it were,
annulled, there are none of those conflicting consciousnesses which, in actual life, destroy each other,
like meeting waves in water. Whatever source of
perception mesmerised persons possess appears rather
to consist in one sense than in many, — in one sense
which can become each of the others, so that he, who
exercises it, seems to become, by turns, a hearing,
seeing, or feeling unity.

It occurs to me here to make a remark connected
with this part of my subject.

Persons in the mesmeric state, however generally
sentient to any injury inflicted on their mesmeriser,
will, if expressly employed by him in any absorbing
effort, appear heedless of those experiments of pain
which may be tried upon him. This is but a natural
result of their singleness of perception; and, moreover, it is evident that all persons, even in the normal
state, will be insensible to pain, if it bear an inferior
proportion to the degree of their mental abstraction.
Yet some persons have triumphantly referred to this
as a proof of the nullity of an experiment, which
only failed because it was not tried under the proper
and necessary conditions.

On the whole, it appears that the increase of perception and of mental power in mesmerised persons, found as it is in combination with a concentrated attention, is in exact accordance with the constitution of our intellectual and physical being.

Yet more than this, the forgetfulness, in the waking state, of thoughts which occupied the mind during mesmeric sleepwaking, is in harmony with other known phenomena of consciousness.

It appears to be a law, that the less of introspective consciousness has been present to a state, the less memory is there of that state.

This is evident; for, when we have been engaged in a reverie, or a series of simple consciousnesses, which die as soon as born, — " the perfume and suppliance of a minute," that we have been thinking is very sure, but of the subject matter of that thinking we retain no recollection. Locke says, " I confess myself to have one of those dull souls, that doth not perceive itself always to contemplate ideas; " and the poet Cowper, evidently copying from the illustrious metaphysician, before drawing an admirable picture of himself in a reverie, confesses

" Fearless, a mind that does not always think."

Every one will probably be ready to make a similar avowal; but we must not forget that this state of apparent no-thought, is, in fact, only one of our modes of thinking. Acknowledging that we do not perceive ourselves always to contemplate ideas, we

by no means affirm that we have not always ideas to
contemplate (though, indeed, I believe Locke meant
his readers to draw such a conclusion), and the
absence of the reflex act of the mind, whereby we
are, as it were, present to ourselves, cannot annihilate
the simple unreflective action of momentary consciousness. Cowper, indeed, by his description of
the objects which he has seen in his winter's fire,
during what he calls " a waking dream," proves that,
in that dream, his mind, though but idly busy, was
still employed, nay even somewhat attentive to its own
operations, for he remembers vaguely the shadowy
ideas that floated across his fancy. It is not every
one, however, who can have a poet's reverie, or turn
it to such good account as Cowper. In general,
when we have fallen into what some call " a brown
study," the fit is succeeded by a total forgetfulness of
the elements which composed it, and the friend who
startles us from our absent mood, by tapping us on
the shoulder, and exclaiming, " a penny for your
thoughts," can obtain no other answer than " I was
thinking of nothing." Still we retain a confused
notion that the apparent blank was really peopled by
the ever-restless mind ; though, not having observed
our thoughts, we feel that we cannot remember them.
This is common reverie ; but there is a deeper
degree of it than this. In some cases, as Darwin has
remarked, " it becomes a formidable disease ;" and
this advanced stage of a mood, which is experienced
in a less degree by us all, is characterised (acording
to the authority just quoted) by " the total forget-

fulness of what passes" during its continuance, and
" surprise on recovering" from it.

Observe, too, how the loftier state of abstraction
indicates the same proportion between the introspective consciousness and memory.

In abstraction, as I have already shown, there is
but a low degree of self-observation present; for
from the moment that we examine our springs of
thought too curiously they cease to play. But it may
further be remarked that, during abstraction, the
retrospective consciousness supplies the place of the
introspective, we stopping from time to time to review our thoughts, and to fix in our own memories
that which originated from our own minds. Only in
so far as we do this, are the workings of our intellect
rendered permanent. Should any one suddenly interrupt our train of ideas, in one of our inattentive
intervals, or before the whole series is connected and
complete, how often we (to use a common expression)
"lose the thread of our thoughts," which by no effort
of memory we can recover.

In abstracted study, also, it should be remembered
that we are often writing down the actual thoughts
of the moment, making our pen the engraver of our
ideas, and our paper the organ of our remembrance.
Thus is preserved much of our brain-work, which
otherwise would have perished in the thinking. This
is more than conjecture; for often, after a few hours,
we have so far forgotten our compositions, as to read
them with what is called "a new eye," and to detect
faults in them which surprise ourselves, they having

P

escaped our observation so long as our effervescing
fancy predominated over our self-consciousness. Dr.
Johnson, it is well known, met with his own Rasselas
many years after its publication, and read it, as he
himself declares, with as strange a feeling as if it had
been the work of another person. In Lockhart's life
of Scott, a similar and more remarkable phenomenon
is recorded. The great novelist composed that grand
story, the Bride of Lammermoor, while suffering
from excruciating spasmodic attacks which caused
him to have recourse to frequent doses of laudanum.
He dictated nearly the whole work under these
painful circumstances; and it is said that, " though
he often turned himself on his pillow, with a groan
of torment, he usually continued the sentence in the
same breath. But, when dialogue of peculiar ani-
mation was in progress, spirit seemed to triumph
altogether over matter — he arose from his couch, and
walked up and down the room, raising and lowering
his voice, and, as it were, acting the parts."

But what is most extraordinary is, that, after the
book was published, Scott assured James Ballantyne
that, " when it was first put into his hands in a complete shape, he did not recollect one single incident,
character, or conversation that it contained ! ' For a
long time,' he said, ' I felt myself very uneasy, in the
course of my reading, lest I should be startled by
meeting something altogether glaring and fantastic.' "
Ballantyne, by whom the anecdote was preserved,
adds, " I do not think I ever ventured to lead to the
discussion of this singular phenomenon again ; but

you may depend on it, that what I have now said is as distinctly reported as if it had been taken down in short-hand at the moment; I should not otherwise have ventured to allude to the matter at all. I believe you will agree with me in thinking that the history of the human mind contains nothing more wonderful."

A similar instance (though on a small scale) of the mind's forgetfulness of its own previous operations fell under my own cognisance.

Some years ago, I met once in society poor Mrs. Maclean — then the admired L. E. L. A song was sung, the words of which struck me as very beautiful, and the young poetess seemed to be of the same opinion, for she expressed her admiration of them warmly, asking at the same time, who wrote them? The gentleman who had sung the song laughed, and said, " Surely you remember your own poetry !" It was a long time, however, before the authoress could be made to recall when or where she wrote the words; which, however, at last she recognised as one of those improvisations which she was accustomed to write down, give away, and never think of again.

That we occasionally remember the dreams that we have had in sleep may be urged as contradictory of the theory I am advocating; for can dreaming, it may be asked, in any way be called a state of introspective consciousness ? I reply, certainly it can and must. Nothing is more evident than that, in dreaming, the mind watches its own fantastic train of thought, since what is the imagery of our dreams but its own creation, and who is the spectator of that imagery but itself? There is in the phenomenon of dreaming an especial duality of action, only to be explained (unless, indeed, we admit, with some wild author, whose name I forget, that all our dreams are the work of spirits,) by that reflex act of the mind which is so swift in its operation as virtually to double our existence. Moreover, it is to be remarked that natural sleepwaking — a state very different from that of common sleep — is almost always succeeded by complete forgetfulness. The exceptions occur when the patient's visions more resemble those of ordinary dreaming, and when the mind has been manifestly more employed about its own creations than about those real external objects which often seem to influence the sleepwaker, though in a wild and unconnected way.

We may then dismiss, as uncorroborated by any fact whatever, the notion that we can remember any past series of thoughts to which introspective consciousness has been altogether wanting; consequently, that mesmeric patients should forget all that occupied them during their sleepwaking is no miracle, but an absolute illustration of nature's laws.

But still further, if we choose to acquiesce in the structure of our own being, and to examine laws rather than frame them, we may perceive that the sleepwaker must naturally resume his mesmeric recollections, when he returns to the mesmeric state; for is it not a principle, that *a former series of*

*thoughts should recur to us then, and then only, when
we are under conditions of consciousness similar to those
in which they were conceived?*

Let any one reflect how he sets to work, when,
having what is called lost the thread of his ideas, he
would take it up again at the point where it was dis-
severed. Does he not endeavour to place himself in
the same state and frame of mind, in fact, to regain
the same conditions of consciousness in which he was
when conceiving the lost idea? Most certainly. And
what is also to be remarked is, that a restoration of
even the same conditions of *physical* consciousness,
under which our intellectual trains of thought were
engendered, greatly promotes our recollection of the
latter. Not only a particular position of mind, but
a particular posture of body, aids us to regain a
frame of feeling to which it has become associated.
When we throw ourselves on our knees, our prayer
of childhood seems to recur of itself. There are some
even who can only pursue their trains of thought
while engaged in some exertion of the physical consciousness which has become familiar to them. The
following anecdote from the memoirs of Sir Walter
Scott will illustrate this truth. It is recorded in
a memorandum of his own. He says, " there was a
boy in my class at school who stood always at the
top, nor could I with all my efforts supplant him.
Day came after day, and still he kept his place, do
what I would; till at length I observed that when
a question was asked him, he always fumbled with
his fingers at a particular button in the lower part of

his waistcoat. To remove it, therefore, became expedient in my eyes; and, in an evil moment, it was
removed with a knife. Great was my anxiety to
know the success of my measure — and it succeeded
too well. When the boy was again questioned, his
fingers sought again for the button, but it was not to
be found. In his distress he looked down for it — it
was to be seen no more than to be felt. He stood
confounded, and I took possession of his place."

Again, we may observe that that action of the
physical consciousness, which relates to the operation
of our senses, is far more linked with memory than
is the intellectual operation of consciousness. That
which we have seen, heard, and felt, we remember
far more strongly than that which we have only
thought. The immediate action of the sensual consciousness is also peculiarly renovative of that which
is past. The smell of a flower, which we knew in
former days, will recall a thousand old forgotten
feelings; and the tinkling of a sheep-bell will bring
before us the wild heath, or the grassy hill-side,
where our childhood sported. Thus, when the poet
invokes memory, he does not summon her to appear
in the pomp of intellectual splendour, but he says —

> " Come from the woods that belt the grey hill-side,
> The seven elms, the poplars four,
> That stand beside my father's door,
> And chiefly from the brook that loves
> To purl o'er matted cress, and ribbed sand,
> Or dimple in the dark of rushy coves."
>
> TENNYSON.

Now let us consider that, to a mesmeric patient, the ordinary sensual consciousness is closed, — that the motive of consciousness has ceased to be intro-spective, — in fine, that all the conditions of conscious-ness are unusual. Can we, then, wonder, either that he cannot grasp mesmeric memories in his normal state, or that he should return to them with ease and perfection as soon as the unwonted conditions of his mesmeric existence are restored? To understand this perfectly, we have only to extend the principle whereby we lose or recover any train of thought whatever. Thus epileptics have been known to finish, in a new paroxysm of their complaint, a sentence begun in an attack which had occurred days or weeks before; and every one must remember the quaint story of the old gentleman who, in passing a particular bridge, answered a question that had been put to him long ago on the same spot. " Do you like peas or beans best for dinner?" his old domestic had in-quired, as he rode soberly behind his master. There was no reply; — but, just a year after, when master and man were jogging on as before, the former turned slowly round, and said, " Peas, John."

Certain facts, relative to the mesmeric state, still further confirm our previous view of the question. Sometimes, when mesmeric sleepwaking is not perfect, it seems to alternate with the natural state; and the patient, occasionally waking, as it were, for a few moments, employs his ordinary senses in the ordinary way. Now, by frequent experience, I have ascer-tained that the remembrance which these patients retain of circumstances that occurred during their sleepwaking is in exact proportion to the usual action of their senses; in other words, to their approximations to their ordinary conditions of consciousness.

The real waters of oblivion — the fabled Lethe — is only to be found where consciousness flows on through channels that are altogether unwonted.

Thus, then, if I have shown, in a former book, that mesmeric sleepwaking is a peculiar state, and therefore worthy of distinct investigation, I have now proved that, though peculiar, it is in harmony with the general laws of our mental constitution; and therefore proper to be classed with other metaphysical phenomena incident to man. I am, indeed, far from saying that it presents no extraordinary appearances; but I must assert that it is by no means one of those subjects in which we cannot arrive at a comprehension of things unknown by that with which we are already acquainted. Every common reverie is a sort of mesmeric slumber; and an absent man lives, as it were, in a mesmeric world. In order to have a perfect idea of mesmeric sleepwaking, as far as regards the patient merely and his mental condition, we have only to imagine a fusion between the two known states of reverie and of abstraction; for the mesmeric sleep approximates to the first in its singleness of consciousness, its deadness to external stimulants, and its complete after-forgetfulness : while it resembles the latter in coherence of thought and intellectual development.

## SECT. III.

### ON MESMERIC SENSATION.

HAVING shown that the metaphysical condition of mesmeric patients is in harmony with general laws and with personal experience, it is now my intention to offer some observations, whereby the phenomena of sensation, in the mesmeric state, may also be brought nearer to our convictions.

In attempting this, I must necessarily sometimes abandon the region of purely personal experience. The real causes of even our *ordinary* sensation (though they may be made mathematically evident to reason) are themselves at variance with our common feelings, and are altogether hidden from the perception of the vulgar. Yet to these real causes I must appeal in explanation of mesmeric phenomena. In order to reveal one and the same base for normal and for mesmeric sensation, I must pierce that which *seems*, and penetrate to that which *is*. Yet will it, I trust, be found that, wherever the subject permits, I retain that best mode of argument— the *argumentum ad hominem.*

Already I have remarked that, sensation being really seated in the mind, a change in its pre-requisites can never imply an abolition of itself. But we may affirm more than this, and adduce facts which shall afford us the strongest presumption that, in the mesmeric state, the pre-requisites of sensation, however changed, are only altered in conformity with established laws ; in fine, that the development of mesmeric feeling is in perfect accordance with nature. That it has not hitherto appeared so, seems to be rather the fault of its expositors than the necessity of its own mode of action. They who would examine into the conditions of mesmeric perception are, in general, so engrossed by its more prominent and superficial wonders, that they rest in external differences of development, instead of going deeper to discover internal identity of origin. They look merely to unusual effects, and do not consider that the causes of these may be such as are actually familiar to us.

The pervading peculiarity of mesmeric perception may be thus briefly expressed : —

Inaction of the external apparatus of the senses, co-existent with the life and activity of some inner source of feeling.

That such an apparent anomaly is not so far removed from the ordinary operations of our being, as on a cursory view of things we might suppose, I trust in the course of the following reflections to demonstrate.

Every fact, adduced to prove the reality of mesmeric perception, has been hitherto met with but one argument, which your established formalist — your limiter of Providence and its resources, deems unanswerable. " Our external organs are the sole appointed means of sensation. How would you then that we should perceive any thing without their aid?"

That there is a flaw in this reasoning is not so difficult to be discovered. It is a mere begging of the question, and proceeds altogether upon an assumption, the fallacy of which I will, at least, endeavour to expose.

Many and striking are the circumstances which assure us that the nerves are the true media of sensation, to which the external organs are only instrumental and subsidiary.

For, in the first place, from the anatomist we may learn that many apparently essential portions of the external senses may be removed, or may exist not, without any perceptible difference in sensation. As regards the ear, we find that "the pinna is not indispensable to hearing; for, both in men and animals, it may be removed without any inconvenience beyond that of a few days." Again: "The membrane of the tympanum may be torn, or even totally destroyed, without deranging the hearing in any sensible degree;" and "the loss of the small bones in the tympanum, with the exception of the stapes, does not necessarily incur a loss of hearing."

With respect to the sense of smelling, it appears that "the olfactory apparatus is but little developed at birth. The nasal cavities, the different convoluted bodies scarcely exist; the sinuses do not exist at all, and yet the faculty of smelling appears to exist." *

* Milligan's Magendie.

P 6

Nay, even parts of a mechanism so wondrous as that of the eye seem not to be essential to vision.

" The contraction of the pupil, though it assists our vision of distant objects, is not absolutely necessary, since Daviel has shown that patients having the pupil immovable see well enough." Also, " Persons who have had the lens removed for cataract, still see sufficiently well." *

Considerable alterations, then, may occur in the external apparatus of the senses without sensation being impaired. But is this indifference, as to losses in their structure, shared by the nerves? Far from it. The smallest abstraction — the slightest change, even, is of consequence here! Whatever deranges the nerves manifestly introduces a corresponding disturbance into our sensations, and, should any of them undergo a substantial change or be entirely annulled, the sensitive functions of the part, to which they refer, cease altogether. Thus injuries to the optic or acoustic nerve will produce blindness or deafness respectively, though the external mechanism of sight or hearing remain entire; and, if the lingual nerve be cut, the tongue continues to move, but has lost its sensibility to savours. " These effects may be produced at pleasure upon animals, by tying or compressing the nerves. When the ligature or pressure is removed, the part then becomes sensible as before."

Magendie also states that, where there is faulty

* Milligan's Majendie.

sensation, it generally originates in defects rather of the nervous than of the organic structure. This dependence of the external senses upon the nerves should diminish our surprise at finding the former occasionally retreat from offices which they only fulfil, as it were, by delegation, while the true powers, as in mesmerism, come forward to manifest their capacities, more palpably, indeed, but not more essentially, than usual.

*Secondly,* The nerves, or portions of the brain with which they are in immediate relation, are capable of gradations in their sensibility.

The proofs of this lie before us in every direction. Though the structure of the external organs be alike in all (of course, we except cases of acknowledged malformation), the capacities to which they minister are, in different individuals, infinitely varied. Thus, of two persons, whose auditory organs are equally perfect, one shall have a far finer sense of hearing than the other. The one, perhaps, may mistake, as did a friend of mine, the tuning of instruments before a concert for the overture; the other shall detect the slightest false note which may occur in an otherwise well-executed piece of music. While some again possess by nature that sense of colour which guided Rubens to all the harmonies of vision; others, with no perceptible defect in the mechanism of their eyes, cannot distinguish between the most opposite colours. A man with this infirmity has been known to order a pair of trousers of the most flagrant red, supposing them all the time to be modest blue.

Others are incapable of even hearing certain sounds, for which

" Nature, though giving two, gave them no ear."

I was acquainted with a gentlemen, who was com-pletely insensible to the chirp of the cricket.  Walk-ing with him and with a large party one summer evening in his grounds, I observed to him that the grasshoppers were more numerous and more noisy just then than I had ever heard them.  They literally seemed to get the better of our voices.  To my sur-prise, my friend refused to believe that any grasshop-pers were chirping, and was only at length convinced of the fact by the united testimony of all the persons present.  In order to ascertain whether this peculi-arity of sensation depended upon any general defect in the organs of hearing, I held a watch at a consi-derable distance from my friend's ear, but he dis-tinctly perceived it ticking.  Again, I placed the watch at an equal distance from him and from myself, at a point where the grasshopper concert completely pre-vented me from hearing it ; yet to him it was as audi-ble as if perfect stillness had prevailed.  He was also cognisant of words spoken in the lowest whisper, which no other person but himself could distinguish amidst the din of the insects.  What can be more evident than that this inability to distinguish a par-ticular sound arose from a defective sensibility in the acoustic nerve, which had not the capacity of responding to aërial vibrations, rapid as those by which a note so shrill as the cricket's cry is produced?

Thus it appears that the nervous sensibility of one man varies as much from that of another, as the mesmeric from the normal.

This again should lessen our wonder when we behold a state of nervous sensibility which surpasses our conceptions even. That we cannot judge of it no more proves its non-existence than the defective hearing of some persons annihilates certain sounds. When we, being in one state, pronounce a verdict upon phenomena exhibited by those who are in another, we act as does the blind man who dictatorially gives his opinion respecting colours.

*Thirdly*, The nerves are subject to varieties of condition, not only in different individuals, but in the same.

That the nerves may vary in their functions of transmission is not only possible, but inevitable, and consequent upon the very constitution of man's nature. Not only does every disease which affects the nervous tissue affect the phenomena of sensation, but the commonest occurrences of our ordinary existence influence them likewise. Viewing the nerves as conductors, along whose sentient line dart those electric impulses which convey to us information of the external world, we may assert that they are seldom in the same conducting condition. In reverie, how deadened is their transmitting power! how enlivened under the excitement of hope, or fear, when we look or listen for the arrival of a friend, or of some messenger of evil ! How differently do they act under the relaxation of fatigue or the tension of vigorous energy ! Sleep alters their capacities of conduction in a thousand varying degrees, exhibiting phenomena only less wonderful than those of mesmerism because more frequently repeated.

In how extraordinary a manner trifling causes may change the nature of sensation, one or two incidents which have fallen under my own notice may demonstrate. A friend of mine, when suffering from slight derangement of stomach, is apt to see every object double (smile not, reader, he is a water-drinker) ; and I am, myself, occasionally subject to a singular affection of the optic nerve. When fatigued by an effort of continued attention, where the light is too vivid (as in a theatre, for instance), I see every object diminished and removed to a distance, as if I looked on it through the reversed end of a telescope.*

At one period, for three months, during which my digestion was slightly out of order, I was entirely deprived both of taste and of smell. All that I ate was as so much earth ; and I could not distinguish between a cabbage and a cabbage-rose : again, a gentleman of my acquaintance lost, and never afterwards recovered, his smell, in consequence of a fall from his horse, which had no other perceptible result.

* Here it is evident that the weaker impression, caused by the exhausted and over-excited state of the nerve, produces the idea of diminution and distance, just as we judge that a feeble sound proceeds from afar.

Phenomena like these, not less than everyday events, assure us that every separate state of man is a separate and peculiar condition of nerves ; and that every peculiar condition of nerves has its peculiarity of sensation. When, then, we learn that mesmerism induces alteration in our sensitive capacities, why should we be as full of wonder as if nothing else could work a change in them? Were we wise, we should, from all we know of the nerves, be prepared at least to anticipate new results from new conditions of man ; and certainly to expect from a state so singular as that of mesmeric sleepwaking some singular variety of nervous action. That the effects of the human influence on the nerves should be new to our experience, so far from being extraordinary, is but in the common routine of things; for the state of mesmerism itself is new to our experience. Did the nervous system present the same phenomena under this peculiar phasis, as under other circumstances, then, indeed, would it violate the laws of nature, and forego its own.

The nerves, then, are capable of changes in their condition : but, change, it may be argued, is not improvement, and the instances adduced of nervous alteration tend to show that their power may be lessened, rather than increased, — deranged, rather than brilliantly exhibited. But mesmerism purports to be an expansion and a gain in nervous energy. Let us then inquire whether the same force, which is susceptible of diminution or disturbance, possesses also inherent capacities of development. That nerves are

sensible to stimuli we know. When their power
has been impaired, — nay, to all appearance perfectly lost, it may, by particular influences, be transiently restored to all its pristine vigour. That the
blind occasionally regain, beneath the warmth of a
summer's noon, a power of vision which declines with
the sun and is again altogether lost at its setting, is
a fact which I need not insist upon to physiologists;
and a friend of mine has assured me that he has frequently known the opium-eaters of the East, after
they had become both blind and deaf from over-in-
dulgence in their favourite drug, regain both sight
and hearing while under the accustomed excitement.
Stimulus, then, may appear to work nervous miracles,
and this plainly indicates a capacity in the nerves to
rise as well as sink — to gain force as well as to lose
it.   But we may go farther still.   The nervous
power, under certain circumstances, may not only be
exhibited as recovering the energy it had lost, but as
transcending itself, and displaying a capacity of
development, the limits of which have not been ascertained.   They who have had opportunities of
studying disease in all its forms will corroborate me
when I assert that, on a depression of the vital powers,
an increase of the sentient capacities is often consequent; and that persons under certain derangements
of the system have manifested a nervous sensibility,
than which, mesmerism can display nothing more
astonishing.   The Chevalier Filippi of Milan, doctor
of medicine, and a most determined opponent to
mesmerism, has acknowledged to me that some of his

patients, more particularly women after their confinement, when suffering from nervous excitement, have distinguished the smallest objects in darkness which appeared to him complete. The same physician related to me the following occurrence : — Visiting a gentleman who had an abscess, he found that the patient had not many hours to live ; this, however, he did not tell him, but answered his inquiries about himself as encouragingly as he could. Taking his leave, he shut the door of the sick chamber, and, passing through two other rooms, the doors of which he also carefully shut, entered an apartment where some friends of the patient were assembled. To these he said, speaking all the time in that low and cautious tone which every one, in a house where illness is, unconsciously adopts, — " The Signor Valdrighi (that was the name of the invalid) is much worse. He cannot possibly survive till morning." Scarcely had he uttered these words, when the patient's bell was heard to ring violently, and, soon after, a servant summoned the doctor back again into his presence. " Why did you deceive me? " exclaimed the dying man ; " I heard every word you said just now in the farther apartment." Of this extraordinary assertion he immediately gave proof by repeating to the astonished physician the exact expressions he had made use of. Subsequently, upon Dr. Filippi testifying his surprise at this occurrence to the servants of Signor Valdrighi, they declared that their master's hearing had become so acute since his illness, that he had frequently told them all they had been talking of in

the kitchen, which was even more remote from the
sick room than the apartment before alluded to.*

The gnat-strainers and camel-swallowers may be
content to accept this story, just because it only professes to relate an instance of exaltation in the usual
senses; but what does this convenient phrase, " exaltation of the senses" mean ? Is it not "Words —
words — words; " one of those formulæ which seem
to account for every thing, while, in fact, explaining
nothing? Considered as a not very precise statement,
it may pass; but when persons would palm it upon
us as a solution, I would ask them whether "exaltation of the senses" (in the common acceptation of
the term) be possible? There may be development,
or, if you will, exaltation in our sensitive capacities —
but how, or where? If we desire to have clear
notions on this subject, we must perforce inquire
whether any increase of sentient power which may
fall under our observation be in the external organs,
or whether it be derived from the nerves which are
masked by those organs ?

In what manner we should resolve this problem
would not be doubtful, but for a certain confusion of
idea, which is extremely prevalent, respecting two
portions of our being, between whose functions and
qualities there can be no parallel. Led by custom,
and blinded by apparent simultaneity of action, we
associate the external apparatus, and the internal
sources of sensation in one common term, — the

* See Appendix.

senses, namely; while, in fact, not only are they not the same, but opposed to each other in every essential particular.

While the action of the exterior organs is always the same, under the same circumstances, that of the nerves is ever-variable. The powers of the one are limited and known; that of the other has not been ascertained.

Is it not evident, then, that the mechanism of sensation is swayed and regulated by something which is not mechanical?

Is it not plain that the external senses are but the letters and ciphers of the scale,— fixed marks, which, by their unalterable character, point out how high may rise, or how low may fall, the variable mercury of our internal sensibility?

These truths, it is to be hoped, are known to physiologists at least; yet how many even of these speak and argue as if they were entirely ignorant of them.

When sensitive power is manifestly increased, where then shall we look for an explanation: there, where all is precise, limited, ascertained; or there, where all is free and boundless, where variety is infinite, and capacity unknown? Can we for a moment make it a question as to where the true power lies? Can we for a moment ascribe it to the external organs, which, far from being capable of development, seem rather to be restrictions upon internal sensibility by their exact and circumscribed action?

Now let us once more refer to the examples of developed sensibility which prefaced the above remarks.

Certain invalids, it was affirmed, upon indubitable testimony, had been known to distinguish the smallest objects in a room, whence light, according to ordinary vision, seemed perfectly excluded, or had heard articulate sounds athwart impediments, which, in ordinary cases, are sufficient to prevent all transmission of acoustic impulses. The usual organs of sense were not in any way mechanically assisted during the exhibition of the phenomena by external aids, such as those of optical glasses, or acoustic tubes ; they were dependent solely on their own resources. Those resources, as we have just observed, are very strictly limited by mechanical and physical laws, which are incapable of change. In the case of the eye, where there is light, there will be a picture formed upon the retina with a clearness and vivacity proportioned to the intensity of that light; or, supposing the phenomenon to regard the ear, wherever aërial vibrations exist, they will strike the tympanum with a force that corresponds with the violence and extent of the original impulse ; but, in the instances cited of sensitive acuteness, the effect is inverse to the known cause. The patient sees in a dark room where no picture can be supposed to be formed on the retina ; or hears in a situation where vibrations of the air, if even propagated so far, can with difficulty be conceived to affect any membrane, however elastic. Under such circumstances, everything is plainly against an increase of sentient power : yet power is gained — how? or whence ? Whatever may be thought of the credulity of a mesmeriser, I have not faith enough to believe in

modern miracles, when they are so interpreted as to
falsify mechanic laws.   I will not, then, load the external organs with the onus of impossible causation;
but that the nerves of sensation may, in particular
states, vibrate to fainter impulses, or even other
media * than those which more grossly move the
mechanism of sense, I may admit to be a probable
explanation of events like the above : — at any rate,
one truth from the preceding observations is clear.
The extraordinary exaltation of the senses, whereof
we have been speaking, was not in the organs, (*that*
is barred by an ex absurdo) but. in the nerves that
minister to the organs.

Thus it appears that, in certain cases, the nerves
may be exhibited in a state of development which
manifestly owes nothing to the mere organs of sense,
and which far oversteps the province of any one portion of our external mechanism.   When, then, mesmerism palpably excludes the external organs from a.
share in the sensitive developments which it elicits,
when it effectuates a revelation of sensibility springing
from an internal source, it may in this, as in many
other respects, be said to detect, rather than to violate, principles.   It sweeps away imagined power,
which a multitude of circumstances had already shown
to be only apparent and superficial : it exhibits power
where power evidently exists, and might be expected
to be found.

But here, perhaps, I may be assailed with the oft-

* See the ensuing remarks on the media of sensation.

repeated question (to which I have once before alluded), " Granting all this to be true, of what use are our external senses ? "

Coleridge, impressed with an idea of the lavish wealth and boundless resources of Providence, when asked of what use were the stars and planets if not inhabited, replied, " I suppose to make dirt cheap." In the same manner I might answer when persons demand, " Of what use is the eye, if we can see without it ? " — " To show us how to make a camera obscura." In graver phrase, how much we learn from the construction of this marvellous organ ; how dependent on that construction are all the discoveries which have been made respecting the physical pro- perties of the luminous medium ! It is a mechanism exquisitely adapted to the mechanical arrangements of the universe. And is its beauty nothing? its power of speaking the soul ? And these advantages, be it remembered, are contingent upon the office, place, and station which the eye now holds in the human economy. But while we adore the Creator's wisdom and goodness in thus suiting our organs to our condition, and to the agents with which they are immediately in contact, let us remember that it is our duty to explore, and to seek out the true grounds of our adoration, and not to rest in a sort of stupid amazement. They who so triumphantly take their stand against mesmerism upon the all-importance of the external organs of sense, would do well if they would condescend to learn the real and most valuable part which they play in our economy. While it is

the office of the nerves to receive and to transmit external impulses with a degree of sensibility proportioned to their ever-varying condition, that of the
external organs is to modify those impulses, with a
regularity as constant as Nature's own. They are the
weights and the balance-wheel in our constitution.
Thus the eye is composed of parts that are refractive,
and of membranes that have each their use in altering
light, such as it acts externally, into light such as it is
perceived by us. The air, again, is modified in the
mysterious labyrinths of the ear; and these organs,
by their form, contribute to direct the impulse thus
modified to its proper recipient — the nerve. Hence
it appears that the apparatuses of the senses are con-
trivances for blunting, not for heightening, the sensibility. They are masks, and careful coverings to instruments of too exquisite delicacy to be bared and
exposed to the outer world. The cutis defends the
nerves from the agony of harsh impact upon their
tender surfaces; the sieve-like mechanism of smell
disarms odour of its over-stimulus, ere we

" Die of a rose in aromatic pain." *

That degree of sensibility which is exhibited to us
in the mesmeric state, as if to give us a permitted
glimpse of the capacities of our being, would be

* The choroid, by its dark colouring, absorbs the light, and
mitigates it to the eye, as is proved by the distressing effect
which daylight produces on those persons called Albinos, who
are deficient in the pigmentum nigrum. An ordinary degree of
light dazzles them into blindness, and they can only see in twilight.

Q

wholly unsuited to the purposes of every-day existence;
and I cannot too often remind our opponents that
man was no more created to pass his days in mesmeric
sleepwaking than in any other state of abstraction
from the affairs of this world. How terrible would
be the condition of a being laid bare and naked to
the irritating influences of all external things, may
be imagined from the sufferings of persons whose
nervous sensibility has become predominant over the
mechanic action of the senses. The Seherin von
Prevorst fell into horrible convulsions if iron were
within a certain distance of her, or if vinegar were
in any part of her room. Even patients in the ordinary mesmeric state are sensible, as I have shown, to
emanations from precious stones, some of which affect
them with very disagreeable feelings. The valuable
experiments of Dr. Elliotson with minerals on mesmeric individuals speak to the same effect. Let it
not, then, be thought that the mesmeriser underrates
the uses of the external senses. That they are as valuable for modifying as the nerves are for receiving
impulses he fully admits ; he is only desirous of exhibiting them in this their proper sphere, conceiving
that he thus renders them service ; for a faculty is
never seen to such advantage as when cleared from
the rubbish which ignorance has heaped around it.
The aggrandisement which is based on error is but
for a time, and God is best honoured where his
work is best understood.

Considerations favourable to mesmerism have now
been drawn —

1st. From the predominance of the nerves over the external organs of sense.

2dly. From their degrees and variations of sensibility.

3dly. From their capacities of development.

And facts have been adduced which may prepare us to resolve the apparent paradox of sensation existing apart from our usual senses.

But the agency of the nerves is capable of still further elucidation which will throw yet more light on the mysteries of mesmeric sensibility. If from the varied *modes* of nervous action we proceed to its simple *principle*, we shall find abundant reason to conclude that mesmerism does never really infringe any established and universal law.

As on the true nature of sensation I ground much that is explanatory of my subject, I must be excused if I endeavour to lead to the full comprehension of my future observations on that head by some preliminary remarks, which will not be deemed needless by those who appreciate the difficulties that surround this branch of our inquiry.

There are evidently but two ways of perceiving objects ;— the one by being present to them in their essential verity; the other by communication with them through the intervention of types or shadows.

The first mode of perception belongs to God alone.

The second is imposed on man by the necessities of his nature.

We cannot go to the object: the object must be conveyed to us.

We cannot penetrate its essence : we must descry it as in a glass, darkly, — by reflection, — by an image which is not itself.

For this reason, in some previous observations relative to this same subject, I likened sensation to a language established between man and his Creator. I now proceed to point out in what that language consists, and how truly it may be called a language, since its symbols no more resemble the ideas they excite, than do words the pictures they raise in the mind, or the printed sentences of a book the thoughts and feelings which they occasion in those who read them.

Limited, and closed in by walls of flesh, with which, for a time, we are almost identified, we cannot but allow that through our body comes all that knowledge, appertaining to external things, to which we have given the name of sensation. This is indeed the tenure on which we hold our individual existence ; for all intelligence that is not limited is God, and in the force of the restrictions which confine the creature (paradoxical as it may sound) consists the independence of its action, and the liberty of its will. The universal ideas, which originate with God, must, in order that they may become proper to ourselves, be transmitted to us through personal experience, and be coloured by the tinge of our own nature. Deceived, indeed, when we think not deeply, by the apparent extremity of effects which we only experience in our own persons, we refer our feelings to the object that occasions them, and speak of sound

or of colour, as if it existed as a thing apart, forget-
ing that aërial vibrations are only sound where there
is an ear to receive them, and that light is only
colour when it is reflected from surfaces to an eye
that vibrates in unison with its rapid scintillations.
Our free communication with external objects is
another source of error to our minds. Borne on the
wings of thought, we seem, as it were, to go out of
ourselves — to annihilate distance — mock at separation, and pervade the universe ; but maturer reflection assures us that sensation does, in fact, never stir
one. inch beyond the corporeal boundaries which
have been assigned to it. Close the ear, and silence
is around us, though to others the brook is murmur-
ing, or the trees are rustling as before : — shut the
eye, and the slight veil of a tender membrane has
blotted out all that glorious creation with which we
so freely conversed, — has for us expunged the beauties of earth, the magnificence of ocean, and made
a blank of the boundless heavens. Were we essentially cognisant of these objects, we could not be
parted from them thus; but it is evident that all our
sentient mechanism (as Newton intimates) is truly
constituted for the purpose of *conveying* the species of
things to our sensorium, and that all that vast expansion to which we seem present is, in reality, present to us within the narrow compass of our brain.

Sensation, then, is personal to ourselves, and man
is strictly confined to his individual sphere ; yet, at
the same time, our consciousness assures us that we
are in sensible communication with objects which

Q 3

are not ourselves, and which are more or less remote from us. Consequently, between us and them there must be media; and thus, in considering sensation, we have three things to take into the account; namely, the external object, or the exciting cause; the medium whereby we are, as it were, linked to that object; and the change in our own corporeal frame, which stands as the representative of the object.

Hereafter I shall have occasion to speak of the media whereby we are brought into communication with the visible universe. At present I restrict myself to considering the signs of things, of which our own persons are, as it were, the living alphabet and recording volume.*

First, that we really do draw our sensations from the signs of things, and not from things themselves, a very simple and well-known experiment may convince us.

Crossing the middle over the fore-finger, let any one rub between their extremities a little ball, and he will have exactly the same sensation as if there were two balls instead of one.†

* Locke mentions "the perception of the signification of signs" as a chief act of the understanding. See his chapter on " Power."

† The reader is begged to remember that experiments like these are not presented as new (as indeed, being appeals to experience, how can they be?) but as known proofs of known propositions, equally necessary to be repeated in a certain train of argument as the problems of Euclid in a course of mathematics.

This phenomenon is capable of but one explanation. Those parts of the fingers which, in the experiment, are brought into contiguity, are ordinarily prevented by their position from touching one object at the same moment. Simultaneous impact on them has therefore become associated by habit with an idea of a duplicate impression. The mind has learnt its lesson, which it cannot forget — its language, which it can only interpret in its own established manner. We touch a single superficies with nervous surfaces which are used to come in contact with two; therefore we think that we touch two objects. The conventional signs are present, and we deem that the accustomed realities are also acting.

An interesting fact, recently observed at Paris, may be adduced in confirmation of the above views.

A celebrated surgeon, M. Blaudin, succeeded in making a new nose for a soldier named Eustache Gressan, who had lost that member from a sabre cut at Waterloo, by means of skin cut away from the forehead of the patient. When Gressan has his eyes shut, if any one pricks with a pin the end of his new nose, asking him at the same time where he has been touched, he immediately lifts his hand to his forehead, evidently referring the feeling to its old place from association. This anecdote, which is taken from a thesis sustained before the faculty of medicine at Paris by M. Chomette of Bordeaux, is extremely valuable, as showing decisively that sensation depends on the nerves and their associated actions, not on their particular localities.

Q 4

Certain signs, then, are really representatives to us of certain objects; and of the nature of those signs we, from the above considerations, may partly judge. It is necessary, however, that on this subject our ideas should be perfectly clear and precise.

Let us, therefore, inquire wherein, individually and generally, consists the language of sensation?

Not in that which is commonly and vaguely called the action of the senses, but in motions of those nerves which are the life and efficacy of the senses. For, let us consider, with Locke, that "all the actions that we have any notion of, reduce themselves to these two, thinking and motion," (thinking being the action of mind, as motion is of matter). Now that the nerves *act* in sensation no one will attempt to deny; and *how* they can act, being material substances, except by a change in the position of their relative parts; in other words, by motion, it is utterly impossible to conceive.

That sensations are the results of changes or motions in some portion of our nervous system does not, however, rest upon abstract reasoning alone. The proposition may be proved, by fact, in a twofold manner.

First, we may, by giving motion to a nerve of sense, convince ourselves that a sensation is the result.

Secondly, knowing that, when bodies are jarred or stirred, their particles continue for some little time in motion, after the impulse which set them vibrating is at rest; we may, by observing that a sensation continues for a few moments after its cause has ceased to act, conclude that the nerve, on which sensation depends, produces its effect by motion.

Each of the senses is susceptible of affording either species of demonstration.

*First.* 1. A blow on the ears will make them what is vulgarly called *ring*, independently of the sound of the percussion.

2. When a sound has been of long duration, we still conceive that we hear it, though it may have been for some time discontinued.

*Secondly.* 1. An electric shock gives such sensation to the nerves of touch as a violent blow usually excites in them.

2. Should a fly have settled on the face, we often raise our hands to brush it off, after it is really gone, being deceived by a continuation of the same feeling which it gave us when actually in contact with the skin. The tickling of a feather produces a similar effect.

*Thirdly.* 1. Let any one place a piece of zinc under his tongue, and a piece of silver upon the upper part of the same organ. At the moment when the two metals are brought into contact, a strong taste will be perceived in the mouth, resulting from the galvanic concussion, and not from any actual flavour.

2. The sensation endures for a short time after the experiment has been tried.

*Fourthly.* 1. A spark, drawn, by means of pointed metal, from the nose of a person who is charged with electricity, will give him the sensation of smelling a phosphoric odour.

2. Musk, however carefully inhaled, so as to prevent any actual particles of the substance from being stopped upon the pituitary membrane, will produce in us sensations of odour for some time after we have ceased to smell to it.

*Fifthly.* To perceive the connection between motion and sensation, with respect to such of our senses as I have already mentioned, is not difficult. But, when we come to consider the eye, it is hard to apprehend that all its beautiful representations of that external world with which it holds such distant communion are but signs and characters, consisting in nervous motions, which the soul, by an intellectual operation, translates into its own glorious and native language. For this reason, I have reserved the consideration of vision to the last, in order to bestow on it more attention. Mesmeric vision also being that part of mesmeric sleepwaking which most alarms the prejudices of mankind, I am most anxious to reduce it to analogy with nature.

If it can be shown that vision does really consist, like all our other sensations, in motion communicated to a certain nerve, I conceive that my task is partly executed, since it is plain that, though other conditions which we deem essential to vision may be wanting to the mesmeric state, the great, the primary requisite may yet be present. There may still be motion.

But let us proceed to proofs.

Sir Isaac Newton asks (Query 12th),

"Do not the rays of light in falling upon the bottom of the eye excite vibrations in the tunica retina, — which vibrations being propagated along the solid fibres of the optic nerves into the brain, cause the sense of seeing?"

And again (Query 16th),

"When a man in the dark presses either corner of his eye with his finger, and turns his eye away from his finger, he will see a circle of colours, like those in the feather of a peacock's tail. If the eye and the finger remain quiet, these colours vanish in a second minute of time; but if the finger be moved with a quavering motion they appear again. Do not these colours arise from such motions excited in the bottom of the eye by the pressure and motion of the finger, as, at other times, are excited there by light, for causing vision? and do not the motions, once excited, continue about a second of time before they cease? And when a man, by a stroke upon his eye, sees a flash of light, are not the like motions excited in the retina by the stroke? And, when a coal of fire, moved nimbly in the circumference of a circle, makes the whole circumference appear like a circle of fire, is it not because the motions excited in the bottom of the eye by the rays of light are of a lasting nature, and continue till the coal of fire, in going round, returns to its former place? And, considering the lastingness of the motions excited in the bottom of the eye by light, are they not of a vibratory nature?"

These observations, which the great philosopher has

so modestly thrown into the interrogative form, modern science has done little else than repeat affirmatively.

It is true, we no longer talk of rays of light as the causes of our visual sensations; we have substituted for these (and, as I believe, with truth,) the undulations of an elastic ether. Still the theory of sensation remains the same; and motions of the nerves, which minister to the eye, are accepted as productive of our ideas of light. The fibres of the retina, according to M. d'Arcet, continue to vibrate for about the eighth part of a second after the exciting cause has ceased; and therefore it is that, if we look at the sun and then turn our eyes from it, or even shut them, we still perceive a bright image before us. The phenomenon is explicable on no other hypothesis.

I might, perhaps, allow the subject to rest here, but it is too important to be abandoned before our misapprehensions concerning it are perfectly cleared up. The great source of error respecting vision has been, and is, the miniature representation of objects on the retina, which was once undoubtingly accepted as the cause of vision — a venerable delusion, with which it is almost a pity to part; for the endeavour to explain how an inverted image (as that on the retina is,) could make men see right end upwards, has given rise to an infinite display of learned ingenuity.

Then, also, the notion of the soul giving a peep into the show-box of the eye is sufficiently ludicrous to be amusing. Still, before we can have just ideas concerning vision, those that anciently found place with us, however captivating, must be altogether swept away. The undulatory theory of light has, it is true, at the present time, already [effected this in part: still, nothing is more difficult than to get well rid of old errors, especially when these are based on a plausible association of ideas. We behold external objects — the same objects are known to be represented on the retina — and we jump at once to the conclusion that here is cause and there effect. Unfortunately, never were two things, apparently connected, so wide asunder. Did not whatever is too near ourselves in general delude us greatly, surely we should not accept the image on the retina as explanatory of vision, only because that image is in the eye, and because we know that, in some way or other, our eyes serve us to see with. For let us remember that the connection between seeing and the picture on the retina has never been *proved*. The theory that links them together is perfectly gratuitous. On the other hand, the connection between seeing and motion of the optic nerve *has* been proved. We cannot indeed see without a retina; — but why? Because the retina is so constituted as to be set in motion by the impact of light, and to communicate motion to the optic nerve. Error continues to be propagated by the loose phraseology employed upon this subject. What is the meaning of the word *impression*, when applied to visual action? An impression gives the idea of something stamped and durable; but the image on the retina endures no longer than while the external object is before it. This is totally at variance with the lasting nature of luminous impulsions on the eye, which consideration, it might be supposed, would alone be sufficient at once to refer vision to a motive cause. Above all, how can the optic nerve *convey impressions?* — a mode of speech, however, which is but too common. What is really conveyed, or communicated to the brain?—a stamped figure, like the impression of a seal on wax — or a picture?— No! a motion, and nothing but a motion. Notwithstanding that this is palpable to reason, and doubtless well known to the physiologist, still it is to be suspected that the multitude are, even yet, misled by a fancied analogy between the act of seeing and taking off the impressions of objects, just as ladies, by an ingenious and almost instantaneous process, transfer engravings to the tops of their work boxes.

This must be reformed altogether.

Taking the question on its widest grounds, let us inquire, Does the image on the retina enable us to see? To see is not merely to have ideas of colour, but to appreciate the relative size and distance of bodies. Now, in spite of the beautiful perfection of the forms represented upon the retina, and the proportion of their distances, many facts contribute to assure us that we gain no perception of form or distance by the eye. We learn from it absolutely no more than relates to light and colour. Thus, though images are formed upon the retina of the new-born infant as perfectly as upon that of the adult, yet, during the first month of its existence, it cannot be said properly to see. It only betokens a certain sensibility to light by shutting its eyes when exposed to a strong glare. That it has actually no idea of the size or distance of objects, even long after it appears pleased with bright and lively colours, is evident from its stretching out its hand to seize objects, however distant, or of whatever dimensions.

Even persons, who, being born blind, attain to sight at an advanced period of life, arrive but slowly at a correct appreciation of external objects. No where is the parallel between language of the precedents of sensation more remarkable than here. Learning to see is precisely analogous to learning a new tongue. It is said of a blind youth, spoken of by Cheselden, "When he saw the light for the first time, he knew so little how to judge of distances that he believed the objects which he saw ' touched his eyes,' (this was his expression) as the things which he felt touched his skin." * It is related also of Kaspar Hauser, who was in the condition of a blind person from long captivity in a dark dungeon, that, when led up to a window, for the first time, whence there was an extensive view, he uttered a cry of pain and hid his eyes with his hands; when able subsequently to describe his sensations, he declared that the landscape had appeared to him as an upright

* Milligan's Magendie.

plane, daubed all over with dazzling and discordant colours.*

Were vision mechanically resultant from a pictured representation, it is evident that effects like these could not ensue.

But, finally, how are we made aware that there is a picture at all on the retina? By its being shown to us in the eye of a dead animal. So that we look upon it externally as we would on any other object. It is we ourselves who create the picture by our previous conceptions of form and distance, which have been drawn, as the metaphysician knows, from the sense of touch or rather muscular action. Thus, we only know that there is a picture on the retina, *because* we are able to see. Can that, then, be itself the cause of our seeing? The absurdity is manifest. It is clear, then, that when we call upon the image on the retina to account for our visual perceptions, the effect does in no way answer to the cause proposed. View it as we will, it appears to be merely the collateral result of certain mechanical dispositions, which in no way regard it as an end. From what has been said, it appears that, in considering the sense of vision, we must absolutely look upon it as no more than the medium of our information respecting light and colour.

To light and colour, then, we have to confine our attention and to limit our argument. But our perceptions of light have been proved to be occasioned

---

* Kaspar Hauser has been accused of deception; but surely this trait of nature alone would suffice to prove the most essential fact in his story.

by motions of the optic nerves, and colours consist so certainly in vibrations, that the number of undulations producing each tint of the solar spectrum has been mathematically ascertained. " The determination of these minute portions of time and space," observes Mrs. Somerville, " both of which have a *real existence, being the actual results of measurement*, do as much honour to the genius of Newton as that of the law of gravitation."

Thus it appears that the real office of the retina is to vibrate in correspondence with the vibrations of light; as they are modified by being turned, intercepted, or broken by meeting with various objects.

Need we observe how much this diminishes the wonder of mesmeric vision? It has been shown that the picture on the retina is not the cause of seeing. In cases, therefore, where a picture on the retina is rendered impossible (as when the eyes are closed in mesmeric slumber) the true cause of vision is not, it seems, on that account, put out of the way. There still may be motion of the optic nerve. If it be inquired, how produced, I must send on my reader to the remarks which will shortly ensue on communicating media. This at least cannot be contested : — the faculty of vision has been proved to result from motion, and is therefore brought under the same law, which, it has been shown, predominates over the other senses. Hence it appears that few propositions, not absolutely mathematical, are more certainly established than this.

Motions of the nerves are representative to man of

external objects. Nor is it less demonstrable that those portions of the nervous system which are central, or proximate to the brain, occasion in us, by their motions, sensations, rather than those portions of the nervous system which are superficial or most remote from the brain. For, in many instances, it may be shown that internal causes, wholly independent of any thing external, produce sensations exactly similar to those excited by external things; their manner of reproduction plainly exhibiting their original mode of production.

In fever there is a repetition of images before the eye, which can only be accounted for by supposing a repetition of the sensations which formerly suggested them. But those sensations are themselves correlative with certain changes or motions of the nervous system. Their recurrence in a confused and incoherent manner is confessedly the fruit of disorder. And what is disorder? An agitation of the frame, occasioned often, and always accompanied by, a deranged action of the nerves. The soul is reading, as it were, an incoherent treatise, of which the disjointed sentences were once familiar to her in their proper order.

Under the influence of strong emotion persons may perceive, as with the actual organs of sense, objects which have no existence but in themselves; nay, even like Macbeth, when marshalled by the air-drawn dagger, may stretch forth their hands to clutch

—— " The false creation,
Proceeding from the heat-oppressed brain."

There are also mysterious affections of the nervous system, in which the patient, retaining all his reason, seems palpably to behold various forms of men or animals. Walter Scott, in his Demonology, gives a detailed and interesting account of the case of a lady who was thus spectre-haunted for months.

I was myself personally acquainted with a physician who told me that, one evening while he was writing, he chanced to raise his eyes and saw, as he imagined, his housekeeper standing close to his desk, with a candle in her hand. " You may go to bed, Betty," he said, and went on writing. A few minutes afterwards he again looked up : there was Betty still. " Did you not hear me ? " he repeated ; " I want nothing more. Go to bed." A third glance, however, showed him the seemingly obstinate Betty yet nearer to his chair, when, raising his hand with a gesture of impatience, he found that it passed through the figure, which, however, retained all its apparent corporeality. With much shrewdness and presence of mind, the physician then felt his own pulse, which indicated fever. He proceeded forthwith to bleed himself, and, as he had anticipated, while the blood flowed, the phantom gradually disappeared — a proof the most convincing that it was the creation of his own disordered state of body.

Nor are such illusions confined to the sense of vision. A lady of my acquaintance, while in a weak state of health, heard constantly a band of music with as much distinctness as if an orchestra had actually been present. As she recovered, the imaginary

sounds diminished, till at length she heard no more
of the aerial performers.

These facts, however, are not within the experience
of all persons. But a proof, as easy of access as it is
interesting, that internal motions of the nerves can
cause in us a perfect idea of something external to
ourselves, is afforded in the following experiment.
Let any one take a five shilling piece, and thrust it
far up between the inside of his upper lip and his
teeth; then let him lay a piece of zinc on his tongue,
and suddenly bring together the edges of the two
metals. At the moment when contact takes place a
faint glimmering light will seem to pass off on either
side his head, which, as nobody else sees it, must be
an optical error, — the mere motion of a nerve,
caused by the galvanism elicited from the two
metals.

Finally, discoveries of recent date demonstrate that
the impulses of sense proceed to a point remote from
the organ in which they originate. It is found that
the optic, acoustic, and olfactory nerves, though remaining entire, may fail to transmit their accustomed
intelligence to the seat of perception, after the severing of the fifth pair. The following experiment gives
still more consequence to the internal action of our
sensitive system. " I destroyed in a dog," says Magendie, " the two olfactory nerves. I presented to
the animal strong odours. He perceived them perfectly, and conducted himself exactly as he would
have done had he been in his ordinary state. It may

then be possible that the olfactory nerve is not the nerve of smell." *

How much these facts throw light upon the internal sensibility displayed by sleepwakers I need not remark. They tend to concentrate sensation, to refer it to a common source. They actually display our senses dependent for their action upon an internal cause.

But, farther still, it may be shown that a quiescent state of the external portions of the nerves is perfectly consistent with an increased and independent activity of their central system.

Nay, the repose of the outer is an absolute condition for the revelation of the inner sensibility.

We all may feel that, in order to call up before our mind's eye the face of a dear friend, or the beauties of a familiar landscape, we must retreat from the obtrusive impulses of the external world.

Would we rise to a yet higher discernment of remembered objects, we must yet more calmly check the beating of our pulses, until we pass into that state of mind so beautifully described by Wordsworth, —

> ——— " That serene and blessëd mood
> In which the affections gently lead us on,
> Until the breath of this corporeal frame,
> And even the motion of our human blood
> Almost suspended, we are laid asleep
> In body, and become a living soul:
> While, with an eye made quiet by the power
> Of harmony, and the deep power of joy,
> We see into the life of things."

---

* Milligan's Magendie.

An instance of another kind of abstraction is to be found in the visionary, who conceives that he possesses the power of really discerning absent objects, or of anticipating such dramas of life as have not yet been acted. These seers, or pretenders to the second sight as it is called, are described as falling at the time of their beholding a vision into a species of trance. Their limbs become rigid, their eyes remain fixed and are insensible to the light. While all within them is in a kind of anarchy, their exterior appearance is that of the dead.

But we need not resort to an extreme case like this, to demonstrate that our greatest enjoyment of the inner mobility of sensation is to be found in the moment when we are most externally tranquil. Dreams, beyond all other phenomena of our nature, prove a central activity of the nervous system. When we have passed into their world of separate existence, the outer organs of sense are closely shut, yet we see, or hear *, or feel, or partake of imaginary banquets. To say that the soul alone reproduces these ideas, were to charge her with unworthy folly, for what can be less exalted or less coherent than are our sleeping visions generally ; yet, were the spirit really free to act unencumbered, would not her operations be distinguished by superior majesty and connection? Rather must we affirm with him, whose

* Dr. Abercrombie doubts that a sense of hearing is ever enjoyed in slumber. It is, however, no uncommon occurrence with me to dream that I hear music, either vocal or instrumental, as distinct as any that voice or orchestra can produce.

every comment on our nature was poetically beautiful
and philosophically true, —

"Dreams are the children of an idle brain."

It is indeed evident that, in sleep, some cause (perhaps the circulation of the blood through the brain)
agitates the nerves of sensation, and gives to our
ideas, by their associate motions, a particular impulse,
form, and feature.    This view of the case is strengthened by the manifest influence which our state of
body exercises over our nocturnal visions, which are
calm or disturbed, agreeable or painful, accordingly
as our digestion, and consequent pulsation, is orderly
or irregular.

A surplus, then, of internal sensibility, co-existent
with a deficiency in the external, is not exhibited in
the mesmeric state alone.    If, indeed, mesmerism
could display an inner development of sensation,
without any sacrifice of the external capacities, we
might have reason to exclaim against its miraculous
deviation from nature's laws; but the increase of
internal power, which it reveals, in connection with
the abrogated activity of the surfaces of the nerves,
appears to be the illustration, and not the infringe-
ment of a principle.

Still are we more and more depriving the superficial extremities of the nerves of potentiality in sensation,— still are we giving more and more importance
to their central motions.    Proceeding in our examination of nervous action, we find proof not wanting,
that like impulses, at whatever part of the nerve they

commence, terminate in like motions, and produce like sensations.

" After the amputation of a foot, or a finger, it has frequently happened, that an injury being offered to the stump of the amputated limb, whether from cold air, too great pressure, or other accidents, the patient has complained of a sensation of pain in the foot or finger that was cut off." *

From this fact we may infer, that nerves have the same conducting capacity throughout their whole length, and that the loss of their extremities or ulterior expansions does not deprive them of their power to convey impulsions.

The common property resident in any part of a nerve, or series of nerves, to excite similar sensations, is also evinced by the difficulty we often experience in referring our pains to the real seat of the disorder which occasions them. Thus it not unfrequently happens, that a person has had a sound tooth extracted instead of a defective one which caused the torment.

But all these phenomena of sensation are, as it were, subservient to one great principle, which, if laws be only the general expression of concurring facts, is most truly Nature's law as regards sensation ; universal — immutable — admitting of no exception. It is this : — similar ultimate motions of the nerves, however produced, excite similar sensations.

*A' priori* reasoning would alone suffice to convince

---

* Darwin's Zoonomia, vol. i. p. 35.

us of this; or rather it is self-evident, as an axiom,
that, where a sign is representative of an object, it
must be so universally and under all circumstances.
It would produce as much confusion did that particular sensation, which now invariably signifies to us
*a tree*, occasionally call up in us the idea of *a man*, as
interpreting vice to mean virtue, or meanness gene-
rosity, does actually bring about, in some instances,
in this our wicked world. God's works are charge-
able with no such imperfections. It is true that the
invariability of our sensations may occasionally lead
us into momentary error (we rashly concluding that,
where the accustomed sign is present, the phenomenon which usually causes it is also at hand); but
this is only the necessary accompaniment of a mighty
benefit. The advantages are great and regular;
the drawbacks from these advantages trifling and
exceptional. That like signs are productive of like
ideas, is a principle which is the groundwork of all
our knowledge. Is it ever violated in mesmeric sensation? Not within my experience. An interesting
proof that Suggestion continues to be, under mesmerism, the law of our thoughts may be found in
the fact, that sleepwakers, with closed eyes, when
asked how they appear to themselves in a mirror,
reply, " as if my eyes were open." Accustomed to
connect the two ideas, of sight and having the eyes
open, in one sequence, the first suggests the second
to their minds.

To prove by general experiment that ultimate
motions of the nerves, *however produced*, are signs

R

which the Mind interprets in one uniform manner, is
but to repeat facts which have been already stated.
The truth of the proposition is involved in that of
preceding ones.  For it has been shown that pressure,
or a blow on the eye, or galvanic concussion, produces, not less than the vibrations of the luminous
ether, a sensation of beholding light; that smell,
feeling, taste may be deceived into action by agents
which are not the proper objects of those senses, and
that hearing may result from any cause that sets
the tympanum in motion or agitates the acoustic
nerve.

And that all these impulsions tend inward, and
that the *last* nervous change is the immediate pre-
cedent of sensation, has been made also evident.

Adopting Dr. Thomas Brown's definition of power,
we may not inaptly call central action of the nerves
the power of sensation.

Wonderful and complex, then, as is the external
mechanism of the senses, their principle is one.  In
vain will a thousand adverse arguments be based upon
the favourite theme of our opponents — namely, the
speciality of office with which each nerve in our
economy is invested; in vain will it be contended
that one nerve is sensible to the touch of light alone
— that another converses solely with the world of
sound.  But however varied their action may appear,
however diversified the ideas which they excite, however distinct and separate the different properties of
matter which they reveal — still, as many rays unite
in one centre, or, rather, as a thousand harmonies

may be all evolved from one fundamental note, so do their multiplied resources all coincide in a single simplicity of plan and operation. Motion is their language — motion is their mode of communicating information to us from without.

This is sensation's one great law, to which all other are minor and attribute; and this, once established, deprives all but the last preparative of sensation of that constancy which is the attribute of power. All other laws relating to our sensibility admit of exceptions, and are not therefore really laws. This alone, throughout the whole history of man, we have never found transgressed. And how beautiful, how wonderful is the contrivance! A few insignificant signs, capable of infinite permutations, are representative to us of the whole glorious universe! The finger of God is here.

Having explored the principle of sensation, we find that the changes which mesmerism induces in its superficial developments are comparatively unimportant. They in no way affect the great internal cause, or throw the mesmeric sensibility out of the pale of nature. We have seen that similar motions of the nerves may be produced in various ways, and that a change in the pre-requisites of sensation can never imply a change in sensation itself. Now, though in the mesmeric state the nervous motions are apparently not *produced* in the normal way, still that there may be normal motions is evident. However altered by mesmerism the mere externals of sensation may appear, the internal efficient change in the nervous structure need not in any way falsify our habitual mode of being. In exposing the secrets and apparent mysteries of a new condition of man, we do but lay bare the fountain of sensation, not change its nature. We pierce beyond external developments only to arrive at their origin and their base more surely even than the anatomist with his scalpel.

Let me, however, hasten to anticipate an objection which may be raised against the mode of argument that I have recently been pursuing. The feelings which I have adduced to prove that similar ultimate motions of the nerves are interpreted always by the mind in a similar manner, are, it may be urged, mere delusions, in no way correspondent to the truth of external things : whereas it has been set forth as the glory of mesmerism, that all its revelations regard realities ; — that, without eyes, it beholds actual objects of sight ; without ears, it appreciates aerial vibrations which do essentially exist.

I answer, that my object was to show the uniform manner in which the sentient nerves perform their functions, and that, in order to judge of any functions aright, we must study them in their derangement. Even their errors point out the truth. Though the mind may be mistaken in its judgment, the nerves do not vary in their action; and this was the great point to establish. In every case, also, of delusive sensation, let it be remembered there still is a distinct perception of each several quality of the external world. Ideas of light, sound, odour, &c., have been perfectly reproduced, and have thus plainly indicated the manner of their original production. It has fully appeared that Nature's mode of conveying to us information is one, and with this one the mesmeric perception need not be at variance.

But is there, it may still be asked, any one acknowledged instance in nature, by which the possibility of receiving correct information respecting actual existences, otherwise than by the usual inlets of sense, can be demonstrated? There is. It is well known that sound may arrive in the tympanum by another way than the external ear. Let any one stop his ears with his hands as perfectly as he can, and hold his mouth open. Let a watch be introduced between his teeth, but without touching them, by another person. As long as there is no contact, he will not perceive the slightest sound of a watch ticking, but, the moment he closes his teeth upon the watch, he will hear it distinctly, — a manifest proof that the sound has reached him by the mouth, not by the ear, in fine by another mode of conduction than the ordinary.

This simple and common experiment is extremely interesting and instructive, for it shows how impulses may arrive at the accustomed seat of sensation by novel ways; warning us how rashly we should judge in deciding that, because the first preparatives of sensation are unusual, the last must be so likewise. They who desire the most to cling to experience in judging of mesmerism may by this consideration be propitiated. It may also be conjectured that an optical impulse may reach the retina by some new route, even as a certain sound was shown to have been transmitted to the tympanum. Should we agree with Magendie, as to the analogy between sight and hearing*, so far as to say, that " sound is to the hearing what light is to the sight," we shall be yet more disposed to admit this. It may, however, be asked, If in the state of mesmeric sleepwaking all the extremities of the nervous system be as dead, does not this form an insurmountable barrier against the transmission of external impulses to the inner seat of sensibility? By no means. In the first place, though the external sensibility of sleepwakers is not manifested in the accustomed localities it does not follow that it exists not somewhere. The truth is, that it appears to be restricted and circumscribed only to be increased in intensity. All mesmeric sleepwakers seem to possess a concentrated sensibility in some one part of their bodies, affording a free inlet for the conduction

---

* " Some experiments have lately taught me that the ear presents physiological circumstances analogous to those of the eye."

Again : " In the same manner as in the apparatus of vision, there are in that of hearing a number of organs, which appear to concur in that function by their physical properties, and behind them a nerve for the purpose of receiving and transmitting impressions." — *Magendie's Physiology.* By Dr. E. Milligan.

Herschel, also, in his Treatise on Sound, says, speaking of those feeble sounds which catch our attention at night: — " The analogy between sound and light is perfect in this, as in so many other respects, &c. The ear, like the eye, requires long and perfect repose to attain its utmost sensibility."

of external impulses to the nerves; nay, if we will, to the accustomed nerves of sensation. Objects of sight (it will be remembered) were transmitted to Mademoiselle M—— by the forehead. Sounds reached her by the same path; for proof of which I refer to an experiment precisely analogous to the just mentioned one of bringing a watch into contact with the teeth. To the slightest touch upon the forehead, even while apparently dead to feeling elsewhere, she was peculiarly sensible. Ammonia held to her nose had no effect upon her, but, when brought before her forehead, made her suddenly draw back with affected respiration; and, moreover, exhibit action in the muscles of the nose; — which last circumstance more especially adds to the presumption that the external impact, in all mesmeric cases, though beginning at an unusual point, is finally transferred to the accustomed nerves of sensation.

Again, even supposing all upward conduction from the extremities to the brain to be impossible in the mesmeric state, it by no means follows that there will be no sensation; for many facts concur to show that the mind can act downward from the brain towards the extremities, and so, as it were, create sensation for itself.

The mere thought of something sour will set the teeth on edge, or of a good dinner will excite action in the salivary glands, while feelings of awe or fear will affect the whole external surface of the body, so as to produce a visible change in the papillæ of the skin  Mrs. Somerville has observed, in her Connexion of the Physical Sciences, that " the imagination has a powerful influence in our optical impressions, and has been known to revive the images of highly luminous objects, months, and even years afterwards."

Again, Sterne says, in his odd book, the Koran, — " I am in possession of a faculty, at any time I please, of communicating a sensible pleasure to myself, without action, idea, or reflection, by simple volition merely. The sensation is, in a degree, between feeling and titillation, and resembles the thrilling which permeates the joints of the body upon stretching and yawning."

Even the motions of the iris of the eye, which are strictly subject to nervous influence, and *apparently* dependent upon the external impulsion of light, have been proved by Magendie to be greatly under the dominion of the will. He says — " I ascertained this in the following manner :—I selected a person, whose pupil was very moveable, for there exist great differences in this respect. I placed one sheet of paper in a fixed position with regard to the eye and the light, and, ascertaining the state of the pupil, I requested the person to endeavour without moving his head or eyes, to read the small characters which were traced upon the paper. Instantly I saw the pupil contract, and its constriction continue during the whole effort." *

Lastly, my own personal experience has convinced me that the mind can act from within towards the

* Milligan's Magendie, p. 43.

external organs of sense. Often I have gone to sleep holding a book in my hand and continuing to hold it. For a few seconds after awaking, I could distinctly perceive that I was unaware of having any thing in my hand, and I have marked the restoration of feeling consequent upon the restoration of perfect thought. My mind has, as it were, proceeded again to the extremities, from which, during slumber, it had retreated.

By these phenomena we establish a truth which all analogy must have suggested to us. The senses act on the mind. Granted ; but action supposes reaction. Reason and fact therefore both concur in proving that the mind has a certain mastery over the senses, even as it has over the limbs, though the apparent passivity of the senses have so far deceived many, even acute, physiologists, as to lead them to assert, that, while the nerves of motion have a downward action, proceeding from the brain, the nerves of sensation, on the contrary, act invariably upwards, or towards the brain. We have, however, seen that this is a fallacy. It may, indeed, be made a question whether the mind can ever be said to be purely passive in any true and philosophical sense. This at least is clear — it can only know that it has been passive, by an act. Moreover, that only in as far as we *attend* to nervous motions we become sensible of them is a positive law of our being. Many an object sends visual beams to the retina, which yet we cannot be said properly to see — many an aerial vibration strikes on the tympanum, which yet excites in us no idea of

R 5

sound. How often, for instance, a clock may strike and we not hear it! The power that we possess, also, of selecting particular objects for our notice, and even of excluding others from any share in our attention, proves that the mind does really — in some cases at least — turn the senses to a higher kind of service than a mere passivity implies. In looking on a landscape, when we are attentive to one pleasing portion of it, how all the rest of the view dies away, as it were, from the sight! How easily from the conversation of a crowd of persons talking around us we can select that information which we most desire to hear! Again, the cultivation of our senses is manifestly dependent on a power which is not their own. Children, till they have learnt to use their eyes and ears, give but feeble tokens that they possess them; — a proof that the mind does not even attend to signs, of which it comprehends not the significancy, and that neither external impulses nor material organs can produce sensation, without the active co-operation of the intellect.

These remarks are of importance to mesmerism — a state in which the mind displays more of its active than its passive character, — casting aside, as it were, the external organs of sense, in order to attend to its own abstracted trains of thought.

Is nature violated or not by this development of internal power?

We have followed sensation inward only to arrive at the mind itself, and to exhibit it as not the play-thing of the senses, but their lord and master.

Is this a truth which is new to us? If so, it is
well that we should learn it now. It is time that we
who talk of the march of intellect at the present day
should rise above the vulgar view of sensation, and,
as Coleridge phrases it, endeavour "to create the
senses out of the mind, and not the mind out of the
senses." Let us no more return to gone-by errors.
Anciently courage was seated in the heart, sorrow in
the spleen, love in the liver, &c.; yet this was not
worse than deeming sensation to be actually in the
organs of sense. We may as well say that modesty
is inherent in the cheeks, because they blush. Again,
when smell is lost through the absence of its external
organ — when hearing is impaired by a collapse of
the external ear — when a person is near-sighted
from convexity of the cornea, art can supply a false
nose whereby odorous impulses are again properly
gathered and perceived — can concentrate in an ear-
trumpet the vibrations of the air — can, by proper
glasses, restore the purblind to perfect vision. Are
we, on that account, to say that sensation is seated in
our false nose, artificial ear, or spectacles? What
better reason have we to suppose that any external
apparatus of sense can actually *create* ideas in the
soul?

A higher philosophy must teach us that the senses
are but the instruments — the mind the power — of
knowledge. The development, indeed, of its immortal stores may depend upon some external touch,
which unlocks the treasures of the casket, and one by
one exposes them to the light; but, as a seed includes

potentially the future plant, leaf, blossom, and fruit, so does the mind contain within itself its own capacities of expansion. Even granting that, till written upon by the finger of the universe, it is a blank, and that all the magnificent endowments which it displays are, until called into action and educated by external existences, as though they were not, still, when once vivified and instructed, it is able to act for itself, and to use its material organs as instruments of its intelligence and will. Till we recognise this truth in perfect clearness, there will be confusion even in our physiological researches. Before, also, we can study sensation aright, we must learn to separate it logically into its two great divisions of general and special — the first relating to us when considered as sentient beings only; the second relating to us, when considered as sentient beings fitted and adapted to a peculiar state of existence and to the mechanical arrangements of this our world. The first is fundamental, the second occasional; the first is a principle, the second the modification of a principle. Our present organisation has reference to our present condition; but sensation is of no time — of no era: it is as old as creation itself. Now mesmerism tends to expose to us the fundamental sensation apart from the organic. Unless, then, we can approach it with a due knowledge of this distinction, its revelations will be spread before us in vain; they will darken rather than illuminate our understandings. Till we thus study man and mesmerism (which is almost another word for man) we shall remain far behind the

German school, both of metaphysicians and of physiologists. But unfortunately we incline to the philosophy of a lighter nation, who have anatomised the body till they see nothing beyond the play and spring of nerve and muscle. With their accuracy of material examination I do not quarrel; I will, if they please, give up the term *soul*, which seems to offend them so mightily; but I will, even from themselves, force the confession that man *thinks;* and whoever does not see clearly that *thinking* has no likeness or relation whatever to any material operation is in no condition to judge any subject, or to argue on any point whatever. There is a radical defect and confusion in his mind, or (if he prefer so to phrase it) in his brain. These remarks are not uncalled for, because it is to be suspected that one of the sins, of which mesmerism is guilty, is that of giving preponderance to mind over matter, and of rescuing sensation from its connection with certain organs with which some persons would absolutely and inextricably identify it. Can it be denied that too many physiologists love to view man as only a *result* of various organs? It has even been affirmed by those who would make us wholly dependant on our material organisation, that the loss of an external sense involves the loss of ideas, which have been furnished by that sense. Nothing can be more absurd or untrue. I have questioned on this very point many individuals, who had lost their sight for years, and they have all concurred in saying that in dreams they had a lively sense of vision.

Milton's beautiful sonnet, beginning —

" Methought I saw my late-espoused saint,"

and ending —

" But oh! as to embrace me she inclined,
I woke — she fled — and day brought back my night,"

is a written record that sensation survives the sense to which it formerly was indebted. Beethoven, it is well known, became perfectly deaf at the age of twenty-eight, and thenceforth his whole world of wondrous harmony was seated in his mind. A celebrated living artist is blind of one eye. According to the theory of some persons he should be only half an artist; but his works are remarkable for correctness of design, and splendour of colouring. These are facts which are in harmony with mesmerism. Let not, then, the determined materialist quarrel with this infant science, as if it alone proclaimed the supremacy of mind over matter. Even should it go to prove that we can see without our eyes, there is no such great cause for alarm. Metaphysicians have told us again and again that we do not see with our eyes, but with our understandings, — and the world is not yet come to an end.

Let us now, by the aid of the principles we have ascertained, combine into one view the peculiarities, and the correspondences with nature, of the mesmeric mode of sensation; pointing out where it agrees with and where it differs from the proceeds of our experience.

I have already likened the nerves to conductors :
the expression is perfectly correct. Setting wholly
aside any theory as to their being the vehicles of a
more efficient force, as the electric, (which, nevertheless, certain phenomena of muscular motion render
extremely probable), it is only a *fact* that they are
conductors of impulses from the external world to the
brain, and thence again to the mind ; and also, as has
been proved, from the mind and brain reversely to
the surfaces of the body. Again, it is a *fact* that
their faculty of conduction varies in many ways,
without any discoverable change in their texture and
substance, or even in the external circumstances by
which they are surrounded, though, from analogy, we
may infer that their alteration does really depend on
causes extraneous to themselves.

Viewing the nerves in a mechanical light, we actually find that, like other substances, they have their
own class of influences, which affect their transmissive
power. In some cruel experiments tried upon half-
dead animals, alcohol applied to the brain was found
to produce action in the nervous system after every
other stimulus failed of this effect; and salt placed upon
the limbs of frogs, long after they have been severed
from the body, will rouse them into life-like motion.
Even such mechanical properties as these, inherent in
the nerves, prepare us to expect varied effects from
submitting them to various conditions. Here, however, our knowledge is so imperfect, that it is safer to
study nervous action as it is connected with our own
thoughts and feelings : the causes are, indeed, less

obvious, but the phenomena more familiar to our experience. We are not acquainted, and probably never shall be, with the conditions by which nervous conduction is varied. That it *does* vary we, however, certainly know.

In the normal state we have the extreme and central parts of the nervous system both in full activity.

In sleep, the nervous capacities die off towards the centre, and the extremities are in no condition to transmit sensation.

In palsy, conduction from the centre to the extremities is impeded, for the brain can no longer transmit the impulses of the will to the limbs.

Nervous deafness, or blindness, afford instances of non-conduction from the extremities to the centre, occurring often without any apparent cause. I have a friend, who used frequently to be afflicted with a complete though temporary blindness. At first, he said, it was as if a black spot were before him, blotting out external objects. This black spot increased till it obscured every thing, and he was in perfect darkness. I once was forced to lead him out of a concert-room at Cambridge, while he had an attack of this kind.

In certain cases of catalepsy, the power of nervous conduction seems to be limited to one particular spot in the body. A lady, whose daughter was afflicted with cataleptic fits, in which she would sometimes fall down, at others be fixed rigidly in one attitude, told me that the patient heard nothing that was said to her in her attacks, unless the person speaking touched the pit of her stomach, and addressed the sound thither. On one occasion the child was seized with catalepsy when standing in the middle of a large public swimming bath, which, for her health's sake, she was ordered to use. The mother, unable to reach her child, was in great alarm lest she should fall, and called out, in hopes to wake her from the lethargy, in vain: she gave no token of hearing. The thought then suddenly occurred to the mother to hold her mouth close to the surface of the water, using it as a sort of conductor to convey the sound to the patient. On this being done, the little girl showed at once that the voice had reached her, and was roused from her catalepsy, as she usually was on her attention being strongly excited in attacks of that malady.

It being then ascertained that nerves vary in their capacity of conduction, we have only to mention precisely what shade of their conducting power characterises the mesmeric state, when it will be found, I imagine, that we have not, in considering this part of man's constitution, to unlearn our previous knowledge, but merely to intensify known conditions, or to combine them anew.

*First.* Under mesmerism, the nerves generally have ceased to conduct from the superficies to the centre; — in other words, from the usual inlets of sensation towards the brain; but it also appears that conduction usually takes place from some one limited portion of the nervous surfaces, whence it propagates external impulsions, by unusual routes, to the brain.

378      MESMERIC SENSATION.

The first branch of the phenomenon is partially exhibited in cases of nervous deafness, or blindness. The second, namely, the limited conduction of sentient impulses, has sometimes been observed in catalepsy. It has, however, been objected that, in cases where the nervous sensibility has been reported to be confined to the epigastrium or pit of the stomach, the communication between this part and the brain has not sufficiently been accounted for, the system of the ganglions, as some persons affirm, scarcely making a part of the general nervous system. This is but a feeble objection. Every part of the system of man is evidently in reciprocal connection ; and if it be asked how the great sympathetic communicates with the nerves of sensation, we may answer this question in resolving another. How does the arterial communicate with the venous system ? By vessels so fine as to have called their very existence into doubt, and to have rendered the circulation of the blood a great discovery. Our senses are not always so subtle as to trace nature in all her links.

The second peculiarity of nervous conduction which the mesmeric state presents, is as follows.

In using, for the purposes of sensation, that portion of the nervous system to which conduction seems confined, the sleepwaker demonstrates more of the active than passive character of the mind. To speak technically, there is more manifestly conduction from the brain outwards than from the superficies inwards. The common process of sensation seems reversed ; for the nerve appears to conduct the sentient power to the superficies, where it *takes*, as it were, the information it seeks, instead of, as usual, conducting the impulsion to the brain, where the sentient power may *receive* notice of what is passing in the external world. More briefly, in the one case the mind *takes*, in the other it *receives* information.

I ground my observation on certain facts, and on this especially : —A mesmeric sleepwaker rarely observes any external object of his own accord. His state is one of concentration, abstraction, and internal thought. If he be conversant with material objects, it is in the mirror of his own mind. Only when the mesmeriser stimulates his volition does he attend to any external thing, and then it is by seeming to apply, as it were, his new organ of perception to the object, with something of visible effort. If I may be permitted the illustration, I should say that he seems to feel out the object, as an insect examines things by putting forth its antennæ and using them as instruments of exploration.

This is more especially observable in the early stages of the mesmeric state, beyond which some persons never advance. There is also a difference which results from native character, some manifesting from the first, some acquiring by degrees, a greater power of independent action than others, — a power less evidently derived from the mesmeriser. When this is the case, there is more attention manifested towards external things, but always, I imagine, in the manner that I have above indicated ; namely, by active exercise of the new organ of sense rather than by passive reception of impulsions through it.

Some, who exhibit a very extended perception in the mesmeric state, seem to possess, to a certain degree, the faculty of shifting the external sensibility of the nervous system, though it continues always restricted, from one spot to another, changing, as it were, the site of their single organ of sensation. Thus E. A., who perceived things by the forehead, would occasionally receive information relative to external objects by the back of his head, the side, or the top; but he seldom did this spontaneously, and would always say it was a great effort and did him harm.   And, in effect, after having induced him to make many such efforts, I always observed that he, on awaking, manifested uneasiness and fatigue.

Other considerations support my view of the case. The mesmeriser can, by stimulating the patient's volition, restore the sense of feeling (but, I believe, that sense alone) momentarily to the extremities of the nervous system.   Thus, Anna M——, when mesmerised, felt nothing if pinched in the hand by an indifferent person ; but if I told her to attend and try to feel, she could transiently perceive that she was touched.   This proves that the mind of the patient can proceed from its inner abstraction for a moment to the extremities of the nervous system, and re-occupy its accustomed channels.   Take this in conjunction with other facts, which demonstrate that conduction *towards* the brain is barred, and I think our point is proved.   E. A. being mesmerised (as will be remembered) at a time when he was suffering from hunger, was no longer conscious of the sensation from the time of his entering the mesmeric state. Again, Anna M—— suffering, when awake, from a gathering in her finger, had no longer a sense of pain when in the mesmeric slumber. In these cases, the nerves of feeling, whether internal or external, did evidently no longer convey impulsions *towards* the brain, but were as inefficient for that purpose as if they had been tied or cut. The reader may also consult, for further information on this subject, the Monthly Chronicle for July, 1838, wherein a most interesting experiment on mesmeric insensibility is detailed. By this it appears that a galvanic shock, which produced a very severe effect on Sir William Molesworth and other persons, had no visible influence on Elizabeth and Jane Okey, patients of Dr. Elliotson. Yet were these girls able to attend to all that was passing on every side of them ; or, in other words, to use their nervous system actively, while incapable of being influenced by it passively. There was no conduction *towards* the brain, but only *from* it.

Again, in the case of Mlle. Estelle L'H——, related by Dr. Despines of Aix, and once before referred to, it is stated that when she was unable, from a species of nervous paralysis, to stir a limb, she could, in the mesmeric condition, into which she sometimes spontaneously fell, actively use all her members as when in perfect health. The transition from a state of active exertion to one of passive helplessness was very marked, and often very alarming ; since if her natural mesmerism dissipated itself (as sometimes would occur) when she was walking, or even riding, she would fall suddenly, and, if not timely assisted, come with force to the ground.

Some of these phenomena are by no means peculiar to mesmerism. The partial restoration to feeling effected through stimulation of the patient's will is paralleled by those cases in which nerves under the influence of strong excitement have been suddenly restored, after years of inaction, to a conducting capacity. Paralytic patients, it is well known, have recovered the use of their limbs for a short time, when they were forced to save themselves from fire or other threatened evils, but, the excitement withdrawn, have fallen helpless to the earth: even those who had been dumb for years have spoken under circumstances where strong emotion burst their bonds of speech. Nor are these the only cases to which mesmeric sensation bears analogy.

The inversion of nervous action has been pointed out by Darwin, and particularly described in his chapter on Reverie in the Zoonomia. A young lady, it is there related, suffering under this singular malady, neither saw, heard, nor felt, during its attacks, any of the surrounding objects; yet did she manifest occasionally in her state of reverie a perception of external things (as when smelling a tuberose and drinking a dish of tea), "*but this was only when she voluntarily seemed to attend to them.*"

Darwin, remarking on the case, observes : —

" In the present history the strongest stimuli were

not perceived, except when the faculty of volition was *exerted on the organ of sense;* and then even common stimuli were sometimes perceived : for her mind was so strenuously employed in pursuing its own trains of voluntary or sensitive ideas, that no common stimuli could so far excite her attention as to disunite them; that is the quantity of volition, or of sensation already existing was greater than any which could be produced in consequence of common degrees of stimulation."

Darwin has well stated this; and this appears to be the history of the sensations in all cases of sleep-waking and states which are analogical to sleep-waking. The mind is deeply occupied on its own train of thought, so as to become inattentive to every thing in which it does not sympathise, and with which it does not co-operate. It is exhibited in the position of *originating action;* and every action which it does not itself originate is to it as a nullity; as, for instance, the natural sleepwalker will be unconscious of the presence of other persons, and will suffer much from them without awaking; yet will he exert — yes, here it may be truly said *exert* — his percipient power on such objects as he pleases; thus manifestly exhibiting a freedom of choice and a capacity of selection which does not belong to man generally. Thus, however paradoxical it may appear, the state from which volition seems the most absent, is really the fullest of free-will.

These things throw light on mesmerism, and approximate it to experience. They are known to be parts of the constitution of man; they are demonstrated to resemble perfectly the state which we deem so anomalous. How comes it that the likeness has not been recognised before, — that mesmerism, in truth, so old, should seem so new — so near, should seem so distant? The fault has been in this: — viewing mesmerism as a strange thing, men have sought strange solutions for it. Deserting the sure path of common life, they have thought themselves obliged by the dignity of their subject not to be too explicit in their explanations, but to wrap the apparent mystery in tenfold darkness. As Arago has well observed, in his life of James Watt, " Les forces naturelles ou artificielles, avant de devenir vraiment utiles aux hommes, ont presque toujours été exploiteés au profit de la superstition." So have we been taught to view with superstitious fear the home-born stranger of our own bosoms. Thus has every effort to approximate it to man thrown it farther off. They who chase it most will find it least. To go far for it is like hunting for happiness —

" Which still so near us, yet beyond us lies,
    O'erlook'd, seen double, by the fool and wise."

The very terms heretofore in use, in speaking of mesmeric phenomena, have been strange and portentous. This is an error. We have undoubtedly startling things to relate; let them not be related in startling language ! Words have much power for good and for evil. On this account I regret that the term " transposition of the senses " should ever have been applied to mesmeric perception; for it is a manifest inaccuracy. The senses, as I trust I have demonstrated, are never by mesmerism transposed or removed from their real seat. Still more unfortunate is it that the early mesmerists should have sinned against precision of language, so far as to talk of their patients seeing *with* the epigastrium — the fingers, &c. To those who might inadvertently repeat such an error, we might suggest that, when we are made to hear the ticking of a watch *by means of* its contact with the teeth, it would be a rash conclusion that we hear with our teeth. Such expressions as these remove mesmerism farther and farther from the safe regions of experience. Our object should be ever to show that it does not alter sensation, save in certain comparatively unimportant particulars of its mode of conduction. But such stumbling blocks as these are fast being removed out of the way of objectors, so as to leave them no excuse for hanging doubts on words and arguments on verbal inaccuracies. Already the obnoxious term magnetism has given way to the unobjectionable name which is well derived from the discoverer or reviver of the agency in question. Once the whole learned world fastened furiously on the word *fluid*, unfortunately applied to this agency. This can occur no more ; for, through the excellent judgment of Dr. Elliotson, mesmerisers themselves agree in exposing the ab-

s

surdity of deciding positively on the nature of mesmeric influence.

But there still remains a stronghold to those who
think that to attack a name is to quash a fact. "What!"
exclaim some persons, "do you assert that any one
can *see* with the eyes shut?" By no means. We will
change the term. Why quarrel about a word? We
will couch the fact of mesmeric perception in any
language that is most agreeable to the objector; we
will allow that to have certain perceptions otherwise
than by the eye is not properly *to see.* Let us say,
then, that it is to *perceive* or *know;* for knowledge, at
any rate, is the end and object of all the senses.
Knowledge is a general expression of the nature of
man. But, be this as it may, the nomenclature of
mesmerism, however faulty, evidently makes no difference whatever in the facts. It is with these we
have to deal, and — unless we actually *desire* to
quibble — it is these we must alone consider. But
the advocates of mesmerism are placed in a singular
dilemma. The world calls out for facts; and, when
we offer such as we alone have to offer, hurls them
back in our teeth. The mesmeric vision, or *clair-
voyance,* especially has been gravely and grandly
pronounced to be "physically and physiologically
impossible." How can we reply to this? Only, I
suppose, as Pascal did to some one who asserted that
it was *impossible* for God, being so great, to busy
himself about our little world, — " Il faut être bien
grand d'en juger." — " To decide such a question,

one must be great indeed ! " Or again (for there is
no lack of answers), " Must every thing be impossible, which our insufficience cannot account for?
Are there not innumerable mysteries in nature,
which accident reveals, or experimental philosophy
demonstrates to us every day? And shall we yet
presume to limit the powers of the great Author of
that very nature?"

But should we still be met with the same wearisome assertion, so as to be provoked out of all philosophical endurance, we may, perhaps, at last be
allowed to exclaim with Young —

> " Impossible is nowhere to be found,
> Except, perhaps, in the fool's calendar."

Did these positive persons know all the offices and
functions of the nerves, they might be thus dogmatic.
But the study of the nervous capacities is itself of
recent date, and the powers of nervous development
are unknown. That the nerves do not always act in
concert with the external organs — that they transcend them — that they can occasionally altogether
dispense with their aid, has been proved. What if
it were reserved for mesmerism to show the nerves
not only acting alone, but acting in correspondence
with external things? Is this a reason for rejecting
an evidence precious in itself — the more precious
from its novelty and rarity? A new science must be
based on new facts. Were all phenomena alike in

the world, knowledge must come to an end, and investigation be circumscribed indeed. Surely, all
that is required to render any class of facts interesting
is, that they should be new, yet not out of analogy
with nature; and those we have been considering
come under both these conditions. Even though
they should stand alone as a class, yet, having been
shown to be uncontradicted by nature's general tendencies, they should be admitted to examination. Yet
here again, though demanding but justice, we are
baulked. We entreat the adversaries of mesmerism to
come and witness our facts. They come not; or, if
hey come, scarcely deign to look at the most important and interesting phenomena; or again try
experiments, not with a rigour, but with an unfairness unknown in other inquiries, expecting the sensitive being we exhibit to them to act as so much
brute matter, and to be in the right when their
sensibility is exhausted.

"But," piteously exclaim the systematic men, "if
mesmerism be true, then all we have hitherto learned is
false." By no means. But even were this so, surely
it is better to get into the right path late than never.
It is better also gracefully to yield than to be held up
as a spectacle of vanquished yet impertinent obstinacy. And light — resistless light is pouring in on
every side to illustrate nature, and to display mesmerism in the first rank of acknowledged truths.

It requires but little sagacity to perceive that they
who now place mesmerism in the category of impossi-

bilities will shortly be in the situation so well described in the following lines : —

> " So fares the system-building sage,
> Who, plodding on from youth to age,
> Has proved all other reasoners fools,
> And bound all nature by his rules ;
> So fares he in that dreadful hour,
> When injured Truth exerts her power
> Some new phenomenon to raise,
> Which, bursting on his frighted gaze,
> From its proud summit to the ground,
> Proves the whole edifice unsound." *

---

\* Beattie.

s 3

# SECT. IV.

## ON THE MEDIUM OF MESMERIC SENSATION.

HAVING shown, from the real nature of the changes in man's personal frame which precede sensation, that an inner sensibility, coincident with an inaction in external organs of sense, is but accordant with the principles of true physiology, I now purpose briefly to consider the pre-requisites of sensation, which are external to ourselves; the media namely, which place us in communication with foreign and distant objects; and I trust here also to demonstrate that mesmerism, instead of violating laws, does in fact bring us to grounds and principles.

Should the problem be given us to solve of conveying to a being, limited in perception and place, knowledge from a distance — how should we accomplish the desired end?

By contact, certainly, and motion.

We will imagine a person, deaf and dumb, placed at a distance from us, with his back towards us. He has been taught to distinguish, like the pupils in the Abbé Picard's institution at Paris, words written on the back. We are too far off to write these with the finger — we take a stick and write; and the person, though distant from us, understands what we mean as well as if our thoughts were actually present to himself. Here a certain motion and configuration convey from our own mind intelligence of what is there existing to the mind of another, through the intervention of *a moving medium.*

So God's intention is manifestly to convey to us certain ideas from his own mind, through intermediary types or shadows; and all knowledge is God's writing on the soul by means of certain touches from afar, prolonged even unto us by communicating media. Should this view of the subject be refused, and should we be required, with cold philosophy, to limit ourselves to material objects, and to their effects upon our physical frame, it will still be perfectly evident, that particles of matter cannot act upon other particles without some means of communication.

When then we are made aware of the existence of distant bodies, it is plain that these are brought into contact with our nervous system, either by a prolongation and extension of the atoms which compose them, or by an impulse continuously carried on from themselves to us by the successive agitation of the particles of some intervening substance. The latter, in cases where the object with which we communicate is extremely remote, appears to be the most rational supposition, and with this modern science is perfectly in accord, no longer considering even light itself as a corporeal emanation from the sun, but as a

vibratory impulse communicated from the self luminous solar body to an ethereal medium, and thence
again to our optic nerves.

By analogical appearances and reasoning, it seems
also to be decided, that " heat, like light and sound,
consists in the undulations of an elastic medium;"
and it may perhaps, in process of time, still further
be proved that odour is not a material emanation,
but a mere action of matter communicated to our
perception by the nerves. This at least would
account for the now incomprehensible fact, that a
grain of musk will diffuse its odour for years without
any perceptible diminution of its weight.

In the admirable language of Mrs. Somerville,
" The human frame may therefore be regarded as
an elastic system, the different parts of which are
capable of receiving the tremors of elastic media, and
of vibrating in unison with any number of superposed
undulations, all of which have their perfect and independent effect."

This view of the nature of our sensations cannot
but greatly simplify our notions concerning them,
and it is astonishing how much is effected towards
the comprehension of a subject, when we put out of
the way all its less important elements. The astronomer, finding it impossible to determine the motion
of each heavenly body when disturbed by all the
rest, takes a simpler problem, and calculates the
mutual relations of three bodies only; yet thence
he learns to judge of the whole celestial system. In

the same manner, the problem of the action of
matter upon mind should be rendered as little complicate as possible; and thus, in considering sensation,
it is desirable to bring into one comprehensive view
its chief conditions, to the exclusion of such con-
tingencies as merely arise from the modifying influence of particular circumstances.  Now we have
seen that the external and mechanical portions of
our senses are not indispensably and fundamentally
connected with sensation, but rather are adaptations
to the exigencies of this our temporary existence.
The two absolute essentials of sensation, without
which it could not take place under any circumstances, in which it bears coincident relation to real
external objects, are — a medium and a system capable of responding to that medium.   Now the
question is, whether there be any fact which renders
impossible, during the mesmeric sleep, the co-exist-
ence of the two essential conditions of sensation.
One of them — that is, motion of the nervous system — has been shown to be by no means of necessity
abrogated during mesmeric sleepwaking; but the
other, that is the correlative motion of a medium,
does, on a cursory view, seem forbidden in cases
where the patient sees through obstacles which inter-
cept the light, and, to our common apprehensions,
break up the continuity of visual impulsions.   On
maturer consideration, however, we must own that
the great principle of sensation — namely, motion
communicated by media — may not only subsist in its

s 5

integrity during the mesmeric vision, but subsist also
analogically to all we know or can discover of nature's
constitution.

It is true that, in the endeavour to reconcile mesmeric sensation with established laws, we are in a
manner forced into hypothesis. But is this a fault?
A great metaphysician has pronounced that " an
hypothesis, in the first stage of inquiry, far from
being inconsistent with sound philosophy, may be said
to be essential to it." Now the case of mesmerism
is one in which we are absolutely obliged from given
data to draw deductions, and to reconcile two facts
by the interposition of a third. With regard to
mesmeric vision, for instance, we have a phenomenon, accurately determined by rigorous experiment,
which involves two things, — a perceiving mind, and
an object perceived ; and these we are forced,
through reason, to connect by a third, namely, a
medium, for we never knew mind to perceive but
through an intermediary, and we have reason to
think that none but the Almighty mind can immediately perceive objects. Yet more, — the sleepwaker
has a nervous system, which, though externally inert,
is manifestly in internal activity; and there is no
ground whatever to suppose that the established
language of sensation, namely, motion of a nerve,
is, in mesmerism, abolished, though it be produced
in an unusual way. If then there be motion of a
nerve correlative with an external object, it is plain
that this motion must be brought about by means,
and by material means. What these means are we

do not indeed exactly know, and we are therefore compelled, in a degree, to form conjectures about them. Certainly the less of this the better, but at any rate we cannot but imagine such a medium as shall meet the necessities of the case; herein, at least, we have a sure guide. If, taking the facts, we proceed, step by step, to deduce from them the characteristics of our medium, fitting always the supposition to the reality, and not the reality to the supposition, it seems as if we could not greatly err.

The first thing of which we may be sure is, that the medium which connects mesmeric persons with external objects is not any of those which we believe to be efficient in ordinary sensation. The mechanism of the senses being null, the media especially adapted to that mechanism must also be useless in conveying information. The ear was made for air, and the eye for light; but, in this case, the ear is closed, and the eye is an abolished organ. That, however, there should be no communication between the sleepwaker and external objects does by no means follow. We have seen that real information respecting the material world can be conveyed through other inlets of the body than the ordinary. In the same manner, it may be affirmed that even our common experience may show us sensation taking place through other external media than the ordinary. Let us adhere to our principle, that similar motions of the nerves may be produced in various ways, and different causes end in the same result; and here also it will not fail us.

s 6

The deaf, who hear and enjoy music by means of a staff connecting their teeth, or their chest, with a musical instrument, do really use another medium than the air for the conveyance of acoustic impulses. In the same way, we have instances to show that real information respecting external actions of matter may reach from great distances by other than the usual media. Water is even a better conductor of sound than air. "According to the experiments of M. Colladon the sound of a bell was conveyed under water, through the Lake of Geneva, to the distance of about nine miles." Yet again: " The velocity of sound, in passing through solids, is in proportion to their hardness, and is much greater than in air or water. A sound, which takes some time in travelling through the air, passes almost instantaneously along a wire 600 feet long; consequently it is heard twice; first, as communicated by the wire, and afterwards through the medium of the air. The facility with which the vibrations of sound are transmitted along the grain of a log of wood is well known. Indeed, they pass through iron, glass, and some kinds of wood, at the rate of 18,530 feet in a second."

Does any thing similar take place with regard to light? There does; and even in a more remarkable manner. The electic medium is a far more swift and subtle messenger of vision than is the luminous ether. " A wheel revolving with celerity sufficient to render its spokes invisible, when illuminated by a flash of lightning, is seen, for an instant, with all its

spokes distinct, as if it were in a state of absolute repose; because, however rapid the motion may be, the
light has come and already ceased, before the wheel
has had time to turn through a sensible space."
Again, some ingenious experiments, by Professor
Wheatstone, demonstrate to a certainty that the speed
of the electric fluid much surpasses the velocity of
light. It is therefore a different medium — yet can
it serve for all the purposes of vision, and even in a
superior manner. After hearing these things, shall
we start at the notion of mesmeric sensation being
conveyed through another medium than that in ordinary action? Even should the sleepwaker perceive
the most distant objects (as some are said to have
done), can we, from the moment a means of communication is hinted to us, be so much amazed? If
his perception be more vivid, there seems to be an
efficient cause in his abjuring the grosser media for
such are more swift and subtle.

But the mesmerised person perceives objects not
only at a distance, but through obstacles which are
complete impediments to ordinary vision. Let us
keep in view our principle. Motions of the nerves,
and something to produce those motions, in correspondence with an external cause, are the required
conditions which we seek to unite. If a mesmerised
person seems to behold an object through an obstacle which commonly bars vision, let us not take
so vulgar a view of this apparent miracle as to suppose that the sleepwaker's sight pierces through
opaque bodies. It is not, in any case, the eye, but

the medium, which penetrates the obstacle. The impulsion on the optic nerve is the cause of vision, and, if we can find a medium to transmit that im pulsion, athwart whatever impediments, vision will take place. Any interruption in the medium would be the *real* obstacle to vision. Cut that off by the finest and most imperceptible barrier, you at once render vision impossible; but, as long as it continues to be transmitted through bodies, the most appa- rently dense vision will not be prevented — an object is brought into relation with us. This is the whole history of seeing; — and is it, in fact, more wonderful that any one should be brought into relation with an object by means of an impulse transmitted through an obstacle, than that the astronomer should be brought into relation with a star by means of light which left it a thousand years ago? We seek, then, a medium, which shall act through obstacles, which is able to transmit impulses athwart them, unimpeded, unimpaired. Now, as Newton observes, " the effluvia of a magnet can be so rare and subtile, as to pass through a plate of glass, without any resistance or diminution of their force, and yet so potent as to turn a magnetic needle beyond the glass." And this effect of magnetism, as we now know, is but one of the actions of electricity, which can pierce not only through diaphanous but through opaque bodies, and which is now supposed to pervade not only the regions of space, but every interstice between the particles of apparently solid matter. The electric force, then, will naturally occur to him who seeks a penetrating and pervading medium, as resolvable of the enigma of mesmeric sensation.

The philosopher also will remember that electricity, when brought to act upon the nerves, in that peculiar form of it which is called galvanism, is capable of affecting all the senses, making us seem to hear, see, taste or feel. Moreover, it has just been proved to equal or surpass light in its capacity of producing real sensations of vision in correspondence with external objects. It may, however, here be objected that electricity never does this save in a manner visible to all men; that if there be vision by means of electricity, there is also an actual electric spark, or flash; — whereas, in mesmerism, it is not pretended that there is any agency cognisable by any person but the sleepwaker himself. This is true. But, then, be it remembered, the luminous medium itself can be manifestly shown to have other actions than those with which men in general are acquainted — actions which can only be perceived by beings in a peculiar state of sensibility, or more finely organised than ourselves. As an author, to whom I am already deeply indebted, has remarked, — " It is quite possible that many vibrations may be excited in the ethereal medium, incapable of producing undulations in the fibres of the human retina, which yet have a powerful effect on those of other animals, or of insects. Such may receive luminous impressions, of which we are totally unconscious, and, at the same time, they may be insensible to the light and colours, which affect our eyes; their perceptions beginning where ours end."

A proof of the soundness of the above conclusions is the fact that the owl and the cat pursue their prey, just when the increasing obscurity renders other animals purblind. The degrees of visual sensibility in different persons are also to be remarked. A lady of my acquaintance performed one evening what might have passed for a mesmeric miracle, deciphering a letter, written in a small cramped hand, at a time when there was so little light that five or six persons of our society (myself amongst the number) could not make out a single word of the letter. That some individuals, in certain states of the nervous system, see in an obscurity which to other persons appears total, has been before observed. A case of this kind was very recently mentioned to me by a medical man, with every particularity of detail. A peasant, who had fallen into a state of illness from having imprudently plunged his head, when heated, into a bucket of cold water, could no longer bear the light of day, but was able to discern objects in apparent darkness. A number of gentlemen visiting him out of curiosity, made him tell the time (in darkness) by their watches, which they altered sufficiently often to be sure of the correctness of the experiment. As the man recovered his health, the susceptibility of his optic nerves gradually subsided to the usual pitch. These facts should lead us to beware under what circumstances we pronounce vision impossible; and should teach us not to limit the perception of others by the boundaries of our own. It is evident that a finer medium and a finer degree of sensibility, co-operating to produce mesmeric vision, are causes perfectly adequate to account for all its phenomena.

There is one remark respecting mesmeric sensation important to be made. In our normal state, the use of our external organs is not only to modify impulses, but to concentrate them, properly modified, on a particular nerve. Now, it may be asked, how, when the concentrating organs are removed out of the way can impulsions be concentrated upon the nervous system, so as to bring the sleepwaker in correspondence with particular objects, it being evident that, in the general motions of a general medium there could be none of that particularity and selection which sensation evidently requires? To satisfy ourselves on this question, we have only to remember that the sleepwaker does not, in general, seem to know on every side, but that the most common mode of mesmeric perception (there being two, as I have pointed out) is by a concentrated sensibility in one particular portion of the frame, which constitutes, as it were, an organ whereby in fact impulses coming from without are concentrated, as they ordinarily are, in another way, by the usual organs. That sleepwakers really do use some part of the nervous system as an organ of concentration, I have had frequent occasion to remark — Anna M. always bringing her forehead in a line with an object she wished to examine, and often starting at the moment when her forehead came opposite the object, as if she were struck by the impulses of some invisible medium. When this was the case, she was always correct in her perceptions. In thus bringing the nervous system into immediate relation with an object by means of one restricted portion of the nervous system, it is evident that the analogies of nature are preserved.

Thus in examining the media, the action of which forms the external pre-requisites of sensation, we seem to arrive at a general ground or principle, just as, in examining the action of the nerves, or the internal pre-requisites of sensation, we formerly arrived at a ground or principle. And the one seems perfectly adjusted to the other. From what we know of the constitution of nature, also, we should expect to come to some ground in sensation, for the discoveries of modern science tend to resolve all forces, even gravitation itself, into varied actions of one medium; and, the universality of that medium being once allowed, it is plain that it must be the ground in sensation, as in all things else. Besides, viewing man as more than the ephemeral being of a day and more especially viewing him in his extended relationships to creation, considering, too, all we know or see as but parts of one great system, we should expect not only to arrive at some ground in sensation, which had not precisely reference to the transitory purposes of this life, but at some general ground, which might be common to all beings, and which might largely pervade the universe.

The media, which now act upon our organs, we should as much conceive to be modifications of the pervading and ethereal medium, as our organs themselves are manifestly modifications of the sentient principle. So, in a new state of man, which takes him, as it were, out of his sphere, for a while, and extends his relations to things in general by removing the modifications of the external senses, we should expect to find a corresponding removal of the modifying media, and a correspondent approximation to the general groundwork of sensation. Something like such a universal principle or agency in sensation has long been suspected to exist.

Newton, speaking of his celebrated ether, says : — " Is not vision performed chiefly through the vibrations of this medium, excited in the bottom of the eye by the rays of light, and propagated through the solid, pellucid, and uniform capillamenta of the optic nerves into the place of sensation ? And is not hearing performed by the vibrations either of this or some other medium, excited in the auditory nerves by the tremors of the air, and propagated through the solid, pellucid, and uniform capillamenta of those nerves into the place of sensation ? And so of the other senses."

Admit this, and the mysteries of sensation in sleep-waking stand revealed. When once we see clearly (as see we must, if we consider nature aright), that the communication between all portions of the universe is continuous and incapable of interruption — that there is a pervading medium, filling all things, permeating all, — the extended sphere of mesmeric faculties appears no longer miraculous. The hiatus once supplied, which seeming obstacles created to our apprehensions, there is no link wanting in the chain which connects us with external objects. All is harmonious — all is clear. And the facts of mesmerism alone would establish this beautiful simplicity in the economy of nature and in our own; proving that Newton, when he suggested a universal agent in sensation, outran his age, and in a moment divined that which to elucidate will occupy future ages. But, alas! he who defends the cause of mesmerism, because he is penetrated with its truth, is " cabin'd, cribb'd, confin'd " in his mode of argument. He cannot support a theory, however noble, by the facts of mesmerism, nor reason from them by the most legitimate induction: for he is called upon to prove that the facts themselves are not absurdities, — that all which he perceives to be so admirably illustrative of Nature's order is not anomalous and monstrous. The foundation-stones of belief for his system are not yet laid; and he has to lay them in a quaking and unstable soil, even as the first builders of Venice had to waste their materials beneath the engulphing sea, long before the ocean city could rise, in stability and beauty, from the waves. A time, however, is at hand, when the facts of mesmerism, being accepted as facts, will themselves be used as the most valuable demonstrations of truths and principles, which the genius of great men foreknew.

If it may be permitted me, for a moment, to antici-

pate that era, I may remark that the fact of the sleepwaker seeing colours, when the usual action of light is
plainly absent, while he does *not* behold objects magnified through a convex glass, is a proof that, as
Newton conjectured, the efficient cause of vision is the
vibration of another medium than light, and is not
dependent on the anterior mechanism of the eye. All
this is in perfect harmony with science. Colours
depend on vibrations of the optic nerves, which vibrations are shown to be more perfectly and permanently
produced by electricity even than by light; for Professor Wheatstone exhibited by electric light colours
distinctly, and as at rest, when they were revolving so
rapidly as to be all mingled together into white under
any ordinary illumination. On the other hand, the
enlargement of objects by a magnifying glass is a
mechanical effect dependent on a modification of
light, which could not take place where the conditions are so plainly absent as in the mesmeric
slumber. The result of the experiment is at once
conformable to reason and to science, and indicative
of the sincerity of the sleepwaker. Here, also, we
may see the difference between sensation general and
special. The peculiar action of light, whereby an
object is magnified, is only an adaptation to our present imperfect organs. By it we gain no new, no
peculiar, no fundamental idea : we only see objects
when magnified, as if they were nearer ; but the
notion of colour is specifically different from every
other, and is common, we may suppose, to all ani-

mated creatures. Thus is mesmerism, in all its parts, calculated to lead us to principles of extensive application. If it seems to contradict our former knowledge, it only contradicts it in its externals and deceptive appearances. But in truth it never does contradict either nature or previous knowledge; it is only in opposition to narrow views. The more we become aware what true experience is, the more shall we find that from experience mesmerism strays not. If men had wished to believe it, half the evidence that is now brought forward to prove it possible would have proved it sure.

I conclude my remarks on sensation by recalling to the reader's mind a few of the most, important reflections on which we have been engaged, taking sight, our chief sense, as the type and example of the rest.

The real nature of vision is as shut to the vulgar as the mesmeric mode of sight is to the learned.

By the eye we appreciate light and colour : only the rest is an operation of the judgment.

Viewed metaphysically, seeing is but a particular kind of knowledge : viewed physically, seeing consists in certain nervous motions, responsive to the motions of a medium. That medium, in our ordinary condition, is light, the action of which seems cut off and intercepted in the case of mesmeric vision.

When, therefore, we hear that a mesmerised person has correctly seen an object through obstables, which to us appear opaque, we, conceiving no means of communication between the person and the object, exclaim that the laws of nature have been violated. But, in all cases where information is conveyed through interrupted spaces, show but the means of communication, and astonishment ceases.

When we know that there is a medium, permeating in one or other of its forms all substances whatever, and that this medium is eminently capable of exciting sensations of sight, and when we take this in conjunction with a heightened sensibility in the percipient person, rendering him aware of impulses whereof we are not cognisant, we are no longer inclined to deny a fact or suppose a miracle.

Finally, all sensation has but one principle. All that is required for its production is that objects should be brought into a certain relation with us by something intermediate; and this is effected by the impulsions of certain media upon nerves, the last changes in which are the immediate forerunners of completed sensation.

Thus things, which are beheld in their primal essences by the Almighty, are to us known in their beautiful results. It is to the *idea*, not to the *contrivance* for exciting the idea, that we are to look; for the unlettered peasant, who has a conviction and a feeling what light is, has in reality as true a notion of this the first effluence of supernal power as he who has divided the sunbeams with a prism or calculated the undulations of the etherial medium. With respect to external things, we have only to consider God and man — not God and the philosopher.

Taking this simple view of sensation, we find nothing in mesmerism contradictory of nature.   Under its influence the human frame continues to be still a system of nerves acted upon by elastic media, for the purpose of conveying to us the primal impulses of the Almighty mind, which made, sustains, and moves the universe.

# BOOK IV.

To every form of being is assign'd
An active principle : — howe'er remov'd
From sense and observation, it subsists
In all things, in all natures, in the stars
Of azure heaven, the unenduring clouds,
In flower and tree, in every pebbly stone
That paves the brooks, the stationary rocks,
The moving waters, and the invisible air."

WORDSWORTH.

## SECT. I.

### THE MESMERIC MEDIUM.

HAVING, as I trust, shown the conformity of mesmerism, in all essential points, with the principles of
nature and the inferences of reason, I now proceed
to exhibit it in connection with such a cause as its
peculiar manifestations indicate and demand.

First, I affirm that, productive of the effects called
mesmeric, there is an action of matter as distinct and
specific as that of light, heat, electricity, or any other
of the imponderable agents, as they are called ; —
that, when the mesmeriser influences his patient, he
does this by a medium, either known already in
another guise, or altogether new to our experience.

T

What proofs, it will be asked, can I bring forward of this assertion? I answer, such proofs as are considered available in all cases where an impalpable imponderable medium is to be considered : facts, namely, or certain appearances, which, bearing a peculiar character, irresistibly suggest a peculiar cause. Let us take only one of these.

Standing at some yards distant from a person who is in the mesmeric state (that person being perfectly stationary, and with his back to me), I, by a slight motion of my hand (far too slight to be felt by the patient, through any disturbance of the air), draw him towards me, as if I actually grasped him.

What is the chain of facts which is here presented to me? First, an action of my mind, without which I could not have moved my hand; secondly, my hand's motion; thirdly, motion produced in a body altogether external to, and distant from myself. But it will at once be perceived, that, in the chain of events, as thus stated, there is a deficient link. The communication between me and the distant body is not accounted for. How could an act of my mind originate in an effect so unusual?

That which is immaterial cannot, by its very definition, move masses of water. It is only when mysteriously united to a body that spirit is brought into relationship with place or extension, and under such a condition alone, and only through such a medium, can it propagate motion. Now, in some wondrous way, spirit is in us incorporate. Our bodies are its medium of action. By them, and only

by them, as far as our experience reaches, are we
enabled to move masses of foreign matter. I may
sit and may will for ever that yonder chair come
to me, but without the direct agency of my body
it must remain where it is. All the willing in the
world cannot stir it an inch. I must bring myself
into absolute contact with the body which I desire to
move. But, in the case before us, I will — I extend
my hands; I move them hither and thither, and I see
the body of another person — a mass of matter external to myself, yet not in apparent contact with
me — moved and swayed by the same action which
stirs my own body. Am I thence to conclude that a
miracle has been performed — that the laws of nature
are reversed — that I can move foreign matter without contact, or intermediate agency? Or must I
not rather be certain that, if I am able to sway a
distant body, it is by means of some unseen lever —
that volition is employing something which is equivalent to a body — something, which may be likened
to an extended corporeity, which has become the
organ of my will?

Surely there is no effect without a cause; and
from actions we may infer the existence of an agent.
We do this a thousand times in other cases, — in that
of mesmerism, for instance. We never behold this
power but in its results. It cannot even be made
evident, like the electric spark, or felt in our own
persons, like the galvanic concussion   The needle
that has become a magnet, has undergone no change
which any mortal sight is fine enough to appreciate,

has acquired no weight which can be detected by our earthly senses. Yet, solely because we are sure that we behold certain phenomena, we allow that there is a distinct form of electricity, to which we have given the name of magnetism. Why should we refuse to mesmerism that which we grant to magnetism? It is true that as yet we have no balance of torsion, whereby the mesmeric force can be measured; but in the human body itself we do possess an instrument whereby its presence may be ascertained; nor would it be reasonable to insist upon separate agencies being detected by the same test. Why, then, but from the force of prejudice, should we call the mesmeric medium a gratuitous assumption? That such a medium exists is not a gratuitous assumption, but an unavoidable deduction of reason. But there is a class of persons who refuse to admit of anything which they cannot see, taste, or handle; with such it is difficult to argue. Should proofs by experiment be exhibited to them again and again, they still return to their cuckoo note — " Show me the agent." One of these practical men, as they are called, actually said to me on one occasion, " I never will believe that what you call mesmerism exists, unless you can put it in a bottle, and submit it to analysis."

To what end, then, is reason given us, if not to judge of things invisible by those which are clearly seen? For what purpose possess we the irresistible propensity to supply deficient links in a chain of causation, if not to prompt us where our senses fail? We move a magnet over a needle; the needle moves in a corresponding manner; and the human mind is so constituted that we cannot behold these two facts, in seeming connection, without uniting them by a third, which we consider as *proved* by them, since it is in truth their necessary consequence. We infer that the effect is produced by means of a magnetic current or medium, — a something which propagates motion from the magnet to the needle. This something we cannot indeed behold; — yet do we believe in it, — and with justice, for that which reason perceives to be necessary is *not* an invention, and can never be superfluous: on the contrary, the *only* immutable and essential truths come out of the mould of the intuitive reason, which, as Coleridge observes, stops not at " this *will* be so," but at once decides, " This *must* be so."

Now, in all cases where motion is communicated from one body to another, the line of communication must be maintained unbroken. The first impulse gives motion to certain atoms, which in their turn propel others, and so on, till the whole series between the active body and the body which is to receive the original impulse is set in motion, and then, at length, the sequence of events is complete, and the body, towards which motion tended, is set vibrating. If the medium that propagates the first impulsion be undulatory and elastic, its atoms only oscillate on either side a fixed point of rest; but, if it be composed of travelling atoms, there is an actual progression of the medium. In either case, motion is propagated by

T 3

a real action of matter till it reach its final destination. This is the history of all communicated motion, and it is plain that this holds good whether we behold the collection of atoms, in a bodily shape, that transmits the motion, as in the case of one billiard ball propelling another, or whether we behold them not, as in the case of sound being communicated to the ear from a vibrating body, by means of the intervening air. I grant that the old axiom, " A body cannot act where it is not," is very properly exploded ; but for it we must substitute another ; namely, " A body cannot act where it is not, save by deputy, or transmissive means." Yet some have overlooked this truth, and in their zeal to avoid theories, when they behold two sensible actions, evidently dependent the one on the other, and yet apparently disjoined, fear to unite them properly by suggesting the presence of an unseen link, which nevertheless cannot but occur between the visible antecedent and the visible consequent ; for motion is not an entity that can go through void spaces independently and alone ; it is merely a property which has no existence out of the subject that manifests it ; and, where matter fails, there motion fails also. It is vain, then, to hold such language as if it were possible for one body to produce motion in another without something intermediate, — that is, miraculously and without means ; yet your good hater of theories will even dare to blame Newton for having suggested an ether to account for that action which one body produces on another, and even, in many cases, from vast distances, and which we call attraction. It is true that Newton may be wrong in

the manner in which he manages his ether, and accounts for impulsion and re-impulsion by differences
of dense and rare; but he cannot be wrong in preserving an unbroken series of atoms between separate
bodies which manifestly influence each other, — between the sun and the earth for instance, — since, in
this case, there is mutual action, and motion communicated from a distance. Extending the principle,
and perceiving that all the heavenly bodies were in
mutual relationship, and the whole celestial system
harmoniously bound together, Newton supposed his
ether to be of universal action, and to fill and pervade
creation, establishing a means of communication between all its several parts. Were this allowed, there
would be but little difficulty in explaining mesmerism;
but a sublime divination of this kind is too vast for
the general understanding. Accordingly, even Newton's name has failed to render the theory palateable,
and men of small views have dared to call even this
suggestion of a mighty mind *gratuitous*, treating its
author with a levity which can only lessen one's respect for the objectors. Have these cavillers an intellect superior to Newton's own? If they have, let
them give us something better than Newton's suggestions (better, not only in their own opinion but in
ours) respecting the great problems of creation;
some theory more solid and sublime to satisfy the
cravings of humanity after pure and lofty generalis-
ation: till then let them, at least by silence, acquiesce
in Professor Playfair's beautifully expressed opinion
of the queries: — " Such enlarged and comprehensive

views, so many new and bold conceptions, were never before combined with the sobriety and caution of philosophical induction.  The anticipation of future discoveries, the assemblage of so many facts from the most distant regions of human research, all brought to bear on the same points, and to elucidate the same questions, are never to be sufficiently admired." In recalling this to the reader's mind, I trust that I seem not to stray from my subject, which is in truth so deeply implicated in the truth or falsehood of Newton's principal suggestions.  But I might leave this great man's defence to time, which already has "brought in its revenges," science being even now occupied in developing Newton's ideas, and in establishing as undoubted truths the greater part of all which he so modestly advanced as queries.  Facts relative to the acceleration observed in the mean motion of comets have demonstrated, to the satisfaction of men of science, the existence of a resisting medium, undulatory and elastic, which pervades the known universe.

How frequently it has thus happened that the deductions of the pure reason have triumphed over the cavils and hesitations of the understanding which, being conversant with matters of experience only, cannot step beyond the sensuous and the known ! Kepler believed that the harmony of our system required a planet between Mars and Jupiter, and the deficiency is now actually supplied by the discovery of the four singular orbs which seem once to have formed but one single body.  My reader's memory will doubtless supply other instances where the philo-

sopher in his closet has outrun experiment, and has
divined what future observation has verified and
facts confirmed. When, then, we find Mesmer (who,
whatever were his faults of conduct, was no con-
temptible thinker) suggesting a universal medium as
alone explanatory of mesmeric phenomena, let us,
instead of unwisely scoffing, inquire whether the circumstances of the case may not possibly render the
existence of such a medium a positive necessity, and
a truth palpable to reason. This at least we know,
that all science seems now tending to refer the apparently distinct agencies of nature to the varied operation of one medium ; to establish, in fine, an ether
such as Newton had imagined, and such as Mesmer
perceived would satisfactorily account for the apparent miracles of his new science.

Now whether mesmerism be a distinct medium, or
only the distinct effect of a general medium, widely
manifested in other offices, I will not take upon myself to decide. We no longer consider electricity,
magnetism, even light itself, to be separate and independent agents ; we call them *effects.* And this is
well, if we remember to refer *effects* to *causes*, and
properties and qualities to real substances and subjects. We must not turn all the goings-on of the
world into mere abstractions. Vibrations imply a
vibrating body ; electric motions or concussions,
something that moves or is concussed. It will, indeed, greatly simplify our ideas, to consider all the
various appearances of nature as so many actions of
matter, but we must beware of supposing that, where

T 5

action is present, matter can be absent. I am very
willing, then, to call all mesmeric phenomena *effects;*
but I not the less contend that they must be effects of
something. I am willing to consider mesmerism itself
as an action of matter, yet still of matter. I cannot
tell whether, in the case of mesmeric agency matter
assumes the form of a fluid or a gas, but I know and
am sure that material agency there is. This agency
may be only one of the modifications of a substance
which operates in other ways, or it may be the single
action of a single substance. But, in fact, the probability is, that there are really various media in nature,
the finer, we may suppose, occupying the interstices
of the grosser, distinct yet interfused, wheel within
wheel, a subtle mechanism. Every one knows that
the atmosphere is the medium through which sound
is conveyed to us. A bell rung under the exhausted
receiver of an air-pump is inaudible; but the crystal
walls, that keep out air, bar not the passage of light
and heat. Newton's experiment of this, and his
consequent reasoning on the fact, appear to me conclusive. He says, Qu. 18. — " If, in two large, tall,
cylindrical vessels of glass, inverted, two little ther-
mometers be suspended, so as not to touch the vessels, and the air be drawn out of one of these vessels,
and these vessels, thus prepared, be carried out of a
cold place into a warm one, the thermometer in
vacuo will grow warm as much and almost as soon as
the thermometer which is not in vacuo. And, when
the vessels are carried back into the cold place, the
thermometer in vacuo will grow cold almost as soon
as the other thermometer. Is not the heat of the

warm room conveyed through the vacuum by the vibrations of a much subtiler medium than air, which, after the air was drawn out, remained in the vacuum? And is not this medium the same with that medium by which light is refracted and reflected?"

The conclusion which such experiments force upon us is, that there really exist in nature different media, related, yet distinct. If, therefore, I am understood literally instead of figuratively, when I speak of mesmerism as an individual agency, I shall not seem greatly to have violated the analogies of nature. This at least I affirm, mesmerism has its own peculiar action; and therefore, for the sake of convenience, I shall denominate matter, as it is developed in this particular way, the mesmeric medium, a term with which, I trust, none of my readers will be disposed to quarrel; the advantage and propriety of referring one class of effects to one cause being manifest. We do this naturally in all cases where distinction is required. The imponderable fluids are still characterised, *pro formâ*, by individual names, though, we believe, that they may be children of one parent. From certain effects we are allowed to presume the existence of a luminous medium. I, therefore, by parity of reasoning, may be allowed, from other effects, to infer the existence of a mesmeric medium.

And, in truth, there is no agency which more manifestly than this may claim to be distinctive, since it is developed under quite other circumstances; and, being developed, presents quite other pheno-

T 6

mena than any material action with which we have
hitherto been acquainted.

It neither results from a union of gases, nor from
chemical composition. It is not developed by the
rubbing of amber or the juxtaposition of minerals.
It is elicited by certain actions of living nature alone.

Viewed merely as a physical agency, it originates
a sleep, *sui generis*, which pervades the external
organs, yet leaves the intelligence free; it brings the
nervous system into a state of exceeding sensibility,
rendering it cognisant of influences by which at
other times it is wholly unaffected.

Now, what is the medium we know not; and therefore all that remains for us, in our ignorance, to do, is
to gather as much information concerning it as we can.
We cannot analyse it in the same manner as light, or
separate it into its component parts, like the atmosphere; but every agent has its own elements, and
consequently a method of analysis proper to itself.
This is clear; in its mode of action it can alone be
made manifest; its sensible action, therefore, is the
legitimate sphere in which it may be studied; and,
till we have all its facts and relations, we have only
to observe its phenomena, and to state the results of
our observation as plainly as possible.

First, then, it is an agency, which has physical
effects on man.*

That any one, who has been conversant with even

---

\* And on brutes also. On this point I *could* state many
curious particulars : but I desire to startle my reader as little as
possible, and to exclude from the *present* work whatever bears
not strict reference to the *human* influence.

the first symptoms produced by mesmerism, should doubt the physical and distinct character of the agency, seems impossible. However marvellous be the train of mental phenomena consequent upon its operation, its primary effects are undoubtedly upon the body of the patient. Those who are neither under the predisposing influence of fear or of imagination, who know not what they have to expect, who, perhaps, close their eyes from the beginning of the experiment, all agree in feeling a weight upon the eyelids accompanied by a slight pricking; then follows the sensation of a cold current of air, streaming in the direction of the mesmeriser's fingers, and of a torpor in the limbs, which gradually increases, until spontaneous motion becomes not only difficult but impossible to the patient.

A lady of much intelligence, whom I recently mesmerised at Rome, was determined to observe, as long as she could retain her consciousness, what would be her sensations under the mesmeric influence. On recovering from the sleep into which she had been successfully thrown, she told me that at first she felt an increasing weight upon the eyelids, and an inclination to close them, which she endeavoured to combat by looking often at a vase containing flowers that stood opposite to her. By degrees, however, a kind of vapour seemed to steal between herself and the object she regarded, until she could distinguish nothing whatever. Her eyelids were then irresistibly drawn together. After closing them she was aware of a tingling in the arms, and currents of air seemed to move around her in various

directions. The strength and velocity of these currents appeared constantly to increase, until the power
of observation failed and consciousness was suspended. Subsequently, when in the natural state,
she begged me to say whether I had not moved my
hands rapidly and violently when mesmerising her,
so as to produce those currents of air which she had
felt. I assured her that, on the contrary, I had
moved my hands slowly and gently all the time.
She, then, still incredulous, begged me to repeat the
same movements which I had used in mesmerising. I
did so, and she declared that she could not at all feel
those currents to which she had been sensible in the
incipient state of sleepwaking.

E. A., (the youth who enjoyed such a high degree
of mesmeric vision,) used to describe similar sensations, and has frequently spoken to me of a kind of
vapour rising before his eyes during mesmerisation.
When questioned about this in sleepwaking, he
declared that the vapour proceeded from me, and
seemed to penetrate and pervade his frame. This he
asserted to be the effectual cause of the mesmeric
sleep.

Nothing is more interesting than to mark the correspondences of sensation under mesmeric treatment,
manifested by different persons, at different times,
and in different places. Every fresh instance of such
analogical feeling is as a new argument in favour of
the reality and constancy of mesmeric influence.

Other facts are not wanting, which contribute to
bring the agency within the domain of physics.

Like the galvanic force it seems capable of aggregation in being transmitted through kindred substances.

I have sometimes formed what may be called a mesmeric pile, by seating five or six persons together in a line, or half-circle, holding each other's hands : I have then mesmerised the first in the rank, who has passed on the influence to the second, who has again transmitted it to the third, and so on, by each pressing the hand held by each, at regular periods of time. Under this treatment I have invariably found that the mesmeric influence was most powerfully demonstrated in the person who was farthest from myself; that is, in the person who received the original impulse through the greatest number of intervening transmitters. The shades of gradation were also in these experiments justly preserved: the first person scarcely experiencing any sensation, the second feeling a more decided influence, and so in progression, till the last was thrown into the complete mesmeric state.

Again : That the mesmeric influence is capable of exhaustion and repair, like any other physical agency, has been frequently forced upon my conviction. That when I am strongest I can best mesmerise, is a fact known to me from personal experience ; also, that my power declines in proportion to the fatigue consequent upon its exertion. This was very strikingly exemplified in some experiments, in which Professor Agassiz of Neufchâtel, in Switzerland, kindly bore a part; thus enabling me to honour mesmerism by associating it with the name of one of the most distinguished natuarlists in Europe.

After mesmerising the Professor himself for an hour, during which I felt considerably fatigued, I made an essay upon a person, whom I usually influenced with full success. This time, however, I only partially succeeded; and a third trial upon another individual, who, in general, was most susceptible to the mesmeric influence, was wholly unsuccessful. The failure was quite unlooked for, and astonished the patient himself, who exclaimed, after I had mesmerised him for a considerable time, " This is surprising! I feel nothing whatever! You produce no effect upon me this evening." After resting half an hour, and taking a glass of wine, I again attempted to influence this patient. I now succeeded better than on the first trial, yet still could bring him no further than the first stage of imperfect mesmeric sleep.

I may also remark that the act of mesmerising is followed by a very peculiar sensation of exhaustion and fatigue, for which the slight muscular exertion which the process demands is quite insufficient to account. Exercise in the open air, and such means as are best calculated to restore the bodily strength in general, I have found to be the best restoratives on such occasions. Yet, though particular acts of mesmerising induce temporary fatigue, it is quite certain that the general mesmeric force of an individual is strengthened, like any other corporeal gift, by just and regular exercise. As a magnet which is laid by has not half the power that it has when kept

in action, so, as I have found by experience that a
mesmeriser who is out of practice has actually less of
mesmeric influence than when he mesmerises regularly, with a due attention to the "ne quid nimis."
This, again, brings the medium within the domain
of physics.

Another discovery which I have been enabled to
make respecting the action of mesmerism is, that it
decidedly bears a relation to the slowness or rapidity
of the mesmeriser's respiration during the act of mesmerising. I have tried this again and again, and
always with the same result. In proportion to the
rapidity of my breathing was the effect produced —
an effect independent of the actual effluence of the
breath upon the patient (though this indeed also aids
the mesmeric processes), for the same result has ensued when my head has been turned away from the
person I was mesmerising. How singularly this fact
connects the mesmeric with the chemical and vital
forces I need not observe.

At present it seems necessary that I should reply
to an objection, which, with some apparent justice,
may be advanced against the claim which mesmerism
puts forward to be considered as a physical agent.

It has been asserted that, when the attention of
mesmeric patients is preoccupied and diverted from
the mesmeric processes; or, when the imagination
has not been previously advised of an effect to follow,
the agency is null.

This statement demands an examination the more
serious, inasmuch as it is one of those half-truths

which Coleridge has denounced as fatal to true philosophy.

It must be allowed that the mesmeric influence is,
to a certain degree, impeded, should it be essayed
upon a person who is determined to resist it, or whose
mind is actively engaged upon other matters; but
does this, can this prove that the agency is null, save
in the imagination of the patient? The only thing
that it demonstrates with certainty is that which every
rational man who has at all studied the subject must
concede — namely, that the force employed in the
mesmeric processes is not sudden, or violent in its
action, but of the nature of those subtle influences
which it requires a certain attention, and indeed education of the sensibility, to perceive. A savage has
been known to track his prey, like a dog, by the
scent alone: and, in doing so, he must, of necessity,
fix all his attention upon the fine exhalations, whereby
he is guided. But how plainly absurd it would be to
affirm on that account, that there was no actual exciting cause of the sensations of the savage ! But, it
is argued, the imagination must be forewarned before
the effects, called mesmeric, can occur; and is not
this sufficient to refer them to the imitative power of
the imagination ? The answer is not difficult.

Imagination, indeed, imitates; but then it must
have something to imitate. The very expression
presupposes a model, and gives real existence to the
subject in debate. Granting that which indeed I do
not concede — namely, that the effects of mesmerism
can be proved without mesmerism, and by the im-

agination solely, it by no means follows that certain effects have not at certain times been produced by mesmerism, and by pure mesmerism. That which is spurious argues that which is true, and many copies of a picture place the existence of an original beyond a doubt.

Again : they who draw strong conclusions against mesmerism by affirming that it cannot take effect unless the imagination be prepared to receive it, should remember that they, who, deeming they shall feel nothing under mesmerism, and do actually feel nothing, are both forewarned and forearmed *against* the influence in debate, and are thus themselves under the predisposing sway of the imagination as much as their opponents, only in a different manner; the one party believing they *shall*, the other that they shall *not*, experience certain effects. Under these circumstances, how the latter can pretend to a more accurate judgment on the point than the former, I confess I cannot perceive. " Imagination does much," say the anti-mesmerists — granted ; and let us have the full benefit of the principle. If imagination be so potent, it may also render insensible to mesmeric influence those who are predetermined to feel nothing of it. Yet more; the force of the mind to resist even the most powerful influences may be easily shown by facts to be great — nay, incalculable. In Lockhart's Life of Scott, an anecdote, proving this, is related by Scott himself, of one of the Duke of Buccleugh's farmers. " His father had given him a quantity of laudanum (writes Sir Walter) instead of some other medicine.

The mistake was instantly discovered ; but the young man had sufficient energy and force of mind to combat the operation of the drug. While all around him were stupid with fear, he rose, saddled his horse, and rode to Selkirk (six or seven miles); thus saving the time that the doctor must have taken in coming to him. It is very curious that his agony of mind was able to suspend the operation of the drug until he had alighted, when it instantly began to operate. He recovered perfectly."

The degree to which a person may resist, yet still be influenced by, the mesmeric agency, when pre-occupied by some counteracting idea, was, on one occasion, demonstrated to me, and, as it were, marked off and measured in an interesting manner.

A friend of mine, at Cambridge, who knew nothing whatever of animal magnetism (as it was then called) but the name, consented to let me try an experiment upon him before some incredulous persons, who had said that they never would believe in the agency until they saw it exhibited upon some one who did not even know that he was to go to sleep. It being as-certained that Mr. H—— (the sincerest of all men) was really ignorant of even so much mesmeric lore, I and my patient sate down in our proper relative positions. At the end of five minutes (though it was the unsleepy time of noon) Mr. H—— began to close his eyes, and to nod; but soon started, rubbed his eyes, shook himself, and went through all the usual formula of a person who wishes to keep awake. This alternate nodding and arousing went on for some time, when, tired at length of such unsatisfactory results, I gave the matter up and quitted my chair. The patient was then questioned as to what he had felt? " Only very sleepy," he replied. " I experienced no electric shock, nor any thing of the kind, though I watched and waited for it." " But if you felt sleepy, why did not you go to sleep?" asked some one. " Oh," answered Mr. H——, " what would I *not* have given to have gone to sleep ; but I thought I must on no account do this, as I was to keep a sharp look out for the electric shock !" My reader may smile at this, but I can assure him that the ignorance of my friend respecting the effects of mesmerism is no measure of his information on other subjects.

It appears, then, that the only concession we have to make to the anti-mesmerist is, that the mood of mind and body which is most favourable to the reception of mesmeric influence is that which Wordsworth has characterised under the title of "a wise passiveness." How completely this refutes the arguments, or rather assumptions, of those, who would resolve all mesmerism into imagination, is manifest. But it is time that this question should be for ever set at rest. How such a cause as imagination could at any time be assigned or accepted as explanatory of mesmeric phenomena, is matter of wonder, and that it should be so, diminishes one's respect for the sagacity of the human species. Here, indeed, is inefficient causation ! Here, truly, is gratuitous assumption ! I have heard of imagination keeping persons awake, but never of its setting them to sleep. This busy power holds no possible alliance with mesmerism, whose gentle influences, like streams, that are only heard when we listen for them in the hush of night, must be attended to with the quiet patience of a peaceful spirit.

But the imagination theory is really too absurd to merit a serious refutation.

A thousand times I have seen mesmeric patients placed under circumstances where the action of imagination was plainly impossible. In proof of this, I have only to refer to the preceding books of this work. Persons, it will there be seen, have been thrown into the mesmeric state when asleep, and wholly unadvised of any experiment to be tried upon them. They have been drawn towards the mesmeriser from a distance, when standing with their backs to him; they have manifested phenomena, coinciding with those displayed by other mesmeric patients at different times and at different places, and which could not have resulted from imitation, since the patients themselves, not knowing a previous type, were plainly incapable of producing a copy.

Surely facts like these imperatively call upon us to acknowledge an agency, which we may call mesmeric, or what we please, so long as we confound it not with imagination, imitation, *et hoc genus omne* of inefficient and inapplicable causes.

But our views of mesmerism, as a distinct and physical agent, would, indeed, be imperfect, were we to stop here. Ourselves would then be guilty of

uttering only half the truth. Physical though the
agent be, its offices all touch upon life and thought
so nearly, that to regard it simply in a mechanical
light is impossible. In order to know it rightly, or
in any other than a partial manner, we must consider
it in its relationship to ourselves, living, breathing
men. If purely physical, why should it produce such
singular mental effects? and why should it not affect
all men alike? These are questions which will occur,
and which demand a satisfactory answer.

The cause must be sought in its acting more especially upon that most mysterious part of us — the
nervous system; — on the apparatus of sensaticn, the
builder-up of all we are, of all we know; — on that
wondrous portion of our being, which occupies the
very confines where mind and matter dimly meet;—
on that which is more intimately ourselves than our
mere external frame, of which it appears to be the
strength and the life: for we may lose any member
of motion and yet preserve in their integrity all our
sources of information; but a trifling injury to the
optic or auditory nerve will cut us off from a whole
world of knowledge, and a slight pressure on the
brain may remove us at once beyond the pale of
humanity, into that fearful region where man's
"thoughts are combinations of disjointed things."
Let us further consider that, in fact, there is nothing
which affects us physically that does not also, to a
certain extent, affect us mentally, nay morally; every
agent in nature being calculated to call into action
all the capacities of our being; thus showing how

exquisitely, according to Wordsworth's beautiful
philosophy,

" The external world is fitted to the mind."

Light, by its presence or absence, disposes us to
courage or fear, joy or sorrow, hope or despondency
Heat, when properly moderated, gives us a sensation
of well-being, of good-humour, and is favourable to
the exercise of our intellectual faculties; while, on
the other hand, cold is apt to extinguish both our wit
and our amiability.  Pope remarks that the benevo-
lence of some persons depends on the state of the
atmosphere, and solves a generous action emanating
from a churlish mind, by suggesting that

" Perhaps the wind just shifted from the east."

Shakspeare's advice not to ask a favour of a man till
he has well dined will also be remembered.  Wine,
and every species of stimulus, affect the mind through
the medium of the body.  Even so mechanical a force
as that of electricity, if made to operate upon us, excites in us, according to our several characters, feelings
of surprise, fear, aversion, or curiosity.  If these remarks be true, as regards such powers as act only
partially on the nervous system, how much more
must they apply to an agency which stimulates not
locally, but generally, the very capacities of sensation!
Other agents affect the nerves circuitously, through
the intervention of modifying organs, but this, removing out of the way, as has been seen, the barriers of

the external senses, proceeds at once and directly to the seat of internal sensibility. Its effects also have been shown to be as durable as they are potent, not operating, like electricity, which gives a shock and is gone, but manifesting itself, when once set in action, for a length of time, so as to demonstrate that the nervous system can be charged with it in a very permanent and peculiar manner. What singular results may not be expected from the operation of an agency like this!

That mesmerism, then, should develope certain mental phenomena in those whom it influences, is no proof whatever that it may not originate physically.

Neither, if it should appear to affect all men diversely, and some not at all, is the reality of its existence to be called in question. All agencies in nature are neither electric nor irresistible, and we cannot but allow that there is a class of powers which act upon man constitutionally rather than mechanically, and so are modified by accidental circumstances of health or temperament. Even should we ultimately be forced to rank mesmerism amongst these, we should by no means have annulled its agency; we should but have ascertained correctly its specific character.

Still keeping in view that its influence is more peculiarly directed upon the nervous system, we shall learn to appreciate it justly, and, even should it appear hopelessly irregular, accept that irregularity as the condition of its existence. How, indeed, can we

U

expect from mesmerism the constancy of a mere
mechanical impulse? Its diversities have their
origin in the very essence of our nature. The agency,
which we have to examine, is almost identified with
man, the ever-changeful — the infinitely varied. No
two constitutions are alike, and the nervous system
exhibits a separate character in every separate person.  If, then, we consider that the action of medical
remedies, or of material substances, even though they
may not directly affect the economy of the nerves, is
greatly modified by the temperament of the individual to whom they are applied, so that certain
idiosyncrasies will convert a means of cure or a whole-
some nutriment into a poison; above all, if we reflect on the varied peculiarities of the senses of various
men, causing that which is a pleasing odour to one to
become an overpowering torture to another (as in
the case of the Roman women, who swoon at flowers
and perfumes), and that which is discord to certain
ears to sound as harmony to those which are differently constituted : thus reasoning, we must perceive
that an influence, which should act immediately and
generally on the nervous system, would, of necessity,
exhibit somewhat of a distinctive character in every
separate case.

It is, above all, of importance to remember that
the variability of the recipients does by no means
prove the agent itself variable.  How would you
that the effects, even though coming from a single
and permanent cause, should be similar, seeing that
the subjects who manifest them are diverse? Does

not the air produce different sounds from different
instruments of music, though it be a single known
and permanent agent ?

Even should a man be altogether insusceptible of
the mesmeric influence, he has surely no just cause
for disavowing its reality.   Is he not, perhaps, insensible to the operation of many other powers by which
a number of his fellow-beings are invariably affected?
May it not be argued that mortal temperaments —
from the phlegmatic to the sanguine — differ as
essentially as water and mercury, and that as little
should we expect men of opposite constitutions to be
similarly affected by mesmerism, as that the above-
named substances should freeze at the same degree of
temperature ?   We should also remember that he,
who is at present insensible to the force, may, at
some future time, obtain a personal proof of its
efficacy.   We are not always in the same frame, nor
always capable of being affected by the same causes.
An exposure to cold, which we brave with impunity
to-day, may be our death to-morrow, and the dose of
laudanum, or of mercury, which, when we suffer, is
but a blessed relief from pain, may, at another time,
be to us deleterious or even fatal.   That these observations hold good as respects mesmerism has been
proved in the recorded case of M. Itard, one of the
members of the French Academy of Medicine, who,
in 1826, experienced no effect from mesmeric treatment, but in the year following was relieved by it
from headach and chronic rheumatism; and is it
always so certain that they, who will not own even to

U 2

themselves that they are affected by an influence they despise, are in truth so proof against it ?    One thing at least I know.    Aware by experience of the external indications of mesmeric affection, I have been able to detect them in some who professed to feel nothing unusual under the mesmeric processes.    It may also be suspected that they, who, like a certain friend of mine, commemorated in these pages, endeavour to solve a problem, during mesmerisation, do in fact experience something of the agency to which they oppose such stout methods of resistance.

Let us but grant to mesmerism that licence which we concede to every agent upon an examination of which we desire to enter.

In all physical sciences, which are based on experiment and observation, it must be granted that the most important phenomena depend for their reproduction on the co-presence of a number of conditions, difficult to be obtained separately, harder still to be assembled and concentrated in one general result.    If this be true as regards physical science, it still more forcibly applies to a science confessedly mental as well as physical.    In mesmerism we have a task incomparably more difficult than to resolve the light into its definite and primal elements, to study the formation of a crystal, or to superintend the action of a voltaic battery ; we are called upon to dissect the mind with all its delicate and fleeting hues, and to unfold the caprices and the varieties of the human constitution.

Under these circumstances, when a mesmeric phenomenon has been observed, how many causes, impossible in the shifting train of casualties to be summoned up at will, must be reunited before it can recur! The instrument with which we have to deal is too delicate to be calculated upon as one would calculate on mathematical certainties. The powers which we have to investigate are amongst the most occult in nature, to be traced only by a few of their scattered, but most extraordinary effects. Add to this that observation respecting the influence in question is yet in its infancy; the relation between its forces, whether negative or positive, is yet a mystery; its statics have not yet been ascertained; but yet, and in spite of all these drawbacks (and this is much to say), the reader may gather from former portions of this work that the irregularity of mesmerism is to be understood with a limitation. It is of degree rather than kind. The diversity consists in the proportion wherein the agent affects various persons, — in the extent of the scale, and the fineness of the gradations, as they ascend from effects which are almost imperceptible to phenomena which exhibit the full mesmeric force. The friends of mesmerism, perhaps, need only claim for it the same experimental patience that is required in examining the properties of light, where a cloud that passes over the solar rays may prevent those beautiful appearances, which are dependent upon their polarisation, from captivating the gazer's eye.

These views of the action of the mesmeric medium are not, be it remembered, presented as complete, but rather as such as are forced upon us by the

recency of the discovery, and the consequently imperfect state of observation respecting it.    I have not forgotten that the great agencies of nature stand in a
two-fold relation to man, namely, physical and physico-mathematical, and are therefore capable of being
considered in a two-fold point of view; that is, either
as they affect all men alike, mechanically, or each
man in a different manner, constitutionally.    But I
contend that the one effect is not less real than the
other, and perhaps that which presents the most variety is not the least interesting to be considered,
though it is with the certain and the steadfast that
science is chiefly occupied.    If, before the analysis of
the atmosphere had been complete, and its laws of
density or temperature established, a man had asserted, because of its various effects on the various
constitutions of men, that there was *no* atmosphere,
how plainly ridiculous would have been the induction !    — or, if a pseudo-philosopher, in a sort of fool-
ish rage and spite against himself, were known
formerly to have renounced as flat, stale, and unpro-
fitable, the study of the constitutional effects of the
atmosphere upon his fellow-beings, because he could
not reduce them to mechanical laws, how manifestly
erring and absurd would his conduct now appear to
us.

Experience of the past should teach us not to despair of seeing the statics and dynamics of the mesmeric force plainly laid down.    Who, in the early
periods of science, would have ventured to predict the
invention of a balance, whereby the magnetic action

could be reduced to weight and measure ?    Look at
the history of magnetical discoveries !   What patience
— what erroneous guesses — what feeble dawnings of
truth — what lessons of hope are there !    The knowledge of magnetical effects is as old, at least, as the
æra of Homer, by whom they are distinctly referred
to; but it is less than a century ago * that Michell
established the true law of magnetic action : and mesmerism has not yet completed its seventieth year, —
a measure of existence which the Psalmist has assigned
as the period of one man's life !   Is this a sufficient
space for the development of a subject the most fertile and the most vast, because the most intimately
connected with man ?

   This, at least, even in the present state of mesmeric
science, may be affirmed, that, considered as a force,
the agency betrays no slight indications of its affinity
with mechanical powers, and that certain circumstances regarding it bring to us nearly a full conviction that its effects are dependent on a certain
invariable proportion between the mesmeric force of
the individual who dispenses, and that of him who
submits to receive the influence.   What that proportion is, once ascertained (and how can this be but by
repeated observation ? ), would reduce mesmerism to
a law, and to a certainty.   Its mental and remedial
effects will indeed necessarily always continue to be
varied according to the character and temperament
of its patients, but its purely physical agency might

* In 1750.

U 4

be brought within conditions, perhaps narrower and more simple than we can now conceive. Supposing (as some now deem) the vital action to be electric, that which is called a man's nervous force, or constitution, would depend on the possession of a certain original measure of the electric fluid. Were this found to be the case (my idea may raise a smile), a neurometer, or instrument to ascertain the nervous power of a person, might give to mesmerism the precision which science requires. Who would have thought, at one time, a measure of magnetism possible?

My suggestion may be deemed absurd; but this will not alter the facts of mesmerism. At present we are only acquainted with the general result, and not with the elements that compose it; but this is no more a proof that it is not composed of elements, than the ancient ignorance what water was could have demonstrated that it was not composed of two airs in certain proportions.

I will state two or three cases out of the many that have inclined me to think that mesmerism is a question of proportional force. A friend of mine at Cambridge was susceptible of being influenced by myself, but transiently and imperfectly, while, on the other hand, he was at once and invariably brought into the mesmeric state by being subjected to the action of a young fellow-student, who (as to the rest) used no art in his manipulations, and merely imitated rudely my proceedings and gestures. Again: the brother of a celebrated sculptor at Rome was always mesmerised by me in a few minutes without difficulty, but to other influence he was by no means so susceptible. On one occasion, before a large party at Rome, I instructed a distinguished artist in the necessary movements, and set him down to mesmerise the gentleman alluded to, precisely as I was accustomed to do. No result, however, at the end of half an hour had ensued. I then took the patient's hands myself, and in a very few minutes he was thrown into a deep mesmeric slumber. Other persons tried the same experiment with a similar result.

E. A., whom I could mesmerise in a few seconds, was operated upon for an hour by another person, who, in other cases, had displayed immense mesmeric power, without experiencing any effect whatever.

An interesting anecdote, related to me by Dr. Chapelain of Paris, corroborates the above.

An English lady, whom he was treating mesmerically, continued, after many days, insensible to his influence. Her son, a child of about ten years of age, who was present while the doctor mesmerised, exclaimed one day, " I think I could make mamma sleep. Will you let me try ? " Dr. Chapelain, curious to see what the little fellow would do, laughingly consented. He was surprised at the gravity and skill with which the boy imitated all the processes which he himself had employed, but still more surprised to behold that in ten minutes the patient was thrown into a mesmeric slumber.

From facts like these we cannot but conclude that it is not the strength of the mesmeriser, but the pro-

portion between the respective strengths of mesmeriser and patient, which ensures success, and that
the less or more on either side would indifferently
prevent a perfect result. To ascertain these relations seems to me to be the great end and object of
research as regards mesmerism at present. In this
labour, let us not be less active and enduring than
men of science have shown themselves for many a
meaner object. What can be so ungrateful a toil as
that of endeavouring to reduce the weather to a law,
and the variations of atmospheric temperature to a
certainty? Yet this has been attempted, and is
attempted still. And how? by the accumulation of
observations, which, since art is long and life is
short, must rather serve as a valuable bequest to
posterity than as a profit to ourselves. All that we
have yet ascertained on this point, and that is much,
is, that the sum of each year's weather and temperature is very nearly alike; that, as Whewell
expresses it, " there is an invariable result of the
most variable quantities." A similar discovery respecting the apparently lawless agency of mesmerism
will, I am certain, reward the perseverance of those
who continue to observe its phenomena with honesty
and patience.

Having shown that the mesmeric agency may truly
be ranked amongst physical influences, and that it
is as real an action of matter as any other which is
made known to us by visible effects, I proceed to
state such additional particulars respecting the agency

as I have ascertained by careful and cautious experiment.

That which I have now unequivocally to affirm of the mesmeric medium is, that it is primarily set in motion by the human mind.

I am aware that, in asserting this, I am committing *lèse majesté* against the French goddess of reason, who has banished the mind altogether out of her dominions; but for justification I must appeal to a higher sovereign, Truth, namely, under whose ægis I would, if possible, take refuge from the storms which are probably brewing against me. "What!" it will be said, "do you render your mesmeric agency dependent on the shifting human mind, on the variable human will, and yet claim for it the character and constancy of a physical influence? What power over physics have the mind and will? When was thought ever known to set matter in motion?" These questions I dare not answer, fallacious though they be, on my own authority. I am a mesmeriser, and my reasoning will be suspected. Let me, then, appeal to the great intellect, which long ago and unanswerably solved the inquiry, "Whence we derive our ideas of power?" From Locke's splendid chapter on this subject I select as much as is necessary for my purpose.

"Bodies, by our senses, do not afford us so clear and distinct an idea of active power as we have from reflection of the operation of our minds; for all power relating to action, and there being but two sorts of action, whereof we have any idea, viz. thinking and

U 6

motion, let us consider whence we have the clearest
ideas of the powers which produce these actions : —
1. Of thinking. Body affords us no idea at at all;
it is only from reflection that we have that. 2. Neither
have we from body any idea of the beginning of motion.
A body at rest affords us no idea of any active power
to move, and when it is set in motion itself, that motion
is rather a passion than an action in it ; for when the
ball obeys the stroke of a billiard stick, it is not any
action of the ball, but bare passion ; also when,
impulse, it sets another ball in motion that lay in its
way, it only communicates the motion it had received
from another, and loses in itself so much as the
other received, which gives us but a very obscure
idea of an active power of moving in body, whilst we
observe it only to *transfer*, but not *produce* any
motion ; for it is but a very obscure idea of power
which reaches not the production of the action, but
the continuation of the passion.

" The idea of the beginning of motion we have *only
from reflection on what what passes in ourselves*, where
we find, by experience, that barely by willing it,
barely by a thought of the mind, we can move the
parts of our bodies, which were before at rest. *  *

" This, at least, I think evident, that we find in ourselves a power to begin or forbear, continue or end
several actions of our minds, and motions of our body,
barely by a *thought*, or preference of the mind, order-
ing, or, as it were, commanding the doing, or not
doing such or such a particular action : — this power,
which the mind has thus to order the consideration

of any idea, or the forbearing to consider it, or to prefer the motion of any part of the body to its rest, and *vice versâ* in any particular instance, is that which we call the will."

If we allow (and how can we deny?) the truth of the above reflections, we must perceive that not merely our very idea of motion is derived from the mind's operations, but that reasoning, as alone we can reason, from what we know, we must conclude that all motion whatever originates with mind and with mind alone. And why start at the notion that the mind, in mesmerism, is the acting force? Do we know any other? Mind and motion, as cause and consequence, are indissolubly connected, the only real antecedent and consequent that can be shown in perfect juxtaposition, and demonstrated to us truly by our self-consciousness. Besides, shall we, who may be said to create to ourselves all we see by thought, doubt of the power of the human mind? One should think that, above all persons, the man of science, who bases all mechanics on the principle of the inertia of matter, and yet cannot but admit " that there is probably no portion of inorganic matter that is not in a state of relative motion," must be aware of the necessity of quite another than a material force to produce all this action in naturally quiescent bodies. Even Descartes (as Pascal says) was forced to bring in God to set the world in motion, though, had it been feasible, he would have shoved him altogether out of his philosophy. The principle once admitted that motion originates with mind, it is plain that such motions as occur in the mesmeric medium, and are thence transferred to a human body, are primarily produced by mind ;—in other words, by will.

Why should we be startled at the expression? Is the will a nullity? Has it no force, no prerogatives? Let us consider what every day and every hour our will effects, merely in controlling and wielding matter. How accurately we move the complicated mechanism of our limbs, through the impulsions of our volition! The least disobedience or want of pliability on their part would mar the slightest of our bodily actions. Let us, with Dr. Maculloch, take the cases of archery and slinging. " It requires little reflection to see that, under the complicated action of so many muscles, the problem to be solved is so intricate that no mathematician could ever hope to assign what was necessary to produce a result, which it as certainly as it is rapidly affected, without any calculation. *    *

" But the precision of muscular action becomes much more wonderful when we refer to the cause, and ask in what manner a definite quantity of cohesive attraction and also various definite and unequal quantities in succession can be transmitted through a nerve, while this too is done by an act of the will, yet of will merely knowing what it desires, not what it performs." *

Our knowledge that the will acts, yet our ignorance how it acts, on matter, should prepare us to receive, without a dogmatic denial at least, proofs of an exten-

* Dr. Maculloch, on the Attributes of God.

sion in its capacities and its sphere. If, in some mysterious way, I actuate another, it is scarcely more wonderful than that I actuate myself. It is true that the latter comes within the range of my every day experience; but can I any more comprehend it?

How great a force the will either has, or wields, may be almost measured off to our senses in a very simple but striking manner.

Let two covered vessels precisely alike, the one empty, the other full of some heavy substance, such as leaden bullets, be placed before a person. Let him first lift the full vessel, and let him then be told to raise the empty one, with an understanding that it is of equal weight with the first. The person doing this will put into the action so much unnecessary force, from the expectation of being about to lift a great weight, that his baffled vigour will, in its re-action, cause quite a painful concussion in the muscles of the arm. Now, could that force, which, as it were, returns upon himself, be directed outwards, it is plain that it might produce a very powerful and peculiar action in the media wherewith we are surrounded. And wherefore not directed outwards? This at least we know, that the will is really the primary agent which enables us to move all bodies foreign to and apart from ourselves. In these cases, indeed, it acts through intermediate agencies which are visible to us. But there is nothing whatever to render it impossible that the mind should act sometimes by unseen, yet even more potent intermediaries than the accustomed. After all, what astonishes us in mesmerism is not that the mind is shown capable of producing motion, but that it is exhibited producing motion in a different way, as we conceive, from that with which our experience is familiar. That the mind should originate a series of motions of which we cannot behold certain of the intervening links (I speak of cases in which the mesmeriser influences his patient from a distance), — this is the true ground of our wonder and incredulity. Yet, in fact, the very same thing occurs in our commonest experience. When I move one of my fingers, I am only acquainted with the first fact and the last in a sequence of events, the intermediate circumstances of which are hidden from my knowledge, and which are, probably, very numerous. Some of the hidden links in the series I know from reason. From the anatomist I learn, that my mind, in the first instance, moves a portion of my brain (for certain injuries to the brain render voluntary motion impossible) ; that this again communicates an impulse to a nerve — (for cut the nerve and the impulse reaches not the muscle) that then again a muscle is moved, and finally the finger. The simplest voluntarily motion, then, is but an impulse, originating with mind in the first place, and thence transferred through a series of atoms. There is a sequence of changes, nothing more ; some of which are known to us — others not. There is (to use the language of Locke) but one real action, yet many passions, or communicated motions. And what is mesmerism but this ? The same definition suits motion whether produced mesmerically, or normally.

The same circumstances attend both. Again, even should we pursue motion beyond the limits of our own bodies, we shall find that there are invisible links in all the impulsions which we communicate to matter. The philosopher knows that we are not in real contact with anything which we appear to touch — that there is always something invisible between us and the object that we handle.

Again: do I not, by every motion of my body, change the relative position of the atoms of the media that surround me? Do I not displace the air, cause various motion in the waves of light, and influence nature to a distance around me, which it were vain to attempt to calculate? The wonder, then, seems to be — not that the mind should produce changes in surrounding objects — but that, being itself " the fountain-head of motion," it should not move matter more forcibly and generally. Doubtless it would be so, had not the all-wise Ruler of creation confined the human mind within necessary limits. Could the will sway the material as it does the immaterial world, what bounds would there be to the tyrant's caprice, to the conqueror's ambition? Mountains would crumble as a dream, and oceans be dried up at our bidding. A Napoleon would " make a sop of all the solid globe." These ideas are not so extravagant as his who doubts of the power of the human will. By that alone we do great things; by that alone we conquer kingdoms, or ourselves; by that alone we achieve the hourly miracle of moving matter, united with, or extraneous to ourselves. Is the will, then, a nullity,

whose influence is to be excluded from our consideration when we treat on any subject which nearly
concerns man ?

Let us now proceed, perhaps with fewer prejudices,
to consider the proofs before us, that in mesmerism it
is the mind which originates the impulse.   It is true
that, in general cases, the mesmeriser employs visible
and physical means to produce his effects.   He is
either placed in contact with the person he desires to
influence, or he makes use of certain looks and gestures, which, to those who regard only the superficial
and the visible, may appear to be the moving causes
of the impression produced on the other person.   But
let us remember that even  the most careless of our
looks and gestures do really spring from mind and
will, acting, indeed, with various degrees of force,
yet still always with a certain force.   Now even as
respects these outward manifestations of an acting
mind, I have invariably found that mesmeric effects
were always in exact proportion to the degree of
voluntary effort I put into the performance.   This
was more especially the case as regarded the action
of the eye.   When my patient's eyes were closed, and
he therefore had no means of knowing what I was
about, I have still found that, the more of *active gazing*
I employed, the greater was the effect produced.   The
necessity of giving an undivided attention to my work
was, of course, most palpable in the first few times of
mesmerising a person ; for it is evident that the more
the patient's sensibility was developed under successive mesmerisation, the less force he would require

to be used in influencing him; consequently the less mental effort would be required on my part. This is a remark to which it is of importance to attend, for some persons in very advanced stages of mesmeric sensibility will be affected by gestures in which the mind of the performer would seem to have little share; that it can have *no* share therein I trust I have demonstrated to be impossible.

Yet even when acting on a patient so accustomed to my influence as was E. A., I found that I by no means mesmerised him so well as usual on days when my mind was preoccupied with other matters. He was himself peculiarly sensitive to any remission of my attention towards him, and would frequently say to me (and always with justice), "You are not thinking of me just now." The poor blind German boy used similarly, and with equal correctness, to complain of my ceasing to influence him properly; and this, from the absence of one of his senses, and that the most observant, is testimony the more remarkable and uncorrupted.

A very certain proof that, when attention remits, there is a remission also of mesmeric power, was afforded me also in the course of some experiments, which Professor Agassiz, of Neufchâtel, permitted me to try upon himself. In these I was of course peculiarly desirous of concentrating all my attention upon the effects to be produced on a man of learning and of science. The very circumstances of the case compelled me to exert my mind in an undivided manner. But one evening the non-arrival of expected letters

from home forced me into another train of feeling,
and, during the mesmeric processes, I could not prevent my thoughts from occasionally straying from the
scene before me into anxious surmises as to the cause
of the silence of my friends.  My patient, although
he had his eyes closed, and his limbs paralysed in the
torpor of the mesmeric slumber, was not slow to perceive the wanderings of my attention, and, although
I was at the time engaged in the mesmeric processes,
to all outward appearance as actively as usual, called
out to me constantly and coincidently with the remission of my thoughts — " You influence me no longer.
You are not exerting yourself."

On another occasion I was convinced in a manner
the most odd, and even comic, how great is the influence of the will in directing the impulses of the
mesmeric medium.  I should hesitate about relating
the story were it not that, in a new science, every
thing has its importance, and the least studied effects
are often the most valuable and sure.  Travelling in
a stage coach in England, with three other inside
passengers (one of them a Cambridge friend, to
whom I can appeal for the truth of the relation), I
fell into the following train of thought: " If mesmeric effects occur only through certain contact and
gestures, why is it that men never mesmerise each
other unawares ?  At a public meeting, in a church,
in a theatre, in all places where human beings con-
gregate, there is plenty of the mesmeric influence
going about unappropriated.  Why is it that this
does not take effect, and that no one should sleep in

such circumstances, unless the orators are prosy, and the play stupid enough to act as a soporific? Surely this is because the influence requires a will to concentrate it and to give it a particular direction, before it can individually operate. Now, here am I, in a most favourable position for mesmerising the person opposite to me. His knees and feet are in contact with mine — in the true mesmeric position. At present he does not look sleepy in the least; and up to the present moment the idea of mesmerising him has not entered my head. Let me see if this want of the idea is the true cause why I have not hitherto affected him. If, by a silent exertion of my will, I can now, from wide awake, bring him fast asleep in a very short period, the relation between cause and effect will be, I think, pretty well established." " Sur ces entrefaites," as the French say, I began mentally to exert my volition, and to fix my attention on my unconscious patient. From time to time I looked at him mesmerically, but watched my opportunity, when his own attention, being turned elsewhere, permitted me to stare at him, without the fear of being challenged. In about ten minutes the charm began to operate. My victim's eyes kept reverting unconsciously, as it were, to my face, which, however, I turned away whenever I saw him looking at me, and at length he began to shut them with that slow and peculiar motion which is indicative of mesmerisation. They did not, however, close, but remained more than half open, becoming perfectly fixed, and, as it were, dead, displaying the whites in a manner which by no means contributed to embellish a physiognomy naturally none of the handsomest.

Behold, then, my man asleep; so soundly, indeed, that the coach stopped, and horses were changed without his being in any way disturbed from his mesmeric nap. When we were again in motion I began to be agitated by some strange doubts, whether my will would be found as effectual to end, as it had been to create, the spell which held the slumberer, who always remained in the same position as at first, with his eyes more dead than ever. As I had soon to quit the coach, I really feared that my sleeper, unawakened, might follow me, attached by mesmeric attraction, to my side. The question of the will's supremacy was now to be tried to the utmost. First, without any effort of volition directed to the end of awakening may patient, I stirred about, spoke loud, and let down the window next me with a rattle; my fellow-passenger did not awake. I then, concentrating my will on the one idea of dissipating the mesmeric influence, leant forward with something of that motion of my hands which, in usual circumstances, accompanied the idea. Immediately my patient began to stir, move his eyes, and rub them, staring still rather wildly, and in a confused manner muttering that he really thought he must have been asleep.

I now proceed to relate some more serious experiments, tried at Antwerp, on Anna M——, relating to the influence of thought and will on the mesmeric medium, and its power of communicating impulses to a distance.

The patient just mentioned was, it will be remembered, in a state of mesmeric sensibility and relation with myself, which successive mesmerisings had carried to a very high pitch. Yet by my mere bodily presence Anna M—— was in no way affected. Unless I mesmerised her with intention, no effect was produced. Never had I led her to conceive it possible that I could mesmerise her in any than the usual way, that is, by contact and gestures. She had never read any books on mesmerism. She had no preconceived notions on the subject. All being thus propitious to the essay of a pure experiment, I concerted with my family what was to be done on a certain evening, when Anna M—— was coming to drink tea with us. According to our arrangements, I pretended to leave the room, while Anna M—— was engaged in conversation with my wife on a sofa near the fire; but taking advantage of a large screen that hid the door entirely from view, and which had always stood in the same place — that is at the very farther end of an apartment twenty-six feet in length — I shut myself into the room instead of out of it. I then tried to concentrate my thoughts on my sleepwaker, and to mesmerise mentally. At first I found this difficult, but at length I succeeded in bringing myself into the same frame and action of mind which usually accompanied my employment of outward means, when mesmerising. Then, and not till then, I suddenly heard Anna M—— (breaking off in the

very midst of an animated conversation) exclaim —
" Oh, where is he ?    What is he doing ? "   Directly
after, the concerted signal, which was to inform me
that my patient slept, was given; and, advancing
into the room, I saw Anna M—— as completely
mesmerised as ever she had been by contact and
gestures.   In the notes taken at the time, by one of
our party, of this transaction, I find, that, after this,
the patient's first words were, —

P. — " Why did you go ?   You only wanted to deceive me."

M. — " Only to magnetise you from a distance."

P. — " Yes.   You can magnetise me now from any
distance."

Anxious to try a second experiment of the same
nature with the above, yet aware that, if I took the
same means, the experiment could not be pure (since
the patient forewarned might, on my apparently
leaving the room, imagine me still in the apartment),
I agreed again on certain measures with my family,
which took effect as follows : —

After the interval of a few days, during which I
mesmerised Anna occasionally in the usual manner,
I, on her coming one evening to see us, was found
deeply absorbed over a volume of Wordsworth's
poems; which, after the first greetings were over, I
continued to read, while Anna was busily imparting
to her kind friend, my wife, all her little plans for
bettering the condition of her family — her hopes
and fears for the coming winter.   While this went
on I was trying to withdraw my mind from the book,

on which my eyes, shaded by my hand, were fixed, and to concentrate my thoughts on my patient. This, however, was no easy task; Wordsworth is not an author to be quitted with indifference. At length, by dint of repeated efforts, I acquired the proper frame of mind for mesmerising; and, coincidently with my success, Anna, as before, suddenly broke off in the midst of a sentence, and exclaimed " Oh, he's magnetising me!" (we called the thing then by its old name) — falling back at the same time on the sofa in a profound mesmeric slumber.

The third trial that I made to mesmerise this patient from a distance was still more remarkable and decisive.

One evening, when sitting with my family, the idea occurred to me, — " Could I mesmerise Anna M—— there, as I then was, while she was in her own house?" to which I knew she was just then confined by slight indisposition. Acting on this thought, I begged all the party present to note the hour (it was exactly nine o'clock), and to bear me witness that then and there I attempted a mesmeric experiment.

This time I endeavoured to bring before my imagination very vividly the person of my sleepwaker, and even aided the concentration of my thoughts by the usual mesmeric gestures; I also, at the end of an hour, said, " I will now awake Anna," and used appropriate gestures. We now awaited with more curiosity than confidence the result of this process.

The following morning Anna made her appearance, just as we were at breakfast, exclaiming " Oh,

x

sir! did you magnetise me last night? About nine
o'clock I fell asleep, and mother and sisters say they
could not wake me with all their shaking of me, and
they were quite frightened; but after an hour I woke
of myself; and I think from all this that my sleep
must have been magnetic. It also did me a great
deal of good, for I felt quite recovered from my cold
after it. After a natural sleep I never feel so much
refreshed. When I sleep for an hour in magnetism,
it is as if I had rested a whole night." These were
the words of Anna M——, noted down at the time
as accurately as possible.

Unless the reader will do me the honour to believe
that I tricked my family, and was in concert with a
poor honest-hearted girl to deceive every one, I know
not what he can say to statements such as the above.
They are facts, to the accuracy of which more than
one person will pledge all the credit of their integrity.

To me, who *think* at least that I behold clearly
the principle of mesmerism, they appear not so wondrous, as doubtless to those who have not yet thought
out the subject they will and must appear. Impulses
conveyed to a distance cease to be miracles the moment there is a communicating medium made visible.
That news should be conveyed from Dover to York in a
few seconds may seem a mighty marvel to the ignorant;
but to those who are aware of the nature of the tele-
graph, a mere common occurrence. Let us not be
so near the vulgar as to be astonished at effects from
causes to us invisible; let us not gape at the ascend-

ing or descending deities, as if they really floated in the air, only because our vision cannot detect the strings and wires by which they are guided and supported. The mechanism once revealed to reason's eye, everything appears easy and natural. But, if we only rest in what we *see*, we must live in wonder and perish in our ignorance.

The following considerations may tend to reconcile us to the phenomenon of thought giving tendency to matter, and propagating its impulses to a distance.

Mental action produces motion in the brain, for the sensible fatigue in that organ, consequent upon intense thinking, proves that during thought it has been exercised in a real and physical manner. The materialist even will confess this, and more energetically, for he will affirm that motions of the brain *produce* thought. Let it be so, if he will; that makes no difference in our argument. Motion of the brain is still, and under all circumstances, allowed to be the accompaniment of thought, either as its antecedent or consequent. Can that motion be possibly insulated amidst the connected mechanism of the universe, where one thing hinges on another, and where the touching of a single spring is but the commencement of a series of actions ("wheel within wheel involved"), which may reach to the throne of Omnipotence itself? Can any one motion live to itself alone, die where it was born, and be bounded by the substance in which it has originated?

Granting that, which few will, in the present day be disposed to deny, that there is one pervading me-

x 2

dium throughout nature, infinitely elastic and impressionable, it follows, as a consequence, that our thoughts must necessarily, in agitating the brain, agitate also the great ocean in which animate and inanimate matter has its being, with which we are in partial relation, with which God is in infinite relation.

Let philosophers pursue their own views to their termination, and subscribe to the consequences of their own theories, and I shall be content. He who adopts the undulatory hypothesis, and contends for the existence of a pervading ether, is already more than half way on his road to mesmerism.

These reflections may serve to introduce another class of mesmeric phenomena, which appear miraculous only when insulated and taken apart from the proposition that Mind must in its every action impress a certain motion on surrounding matter.

It has been said that persons in certain states, either mesmeric or akin to the mesmeric, can become aware of the thoughts of others without the usual communication of speech. Bertrand, who (be it remembered) wrote against mesmerism in the later years of his life, relates an amusing story to this effect, a story in which he professes his full belief, for he knew the parties concerned in it : — A little girl of about ten years of age fell into a singular state of abnormal sensibility. In her fits of auto-mesmerism, she alarmed her family by proclaiming aloud to them all the subjects of their thoughts. She would say to her sister, " You are now meditating whether you should or should not go to such a place, to meet such a person; I advise you to stay quietly at home;" or, to her mother, " Do not ponder why papa stays out so late; it will do you no good." These revelations were at times not a little awkward and mal-apropos, and so the poor little girl was not thanked for her discernment, but voted to be under the influence of a deluding and wicked spirit. For the purpose of exorcising this familiar, so much more malevolent than that of Socrates, the young patient was committed to the care of a pious community of nuns, with directions that much prayer and holy water should be spent upon her; but, in the convent, matters went on much worse. The holy water threw the patient into convulsions, and (still more horrible) whenever a metal cross was laid on her breast, she threw the precious ensign of redemption from her with symptoms of the greatest aversion. The pious sisterhood, though not aware of the fact that the touch of metal powerfully influences persons in certain states of sensibility, happened, however, to exchange the metal cross for one of wood, which, having been blessed by the pope, was supposed to be of peculiar efficacy; and, lo! in proof of that efficacy, the little girl allowed the relic to remain quietly on her breast. This was a most favourable omen; but, alas! the evil spirit was not to be thus tamed! — the great, the terrible symptom of diabolic agency broke out in fresh vigour; for the patient began to proclaim the thoughts of those around her. When irritated by the kind but mistaken officiousness of the nuns, she was especially formidable in that way, so formidable, indeed, that at length she

x 3

completely controlled and governed the saintly community. " Sister Agatha," she would exclaim, " you
had better not bring that cross here, or I 'll tell why
it was you nailed your ear so close to the key-hole of
the abbess's parlour." " Sister Ursula, do not force
me to say any more paternosters, or all the world
shall know what you were thinking of in your cell
last Tuesday."

Now this phenomenon of thought-reading has been
observed by me, to a certain extent, in some of my
mesmeric patients.

The first occasion of my remarking it was in consequence of an accidental occurrence — for, in truth,
I have never *sought* for marvels in mesmerism; if
marvels I have to relate, they have *presented* themselves naturally to my notice. E. A., when in his
normal state, used sometimes to exhibit a trick on the
cards, which consists in a long and rather tedious
process of arranging a certain number of cards according to the letters of certain words, and then, by
the correspondence of the letters, discovering any two
cards which may have been chosen by another person. I was curious to observe whether my sleepwaker retained enough of his calculating and com-
bining powers in the mesmeric state to go through
this trick as usual, and I therefore asked a lady, who
happened to be present one day when E. A. was
mesmerised, to choose two cards from amongst the
little packets of pairs, which the sleepwaker had
already placed in proper order on the table. The
lady chose the cards by her eye only, in perfect

silence and standing behind the sleepwaker, so that there was absolutely nothing of word or gesture to guide him in his subsequent discovery.  He began to gather up the cards as usual preparatory to completing the trick, when suddenly he threw them down with an air of disgust, saying words to this effect :— " Why should I go through this farce? I know already the two cards which the lady thought of. They were so and so." He was perfectly right. Another time, a little basket, brought in by my wife during his sleepwaking, was standing on a table near him.   He took it up and considered it (always with his eyes shut), and said to my wife, " Ah, you are thinking now of making this a present to Mr. V. O.'s little girl." This was quite true.  Again: I asked the same sleepwaker in his state of mesmerism, if he knew of any application which would strengthen his eyes — then weak.  " Yes," he said, " something of which you have been thinking — a wash, for the prescription of which you wrote to a physician in Paris not long ago for the use of a friend of yours." He spoke correctly; and I declare most positively that the sleepwaker had no means whatever of knowing the facts he stated, except through the medium of my own thoughts.

Anna M—— occasionally manifested phenomena similar to the above, and (woman-like, perhaps) accompanied them by a far more delicate and accurate perception of the state of my feelings, at such times as she was placed in mesmeric relation to myself. Frequently she has surprised me by the manner in

x 4

which she anticipated, on these occasions, my wishes with regard to any little experiment I desired to make. "Come," she would say, "you are now wishing that I should do a piece of needle-work. I own just now it is rather a trouble to me, but I will do it notwithstanding."

In my notes relative to the same period I find the following fact mentioned. A musician, with whose name or person Anna M—— was not in the least acquainted, was in the room during her sleepwaking — February 16th, 1838. Not a word had been spoken about music; all present were engaged in observing the sleepwaker, when suddenly she said to me, "You know there is a gentleman here who sings and plays very well. You were wishing a while ago to ask him to sit down to the piano. You must ask him now, and, when he is playing, lay one of your hands on his shoulder, and the other on mine, and I shall hear the music too."

These phenomena, as I have said, were *manifested*, not *forced*, and I had occasion to observe that, like all other phenomena pertaining to mind, they were best exhibited when most spontaneously exhibited. Attempts to elicit them in any other manner generally failed. Once, standing near Anna M—— I addressed to her a sentence mentally, but she did not comprehend it, though, that I wished to say something to her and that there was an action of my mind, she manifested knowledge. Her words on this occasion were remarkable. "Why do you *speak* so low, sir? *Speak* louder, that I may hear you!"

Now, be it remembered, I had not spoken at all, nor given her to understand, in any way, that I was about to address any thing to her.

One experiment, however, of this nature, was almost invariably successful. If I mentally ran over a tune in my head, Anna would immediately begin to beat time, and sway her head about in the mea- sure of the air.  Anxious to have a correct witness of the experiment, I agreed with a musician that at a certain silent signal I should begin mentally to re- peat an air, and, at another signal, change the air and measure from slow to fast.  I made the musician acquainted with both the airs before-hand, in order that he might accurately judge whether the sleep- waker kept to the time.  The experiment answered perfectly, both as to beating time in the first place with accuracy, and then as accurately changing the measure.

I could relate a crowd of other circumstances of a similar nature, and some perhaps even more remark- able than the above; but I forbear: the above are sufficient to illustrate the principle which I am en deavouring to educe from facts.  Should the prin- ciple itself remain unacknowledged, I shall have already drawn too largely on the faith of my reader. Admit that thought communicates action to sur- rounding media; admit that the mesmerised are sen- sible to that action; and all that may seem wondrous in my statements vanishes.  It may be doubted, indeed, whether the prejudice that exists with respect to such phenomena may not have been wholly

x 5

caused by their having too often been brought forward, not only unaccompanied by a proper explanation and in disconnection with a sufficient cause, but
actually expressed in such a way as to deepen mystery and to shock the scientific. The power which
sleepwakers, in certain states of sensibility, undoubtedly possess, of perceiving the thoughts of
others, has been called divination; a term which
approximates the faculty in question to witchcraft
and all its vulgar associations. The sleepwaker does
not *divine* what passes in the minds of others; he
*reads* it there by means which are in perfect accordance with the economy of our being. Let me endeavour to set this in the strongest, clearest light. Should
I even repeat myself, it will scarcely be a fault on a
subject like this, where indeed a certain iteration
seems necessary to familiarise our minds with themes
so new and of such grave importance.

As we are more acquainted with that corporeal
change in ourselves which accompanies sensation,
than with that other more hidden which accompanies
reflection, let us reason by analogy from the one we
best know to the one we know the least. Both are
mental operations, but the one, as we are sure from
facts, is connected with a series of material movements which stand as actual signs and representatives of itself. There is sensation, and there is also
the language of sensation; in other words, nervous
motions, which are equivalent to a language. Now,
were we enabled with equal certainty to say, when
speaking of our intellectual operations, — " There is

thought, and there is the material sign of thought : "
could we affirm positively that, as every sensation has
its peculiar nervous motion, so also every reflection
has its peculiar action of the brain, we should be
greatly helped to a comprehension of the point in
question.   For it is plain that, could another person
be made aware, in any way, of the motions of my
brain during thought, and were he also properly
instructed in the significancy of those motions, he
might read from " the book and volume of my
brain" as readily as from any other collection of
symbols.   For it is not the nature of the sign, when
once its meaning is fixed, that can make any difference or create any difficulty.   The deaf and dumb
converse as well by motions of their fingers as persons in general by giving impulse to the air.   Now
the presumption certainly is that every thought really
has its particular configuration in that texture of our
brain; for, 1st, we know that thought does indeed
move the brain generally, whence we might infer
particular motions for particular thoughts.   2dly, We
are sure that, in some instances, individual thoughts
do move the brain and thence certain nerves in an
individual manner ; as, for instance, in cases where
a thought reproduces a corporeal phenomenon, generally dependent on sensation, such as a distinct
visual image, or a certain feeling in the nerves of the
teeth.

From these phenomena it seems that we only
draw a legitimate conclusion, when we say that every
thought moves the brain in its own appropriate
x 6

manner. Our personal ignorance of these specific changes, though they occur in our own persons, is no proof whatever that they do not take place. We are perfectly unaware that sensation depends on a corporeal sign; yet a corporeal sign there is.

If it should be asked in what manner the sleep-waker has come to be so well acquainted with the signs and characters of thought, I might think it perhaps a sufficient answer to reply, " The soul is wise — yes, wiser than we know." There are more intuitions, or (as Kant calls them) " cognitions *à priori* " than we suppose. What if motions of the ethereal medium were the native and universal language of the mind? Or, to answer one question by another, how, in sensation, have certain motions come to be representative of certain external objects? I may be told, by experience; but again, experience itself must have a basis. What *first* taught the mind the connection between the sign and the object? Plainly an intuition; for it is self-evident that we cannot advance to new knowledge but by the aid of previous knowledge. But where is this previous knowledge to begin? It must begin somewhere, or we shall never know anything. When we grant that we do know something, we also imply that this our knowledge had a beginning, and this beginning could only be a truth so clear as to be apprehended at once without the aid of anterior knowledge; for, to suppose a knowledge anterior to the beginning of knowledge is manifestly absurd. Thus it appears that we cannot but concede the existence of a first knowledge, which came to us in a totally different manner from any other, about which we could neither have pondered nor reasoned; a knowledge which, luminous itself, required no light for its illustration; a knowledge which is the guiding torch to all other knowledge whatever; in fine, an intuition or cognition *à priori*.

This reasoning, applicable to all knowledge whatever, is more especially so to that peculiar knowledge which we call sensation, and which Locke has satisfactorily proved to be the avenue at least to the greater part of all we know, of all we are; the occasion of development to the mind, the awakening touch that sets in motion the germ of all our faculties. Locke, indeed, by his unfortunate and misconceived comparison of the mind to a sheet of white paper, has obscured his own beautiful system; but to him who studies the work on the " Human Understanding" as a *whole*, it will be seen that the reflection of Locke is nearly identical with the pure reason of Kant, Locke allowing clearly that sensation is only (to use his own words) " *one* of those ways whereby the mind comes by its ideas."

Now, let it be remembered that the mind in the complete mesmeric state has been proved to be more rational, more reflective, than in its normal condition; consequently nearer to that which Wordsworth so finely calls —

——— " The fountain-light of all our day,
The master-light of all our seeing."

Is it then wonderful that its first-developed intuitions should be recalled with superior ease and clearness to its perceptions ?

This is the view of the subject which searches it most deeply and entirely; but, while I maintain that the mind's knowledge of the mesmeric signs must in the first instance be intuitive, I am also enabled to modify the statement so as to render it more analogous to our ordinary views and common experience. Like our other intuitions, this particular one is developed manifestly through time and circumstance. The phenomena of silent perception are, let it be remembered, not only rare in occurrence, but slow to occur. As far as I have observed, they only take place when a person has been brought by degrees into a very exalted condition of sensibility, only when he has learnt, and, as it were, perfected himself in his new mode of being, and in the language appertaining to that mode of being.

In the chain of our argument, then, there appears to be no link wanting. It is thus connected. Every thought moves the brain in its own appropriate manner.

A pervading medium being allowed to exist throughout nature (such as the electric), it follows as a consequence that every thought which moves the brain imparts motion also to the ethereal medium.

Mesmerised persons, being in a state of extreme sensibility, are cognisant of the motions of finer media than common.

The motions, created by the thoughts of other persons, being transferred through the brain, and through a certain medium, to the sensorium of a mesmerised person, are to him intelligible signs of thought — a language which, though new to him at first, he, by a gradual process of association, gives meaning to and learns to comprehend.

We may further remark (in order to elucidate certain other phenomena of mesmerism) that a mesmeric patient is in such relation with his mesmeriser as to be peculiarly attentive to the workings of his mind; with these he is evidently far more connected than with the intellectual operations of any indifferent individual. Hence the thoughts and feelings of the mesmeriser are transferred to the patient more vividly, and with a superior degree of accuracy, so as (if I may so express myself) to bring into sympathetic vibration the very brain and nervous fibres of mesmeriser and patient. Facts coincide with this supposition. In various parts of this work it will be seen that the mesmeric patient takes, *pro tempore*, the tone of his mind from that of the mesmeriser's mind. It may farther be asserted that he shares the knowledge of the latter — a circumstance which accounts for many apparent miracles regarding the information of sleepwakers on subjects with which they were not previously acquainted. When I first began to mesmerise, I used to consult my sleepwakers on dark and dubious points, with something of the blind faith of a novice in a new and wondrous science. Their answers to such inquiries were calculated to bewilder me by the pure influence of astonishment; for the simple had become theorists — the uneducated were turned into philosophers. At length I was awakened from my dream of somnambulic knowledge by finding that my patient's idea shifted so visibly with my own, and were so plainly the echo of my own thoughts, that not to have perceived the source whence they originated would have been pertinacious blindness indeed. I was but taking back my own, and receiving coin issued from my own treasury. I particularly observed that what I had last read, or most recently reflected on, was most vividly returned to me by my sleepwaker. It was a vexatious discovery. Would that I could present to my reader, as oracles, the speculations of my patients on the ethereal medium! Would that I could enlighten the world by displaying, as authenticated by a preternatural illumination, the harmony and mutual relationship of all things! But, alas! I should only be repeating my own feeble explorations of the regions of truth; I should but be retailing my own mortal and uninspired surmises!

A few particular instances of the coincidence between my own thoughts and those of my sleepwakers may interest the reader, and prove that which, as yet, I have only generally advanced.

E. A. was perfectly ignorant of the mechanism o. the eye, or of the fact that images on the retina are reversed representations of external objects. I had been reading some works on this knotty subject, in which needless pains had been taken to turn the image the right way upward, when I questioned my patient as to our mode of seeing. To my surprise, he entered immediately upon the topic of the reversed image on the retina, and—though confusedly, indeed — repeated partly some mystical stuff which I had been reading in the " Seherin von Prevorst," (he did not understand German,) upon that head — and partly a scientific statement of the mere fact, which latter he illustrated by drawing on paper the very same mathematical figure which I had been contemplating that morning, and which represented the intersection of the rays at the focal point, which causes the phenomenon in question.

The following coincidence was still more remarkable and perfect than the preceding.

In a work by M. Despines of Aix, in Provence, I had been reading some speculations on the (supposed) vital fluid. It was there conjectured that some of our maladies take their rise from the unequal distribution of this fluid along the channels of the nerves. I declare most positively that the sleepwaker had never seen the work in question, which had just been forwarded to me by Lord N——, from Geneva. Nevertheless, when in the mesmeric state, he àpropos of something said respecting a person's illness, thus expressed himself : —

" Il y a des maladies qu'on ne peut pas voir. Ce sont les dérangemens de la fluide nerveuse, qui est accumulée ou par ci, ou par la ; et c'est cela qui fait les maladies."

I copy the words exactly from notes taken at the time they were uttered.

It must not, however, be supposed that this reflection of thought from the mesmeriser to the patient destroys the individual character of the latter. The knowledge returns to its originator, tinged always by the medium from which it has reverted. Thus Anna M—— did not philosophise like E. A. Her speculations were influenced by the synthetical habits of woman's mind. In her there was more of the heart, in the other more of the head. In every case I found that sleepwakers do not retail things like mere parrots, but as thinking beings, themselves elevated to a higher intellectual region than the ordinary. Besides, there is always in the mesmeric state, a certain independent power of thought, proper to the patient, the degree of which varies with the natural independence of the character. Those ideas I was never conscious of in my own mind might be fairly attributed to my sleepwakers as their own property. From some of these I have taken hints, at least; especially in all that regards the mesmeric state, of which mesmeric individuals may be supposed to have the same personal knowledge that we have of our normal existence. Let us now pause to generalise the more important of our facts in the spirit of the following remark.

" The great object of philosophy is to ascertain the simple or ultimate principles, into which all the phenomena of nature may, by analysis, be resolved."

Striving after the philosophic unity recommended by Dr. Chalmers, yet, keeping within the verge of safe analogy, we may affirm that we have been considering — relative to mesmeric action — a medium

which, by its character and offices, is capable of being
identified with that other medium treated of under
the head of " mesmeric sensation."

For that which is predicated of the one may be
said equally of the other : —

Both have been shown capable of communicating
impulsions to the human system from a distance.

Both act through obstacles which are impediments
to grosser media.

Both bear an especial relation to the human mind.

In the one case, indeed, it appears that the medium
acts upon mind, while, in the other, it is manifest
that mind acts upon the medium ; but this distinction
will, I imagine, vanish, when it is remembered that
we have already proved sensations to be in reality
actions, and not mere passive performances of the
sentient principle, and that it is more probable that the
mind always consults actively its book of signs (as the
nervous system may be called), than that it is swayed,
in any case, inertly by the impulses of the external
world.

More briefly, sensation and reflection are, in truth,
both *actions*, though different actions, of one thinking
substance.

This being allowed, a pervading medium being
allowed, and it being manifest, as a consequence, that
this mere ethereal medium must respond to every action of the mind and brain, it will be evident that the
mind impresses motion on the medium by the very
act in which it seizes the meaning of the motions to
which it had been passive. Taking the proposition in a

material point of view, it may be said that, in the undulations of the nervous system, and the elastic media,
of which it is the centre, there must be reaction not
less than action.   In either case, the motions of the
medium have been shown to be symbolical of something beyond themselves, and to compose a language which the mind recognises and interprets.

Other points of resemblance between the sentient
and the mesmeric medium may be found.

It appeared probable that the former, from its
action from a distance, and the rapidity of its com-
munications, was, like the luminous ether, of an
elastic and vibratory nature.   Now considerations of
a similar kind induce us to come to a similar conclusion with respect to the mesmeric medium.   We have
seen that physical effects have been produced by one
person on another from distances and through obstacles which render the supposition of a material substance emanated from the body absurd.   Anna M——
was by me brought into the mesmeric state when I
was divided from her by a screen, and by the whole
length of a large apartment ; and again when I was
in a house a quarter of a mile distant from that in
which she was sensible to my action.   Facts like these
intimate not doubtfully that the mesmeric medium
has not a real transitive motion, but only a motion
of oscillation ; that the mesmeriser does not project a
means of communication to his patient, as one would
toss a string to a person at a distance, but merely
agitates an already existing means of communication,
as one would impart vibration to a string already

stretched from one's own hand to that of another person.

The phenomena of sympathetic sensation between mesmeriser and patient come in aid of the above conclusion. It has been shown that, in certain cases, where a mesmeric relation is perfectly established between two persons, the impressions of tact and savour are transferred accurately from the nervous system of one to that of the other. Thus Anna M——, when any one gave pain to my shoulder, foot, &c. immediately began to rub the corresponding part of her own person ; and she could also, from a real impression on her nerves of taste, correctly state what I was eating or drinking. Miss T—— likewise, through nervous motions in her own hand, became aware of the size and form of objects which I was secretly handling.

These phenomena extremely resemble those of sympathetic vibration produced by the tremors of the air. " It is thus " (says Mrs. Somerville, speaking of the latter) " that sympathetic undulations are excited by a body vibrating near insulated tended strings, capable of following its undulations, either by vibrating entire, or by separating themselves into their harmonic divisions. If two cords equally stretched, of which one is twice or three times longer than the other, be placed side by side, and if the shorter be sounded, its vibrations will be communicated by the air to the other, which will be thrown into such a state of vibration, that it will be spontaneously divided into segments equal in length to the shorter string." Again, when sand is strewed on paper or parchment

stretched over a large bell-shaped tumbler, with pro·
per precautions, " if a circular disc of glass be held
concentrically over this apparatus, with its plane
parallel to the surface of the paper, and set in vibration
by drawing a bow across its edge so as to make sand
on its surface take any of Chladni's figures, the sand
on the paper will assume the very same form, in consequence of the vibrations of the disc being communicated to the paper by the air."

" When a slow air is played on a flute near this
apparatus, each note calls up a particular form in the
sand, which the next note effaces to establish its own."
It is plain that this visible manifestation of an invisible agent might be made a language, and might
suggest to a person versed in its symbols the air
played on the flute so perfectly as to enable him to
write it down, even though he neither saw nor heard
the musical instrument from which it emanated, each
form in the sand indicating to him a particular note.
The analogy between this musical language and mesmeric sensation is manifest. Just as the vibrations of
strings, or of vibrating plates are reproduced in other
strings, or visibly represented in the sand by the
agency of the intervening air, so are the nervous motions of the mesmeriser's system reproduced and faithfully represented in that of his patient by the agency
of the mesmeric medium. It is true that the aërial
vibrations are only manifested in tense substances,
and that the nerves are not, as generally supposed, in
a state of tension, and herein our parallel is not complete; but then, in the latter case, if the undulating

body is not the same, neither is the medium that excites the undulations the same. The true and important point of resemblance is this; — the nerves
have a capacity to vibrate in unison with a certain
medium, just as tended strings have a capacity to
vibrate in correspondence with the air, and sympathetic vibrations may undoubtedly be just as well
propagated from one system of nerves to another by
their own proper medium, as from one string to another, or from the glass disc to paper, by the air.

This being allowed, is it any marvel that the transferred vibrations of the nerves should be an intelligible language? Even as regards the mere tremors
of the air, we find that " by the vibrations of sand
on a drum-head, the besieged have discovered the
direction in which a countermine was working." If
personal interest can thus awaken the intellect to a
comprehension of mere external motions, how much
more keenly must personal feeling rouse the sentient
being, who is himself the subject of sympathetic nervous action, into a perception of every change in the
system with which he is in unison? Besides, motions
of the nerves are more than a language; they are the
language of sensation; each of their changes is not
merely a sign of something occurring externally to a
man's self, but in his very frame and being: thus,
when a mesmerised patient has a nervous vibration
transmitted to him from another, he has also an actual
corporeal feeling, not only coincident with that vibration, but also perfectly corresponding with the
original feeling in the other person. It is, however,

of importance to remark that to all these phenomena the will of the mesmeriser bears more than a relation; — it has an evident share in their very existence. The degree of attention that he bestows on producing the phenomena of sympathetic sensation has a great influence on the time and manner of their occurrence; neither in this instance, nor in any other relating to mesmerism, can we leave the will out of our calculations; it is, indeed, the focus, where all the rays of our information meet, — the centre without which there would be no general relationship — no union between our forces: this premised, we proceed to our conclusion, that the relation between mesmeriser and patient is established by an elastic medium is to be presumed from the nearly simultaneous and correspondent action of the nervous systems of both. Here, as in former cases, admit a vibratory medium, and that the mesmerised are sensible to that medium, and everything is clear and satisfactory; without it all is dark.

They who watch for my halting, and are on the look out for discrepancies, may here remind me that I have in other places advanced facts which are at variance with the supposition of a vibratory medium in mesmerism, and which rather tend to establish the existence of a material emanation, of which the mesmeriser's body is the original source. Contact, the breath, motions of the hand near the patient, have been described as powerfully aiding the transmission of the mesmeric influence. To this I reply, that it is now very generally admitted that heat is but a modification of the same agent which produces the sensation of light, — in other words, no real entity, but a peculiar action of matter. Thus to heat a body is only to bring the atoms which compose that body into a particular state of vibration. Yet in order to effect this, in order to induce that peculiar action of matter which we call heat, we employ certain means, which, being themselves material, tend to invest the agent with a material character, but which are in reality only the occasions of bringing certain atoms into a particular state.

It may well, then, be allowed by those at least who adopt the undulatory theory, that the material means employed in mesmerism, to charge a patient with the mesmeric influence, are no proof whatever of the material character of that influence, but are simply methods whereby a medium may be thrown into a particular state of action. That there is no just reason why that action should not be vibratory is also manifest.

Pursuing the analogy between heat and the mesmeric agency, we know that, with regard to exciting the former, the means are not only various, but that they may be altered according to circumstances. Occasionally they may appear very complex, as where chemistry employs its resources to develope heat by the union of two liquids or gases; and occasionally nothing can be more simple, as where, by mere friction, we produce heat. This should warn us not to be surprised that mesmeric results should accrue from different processes, simple

as well as complex; that at one time the mesmeriser
should employ all the intermediate aids of gesture,
look, and respiration; yet at another, with equal
success, influence his patient by the mere action of
his mind and brain.

I can assure my reader that I am as anxious to
give the subject a fair consideration as he himself
can possibly be; and I can suggest to myself objections to a vibratory cause in mesmerism, which, perhaps, might not have occurred to his own mind.
Thus I found on one occasion, when I mesmerised a
person out of doors, that the wind had a manifest
power to disturb, and, as it were, bear away the
mesmeric influence. This, on a cursory view, goes
very near to investing it with a transitive character;
but, on a more accurate examination, does not in any
way affect our previous conclusions to the contrary.
We have only to remember that even the transmission
of light depends on the condition of the atmosphere,
and that wind disturbs that vibratory property of the
air, which produces sound, in a marked and acknowledged manner. All then that is really proved, when
the wind is found to disturb the mesmeric influence,
is that wind has the power to affect the vibratory
properties of the medium.

Again: substances touched or breathed upon by
the mesmeriser will, when tested by such patients as
are extremely sensitive, be found to retain something
of a mesmeric virtue for a considerable period, as if
a material emanation clung around them, and at
length became dissipated in the surrounding atmo-

sphere. Nay, I must avow a circumstance, that, more than any other, seems to shake my hypothesis of undulations. I have mesmerised Anna M——, by sending to her from a distance a note, which I had previously carried about me for some time. But what do these phenomena in reality prove? Merely that the mesmeric medium, like other media, has the power of imparting a change to the particles of the substance on which it acts.

The following reflections will at once support my hypothesis, and set my meaning, as I trust at least, in a clearer light.

Whatever is corporeal is ponderable. When, therefore, we pronounce that an agent is imponderable, we at once pronounce that it has no existence (that is to say, virtually, and as far as our senses are concerned), but in motion — that it is only an effect or action of matter, and not an individual substance. Such agents, then, as affect bodies without adding to their weight, can only be conceived to act by introducing a change into the relative position of their particles. Thus, when a grain of musk perfumes a sheet of paper, without either losing in its own weight, or imparting additional substance to the paper, what can it prove more than that the action of one thing can be imparted to another? Or, when what we call infection is conveyed from one person to another by wearing apparel, or by a letter; and again, when fumigating the apparel or the letter properly is found to restore those substances to their former innoxious condition, is it not because, in the one case, the substances had acquired the property of imparting a change to the atoms of the body with which they should come in contact; and, in the other, had this new mode of action abstracted from them? This at least we know, that colour, which appears to be a material body, and is transferable from one body to another, is actually nothing more than a chemical change in the position and arrangement of those atoms which by their varied manner of cohesion produce what we call the different textures of substances. Again : science has ascertained that the mere action of the luminous medium can produce an internal change in the texture of crystals, and can develope magnetic properties in iron ; while our own observation may assure us, by daily examples in the fading of vegetable colours, that light alone can so alter the atoms of bodies, as to change their capacities of absorption and reflection. The recent discovery, whereby a landscape is transferred, in all its natural hues, to paper prepared with nitrate of silver, by the simple agency of light, speaks more strongly still to the same effect. Nor is it by any means fatal to our cause, that the change which mesmerism induces in bodies should be but temporary. Electricity induces temporary alteration in bodies ; the magnetic virtue in iron may be diminished or reinforced by light, by heat, and other external agencies; while to or from such metals as only become magnetic by induction, the properties of magnetism can be given or taken away at pleasure. Heat, also, can impart a temporary change to the molecules of matter, giving them a new vibration and impulse, which gradually ceases and leaves the body they compose in its former condition. The analogy in this case to mesmeric action is particularly striking, and would alone be sufficient, one might imagine, to rank this agency amongst the imponderable media. It renders perfectly comprehensible not only the transference of mesmerism from the mesmeriser to an inanimate substance, but the subsequent and further transmission of the action from the inanimate substance to a mesmeric patient.

It may further be conceived that this particular action of matter, when weakened by transference to a mesmeric billet, for instance, can only affect such frames as are predisposed to receive the influence, just as the contagion of illness conveyed by letter will act most readily on persons predisposed to take infection. In every case, where an agency so subtle as that of mesmerism is to be appreciated in its minute gradations, there must exist a previous acquaintance with its effects, and an acute degree of sensibility both natural and acquired. Beings possessing such qualifications are indeed rare, and, as detectors of nature's finer influences, most interesting to the observer. They alone, it is evident, can throw new light on the action and properties of the mesmeric medium. If we only regarded them as instruments of extreme delicacy, we should value them as the chemist, his test and reagents, or as the student in electricity, his balance of torsion. Dr. Elliotson, by means of two young females, who from their extreme sensibility may be called mesmerometers of the highest precision, has ascertained particulars which bid fair to develope the laws of this remarkable agency. Amongst other things he has discovered that mesmerisable substances may be mesmerised by the eye alone, without contact of any kind. This is of importance; for, to what can we refer such an effect, but to a change in the atomic relations of the affected substances, which change cannot but be brought about by impulsion on some medium subsisting between the eye and the objects? Thus, if some phenomena seem to give a material character to mesmeric agency, others again more truly invest it with all the finer attributes of that subtle, ethereal, and elastic medium, which pervades creation.

Whatever objections, then, may be urged against the undulatory character of the mesmeric medium are shown to be null; while, on the other hand, we should be forced into absurdities, did we attribute to it any other mode of action.

Thus proved to be undulatory, elastic, pervading, in peculiar relation with the mind of man, it should seem clearly to be identified with the medium spoken of in sensation.

But of this we have a yet closer method of proof.

*First.* The phenomena of sensation proper to the mesmeric patient proved to us originally the existence of a certain medium, which circumstances led us to suspect might form the groundwork in all sensation whatever, since it acts pervadingly, and by simple impulsion on the nerves, which are the real agents in sensation.

*Secondly.* The phenomena of sympathetic sensation, originating with the mesmeriser, and from him transferred to his patient, bring the sentient medium into relation with the mesmeriser, and presumably identify it with the mesmeric medium; especially as it is found that the mesmeriser's will and attention aid the production of the phenomena.   Besides, it is fair to infer that all the relations between mesmeriser and patient, being connected in one series, and bearing one character, are brought about by one medium, which we call mesmeric.

*Thirdly.* That such a medium is really the great medium in all sensation we may again deduce from this.   It is found capable of transferring sensations from the mesmeriser to the patient; we may therefore conclude that it is capable in general of producing sensation,— a task to which it seems peculiarly adapted from its acting (as has been shown) with such exquisite precision on the nerves.

The chain of evidence seems thus complete.

In the first place we have a particular medium of sensation indicated to us by the phenomena of mesmeric perception ; in the second place that medium is connected with the mesmeriser, by his capacity of using it so as to produce sensation in his patient. Finally, it is again reciprocally connected with sensation in general by its capacity of acting on the nerves of a human being, so as to produce sensation of various kinds.

Having thus assimilated the mesmeric medium to such a medium or to such media (if there be many) as form the groundwork in the external preparatives of sensation, I would endeavour still further to associate it with other media, now generally considered to be of an undulatory nature and universally allowed to be proper objects of physical and certain experiment.

If we can succeed in this, we shall at once give to our agency a claim on science ; connecting it with calculation and with mechanical analysis.

It is plain that this can only be effected by detecting and bringing to view similarities between its action and the action of such other forces as are subjects of the physico-mathematical sciences, — forces which are now considered to be mere modifications of the electric.

Already we have seen that, like heat, it can be imparted by one body to another, and that, like heat, it gradually leaves the body to which it is communicated.

Some interesting experiments by Dr. Elliotson prove that, like light, it may be reflected from mirrors. A celebrated sculptor at Rome assured me that he had seen the experiment tried and verified on his own brother by a nobleman, whose name it might be thought an impertinence in me to mention.

These are striking analogies between the mesmeric and other media ; — but to electricity — now considered as the great parent of them all — it bears a yet greater resemblance.

*First.* We know that electricity is capable of all that modification in its action which the necessities of our case demand. Sometimes its effects are sudden and energetic; sometimes of indefinite and uninterrupted continuance. It is "capable of moving with various degrees of facility through the pores or even the substance of matter;"* and is not impeded in its action "by the intervention of any substance whatever, provided it be not itself in an electric state."

This capacity of varied action and of pervading influence has already been shown to characterise the mesmeric medium.

*Secondly.* Electricity may be called into activity by heat. Here the resemblance between the electric and the mesmeric medium is still more marked.

I have found that I could not successfully mesmerise in a cold room, or when I was in a low state of bodily temperature; while, on the other hand, a warm atmosphere, and warmth in my own hand, have always intensified and expedited the transmission of mesmeric influence from myself to another. The first time I ever saw a mesmeric experiment, the mesmeriser made it an express condition that he should have a warm room for his process. The experiment, as will be remembered, was quickly and successfully performed. Now, of electricity we know that it may be called into activity by heat, and that the heat of the

* These and the following characteristics of electricity are taken from Mrs. Somerville's admirable work on the Connexion of the Physical Sciences.

air is intimately connected with its electrical con‑ dition.

*Thirdly.* " Electricity readily escapes from a point; and a pointed object receives it with most facility."

I have found, by experience, that the mesmeric medium is most powerfully conducted by the tips of the fingers, and that mesmeric actions, directed also towards the tips of the patient's fingers, have a more remarkable effect than when simply directed over the surfaces of the body.

*Fourthly.* " Iceland spar is made electric by the smallest pressure between the finger and thumb, and retains it for a long time."

How analogous is this to the transference of mes‑ merism to certain metals by human contact, and to the manner in which they slowly part with the com‑ municated property !

*Fifthly.* Electricity may be transferred from one body to another in the same manner as heat is com‑ municated, and like it, too, the body loses by the transmission.

This resembles the exhaustion and repair of mes‑ meric force, of which I have, in another place, taken notice.

It is true that, in the case of mesmerism, a human body is the electric machine, another human body the recipient of the force, and that neither are repre‑ sented to be in a state of electrical insulation ; — but, be it remembered, we are contending — not for the identity of the agency with the electric, but for the propinquity. While we affirm that there are analogies too great to admit of a separate source, we not the less contend that there are differences too striking not to demonstrate a varied action. Their powers not being identical, it is evident that the laws of their retention and escape may vary, as do those that regulate the transmission or accumulation of the galvanic or magnetic forces. Thus, a body may be in a state of mesmeric though not of electrical insulation. Magnetic or electric agency seems to aid the mesmeriser in influencing his patient. I found, when mesmerising Professor Agassiz, that a magnet held in my hand assisted me to raise his limbs by that species of attraction which may be called mesmeric. It is singular, likewise, but true, that, whenever I took the magnet into my hand, however secretly, he, having his eyes always shut, invariably com plained of a palpitation at the heart.

It has also been told me, by Dr. Elliotson, that a lady, who is remarkably gifted with mesmeric force, gains an additional share of that power when she mesmerises standing on a glass stool and in a state of positive electricity. This fact aproximates the two agencies in a very obvious and striking manner.

A few additional remarks on the connection of the human body with electricity may be pardoned.

Much needless pains have been expended to prove that the force in question is an agent in our economy. It cannot but be so. Its presence in our own frames is only one of the consequences of its universal pre-sence — now accorded by all philosophers.

Recent discoveries tend to prove that gravitation

itself is but a residual and comparatively feeble portion of the electric energy; and the power which electricity displays in effecting chemical combination, or separation, demonstrates that it really binds or looses the atoms of which material bodies are composed. To separate one drop of water from another, it has been calculated that as much electric force must be employed as would be called into activity during an ordinary thunder-storm. How, then, can the electric principle be absent from the frame of man, which is itself but a collection of material particles? Facts moreover prove that our bodies are electric, and that the degree of electricity varies in different individuals. The silk stocking drawn off hastily at night rustles with electricity; the hair when rapidly combed, especially in frosty weather, will crackle and emit sparks; and these phenomena are more or less evident in different persons. All this is true and undeniable. The fault has been in theorists to assign specific offices to the electric agency in the human frame, — offices which it may indeed fulfil, but which the march of experiment has not yet sufficiently demonstrated that it does fulfil. The inquiry has not, as it seems to me, been conducted rightly. We have sought, as I said before, to demonstrate the presence of electricity in the human body; and that was needless. The only sure way of demonstrating that any agency in the human body is akin to the electric is to identify the laws of action of the two forces.

I think, then, at present, that the most striking fact

of which I have heard relative to the identity of the
nervous and electric agencies is the discovery of Des-
moulins, that the transmission of sensation and motion
is made by the *surface* of the spinal marrow, and not
by its central parts. This is exactly parallel to the
action of electricity, which is developed only and
transmitted along the surfaces of bodies. That nerves
really do conduct a matter similar at least to the
electric has been also proved by the fact, that a magnet held between the two sections of a recently divided nerve was observed to be deflected as by an
electric current.

But the kind of electricity which is in the human
frame is probably a modification of the original principle. In many particulars it bears more resemblance
to galvanism, and it is really ascertained by experiments on dead animals that the nervous fibre has a
property of being galvanically affected, which (though
varying of course like other properties with the condition of the substance in which it resides) may be
called inherent. Some most interesting experiments,
by Dr. Elliotson, in which patients, by a reinforce-
ment of mesmeric power, were shown capable of
swinging round large weights, impossible to be even
lifted by them in their ordinary condition, prove
again the intimate connection between the mesmeric
medium and the muscular force, which, as every one
knows, is dependent on the state of the nerves and
by them conducted from the brain. If personal evidence may be allowed to have importance, I may add
that I am of an electric temperament, so much so, that

long ago when a child I used to amaze and even
alarm my young companions by combing my hair
before them in the dark and exhibiting to them the
electric coruscations. Of course, also, this phenomenon takes place most remarkably in a dry, and, therefore, non-conducting atmosphere. Now between this
electrical endowment and whatever mesmeric properties I may possess there is a perfect relationship and
parallelism. Whatever state of the atmosphere tends
to carry off electricity from the body hinders in so
far my capacity of mesmerising; and whatever state
of the atmosphere tends to accumulate and insulate
electricity in the body promotes greatly the power
and facility with which I influence others mesmerically.

My feelings of bodily health also vary with the
plus or minus of electricity; and, perhaps, did persons
oftener attend to such things, a similar phenomenon
might not uncommonly be remarked. This at least
we may admit, that the welfare of the human body
depends on the equilibrium or proper distribution of
its forces, and that the electric is one of these just as
much as heat or oxygen. The mesmeric force has,
more than any other, been shown to be inherent in
man; and, taking all the above facts into consideration,
it is by no means a strained conclusion that it actually
is that particular modification of electricity which is
appropriate to the human constitution. When, then,
after having mesmerised a person, I have a peculiar
feeling of loss of strength and general uneasiness,
which can by no means be traced to usual causes, I

am compelled to consider this as a proof that I have suffered by a temporary destruction of equilibrium in that medium wherewith I have charged another person,—that medium, namely, which we have agreed to call mesmeric. That which greatly adds to the presumption is the fact that there is gain in the patient as there is loss in the mesmeriser. The tendency of mesmeric influence to restore equilibrium to the bodily forces is manifest. Under its beneficial action I have seen headachs cured, fatigue dissipated, and trifling bodily ailments removed in a short time. How much it calms and equalises the circulation was shown in the case of Anna M——, whose pulse, agitated by fever, was rendered even and natural after the occurrence of the mesmeric slumber. I have had occasion also to remark that my own sense of exhaustion bore a perceptible relation to that which I was called upon to cure. In mesmerising the weak, even when using very slight exertion, I have found myself more than commonly fatigued. A young lady, whom I mesmerised occasionally at Antwerp, and who was in delicate health, remarked once to me, with that quick perception of every thing relative to mesmerism which persons under that influence display, — " I am afraid it is a great fatigue to you to mesmerise me, because I feel so much strengthened." It was perfectly true.

A medical man at Berne told me also that he once undertook, by mesmerism, to treat a lady who was suffering from periodical convulsive fits. By mesmerising her every day a little before the hour

when the fits usually came on, he first deferred,
and finally altogether prevented, the recurring attacks of the malady; but he himself suffered in a
singular manner. Sometimes, after mesmerising his
patient, he would faint away from the consequent
exhaustion.

From these facts, as it seems to me, there is every
reason to conclude that the principle of mesmeric
action is a disturbance of equilibrium, and that its
mechanical effects may be all explained by differences of positive and negative, of plus and minus, in
the mesmeriser and the patient.

Add to these features of coincidence between electric and mesmeric agency, that both are subject to
actions and reactions, and that the one, like the other,
may be accumulated in bodies to a great extent without producing any sensible change in their properties; and it is impossible but that the analogy between
the two media should appear remarkable.

Thus generalising, we have arrived at the point to
which all science seems tending, — namely, the demonstration of an ethereal medium, of which all other
agencies are but the effects. Should I have been
successful in proving the existence of a mesmeric
agency and its connection with other actions of matter,
I trust that I may be allowed, not to have annulled,
but to have extended the sphere of former knowledge.
I have only exhibited that ether, which Newton imagined and subsequent philosophers have proved to
exist, in a new point of view and engaged in new
offices. I have shown that that medium is the proper

organ of the mind, and the transmitter of all sensation, — that its motions are identical with thought, and are not the accidental, but the permanent, language of sentient beings. Let our contribution be received. The question, — Can mesmerism take its stand amongst the sciences? is even now decided. Even now mesmerism is a science of observation. They who have always rejected the undulatory theory may be pardoned for rejecting it on the present occasion : but they who allow it will not be excusable, should they refuse to trace it in a new and most interesting department of its agency ; — at least, they cannot but confess it to be capable of the effects now ascribed to it. What then will be thought should the very supporters of the undulatory system desert us here — reject our facts as absurd, and proscribe, in this particular instance only, the hypothesis that would render them credible, nay, even necessary?

An elastic ether, modified by the nerves, and the conduction of which depends on their condition; which can be thrown into vibration immediately by the mind of man, and mediately by the nervous system ; which manifests itself when thrown out of equilibrium, and produces mental effects through unusual stimulation of the brain and nerves, cannot but be allowed to be a cause which answers to all the conditions that we desire to unite, and which is sufficient to account for the phenomena that we have been considering. And, be it remembered ever, we have not created the medium for our purposes, — the necessities of the case have created the medium for us, as alone adequate to the production of certain effects which do really and visibly exist.

Even should my theory of undulations be rejected, we cannot but conclude that in mesmerism there is an especial agency. Not one argument against the reality of its existence is available, while, on the other hand, a mass of positive evidence demonstrates that it does exist. Should then the learned at once refuse to examine and to believe, I own that I shall feel surprise. But, that the world in general should reject mesmerism, till the great leaders of opinion have marshalled the way to its precincts, I not only expect, but am ready to forgive, since there are so many received truths (as they are called) from which it seems incorrect to differ, and which the most part of mankind carry about, as Foster says we do our characters and our watches, without scrutinising of what they are composed. In the mean time let me, at least, in drawing to a conclusion, set forth the claims of mesmerism to general attention — to the philosopher as a philosopher — to man as man. My own debts to this despised science are too many and too various to permit me to leave undischarged this duty towards others and myself.

First, in a philosophical point of view, mesmerism may be affirmed not only to touch nearly, but to throw light on, questions which have long agitated the intellectual world.

One of the most interesting of these is whether there may not exist a special though invisible agency appropriate to the nervous system, through which

sensation is accomplished and animal motion performed.   Charles Bonnet of Geneva, a century ago,
suggested that there was an elastic ether resident in
the nerves in a manner analogous to that in which
the electric fluid resides in the solid bodies along
which it is conducted.   He says, " Should we admit
that there is in the nerves a fluid, which in subtlety
and elasticity resembles light, or ether, by the help of
such an agency we may easily explain the celerity
with which external impressions are communicated
to the mind, and that with which the mind executes
so many diversified operations." (*Essai Anal.* chap. v.)

This suggestion has been variously received, or
combated, but never absolutely dropped or decided
one way or the other, so recent a writer as Mrs.
Somerville (who may be supposed to represent the
prevailing authorities of the day) allowing that the
galvanic fluid " may ultimately prove to be the cause
of muscular action."[*]

Now it appears to me that mesmerism is more calculated to decide this point than galvanic batteries
or experiments on the divided nerves of dogs and
rabbits.

We have seen that, in cases of mesmeric sympathy,
the actual sensation of one person is transferred accurately to another : so also the mental action of the
mesmeriser can — so to speak — perform motion in
another.   I have actually seen mesmerised persons
make the most laughable faces simultaneously with
their mesmeriser, though the latter was standing quite

[*] Connexion of the Physical Sciences, p. 331.

behind them.   If he put out his tongue, the patient
also protruded his ; if he wrinkled his brow, the patient wrinkled his brow likewise.   These things may
appear ludicrous — to those who view but the outside
of things they are purely so ; but to the thoughtful
inquirer every trifle connected with a grave subject
is of importance.

Be it remembered that the phenomenon of which
we are speaking does not take place indifferently in
all cases of mesmeric relationship between two persons ; but only (with few exceptions) when the mesmeriser exerts his will and attention to that particular
end.  Where, indeed the relation between the parties
is very strong, the patient is apt, as it were involuntarily, to adopt the gestures and mode of walking of
his mesmeriser.   He is as a machine swayed ever by
the volition of the latter.   But in general the connection between particular acts of will and intention,
originating with the mesmeriser, and particulur motions occurring simultaneously in the patient, is
marked and decisive.   The inference is irresistible.
One mind originates motion in two bodies ; one act
tends to two performances.   The medium, set in
motion between mesmeriser and patient, does not, in
this instance, move the latter in the manner of a
mere stick or rod ; it is plainly impulse as well as
guide, and conveys to the motory nerves of the patient the exact impulsion which it takes from the
will of the mesmeriser. Whatever be the force which
has moved the muscles of the one, precisely the same

degree of that force is meted out to the muscles of the other.

I may now ask whether the mesmeriser, having thus in the truest sense performed muscular motion in another, and evidently by a medium (for actions are not propagated save by transference from atom to atom), may not legitimately infer that what is so efficient to influence the nerves and muscles of another person is also adapted to act upon his own and is capable of the production of that very effect in himself which he has remarked in another? So also, when he extends, as it were, to his patient, or rather gives action to, a medium of sensation the finest and most delicate imaginable, may he not at least conjecture that a similar medium of sensation is employed in conveying to himself impulsions from the external world?

But to consider the matter more closely. The first effect of mesmerism is a bodily sleep, from which volition seems to be absent. In such a state, that it is not natural for persons to get up and walk must be conceded. Constitutional sleepwaking is the affliction of but a few, and of those who are mesmerised scarcely one perhaps has been addicted to sleepwaking. In accordance with this, it will be remembered that, in my cases of mesmeric sleepwaking, the patient always manifested at first the strongest reluctance to move, nor indeed would stir until compelled by the strong command of the mesmeriser. There is, then, at such moments as the patient is moving, a force in operation entirely opposed to a natural desire, and of sufficient strength to prevail over that natural desire.    But this is not all ; we have not here a common case of sleepwaking, whose only phenomenon is that motion ordinarily performed with consciousness is performed unconsciously ; but we have sleepwaking directed by an external force ; we have even, be it remembered, motion cosentaneous with the motion of another, as often as the mesmeriser wishes to produce the phenomenon.    We have the nerves of motion inert and incapacitated one moment, yet brought the next into involuntary action.

Professor Agassiz, who could not of himself stir a muscle when mesmerised, moved, when impelled by me, like an automaton across the room.    Even while retaining his consciousness enough to resist my efforts, by mere gestures, without contact of any kind, to move his limbs, he subsequently owned that he was actually compelled into such motions as I wished him to perform.    Surely a phenomenon like this can be interpreted but in one manner.    The mesmeriser restores to his patient's nerves of motion that force, the previous abstraction of which rendered them incapable.    Whether that force be galvanic or not is a different question.

Again, pins have been thrust into the flesh of healthy persons during the mesmeric slumber, and they have betrayed no mark of feeling.    Now, in such cases, it is plain that the nerves, which ordinarily convey sensations of touch to the mind, are as inefficient to propagate impulsions upward to the brain, as if they had been tied or cut. — But wherefore ? The nerves have been neither tied nor cut ; they have undergone no alteration that we can perceive. What are we then to conclude but that they have ceased to become conductors to some invisible force which is indispensable to their action? For, lo ! the same person who does not feel pins thrust into his own flesh betokens sensibility to the slightest injury which can be offered to his mesmeriser, and that not merely by a simple idea conveyed in some way to his mind, but by an actual impulsion on a nerve of touch ; for he will raise his hand to the part of his body corresponding with that which suffers pain in the mesmeriser. Seeing this, what can we surmise but that the nerve which was so lately inactive has been restored to action by the very force whose absence marred its efficiency ;—that the mesmeriser communicates to his patient not merely a conductor, but a means, of sensation? And impressions of this kind are conveyed in no other way to the patient; for, unless the mesmeriser act, his nervous extremities are in general as dead.

Since then we see the motory and sentient powers of the mesmerised person as active through the mesmeriser as they were shown to be powerless in themselves, we infer that the latter restores the very force through the obstruction of which they become inefficient.

And thus a new antecedent to sensation and palpable motion becomes manifest: for, however uncertain a person may be that there is an intermediate agency between his own will and the motions of his

own body, he cannot doubt that there is an intermediate agency between himself and the motions he
produces in another; and again, however doubtful
he may be that his own nerves, in sensation, are but
conductors to some more efficient force, he must be
certain that he produces sensation in another person,
not by nerves merely, but by some invisible medium
acting on those nerves and restoring to them their
capacities.  And that this is all effected by one medium
seems certain, for the mesmeriser can concentrate its
motory and sentient efficacy into one act; as when
he handles a ball, or other object, and the patient
(with his back turned and at a distance) goes through
the same motions; and, by sympathetic feeling, becomes aware what it is that the mesmeriser holds in
his hand.    It would be perhaps rash to say that
hence he may decidedly infer that motion and sensation are performed by a medium in himself; yet
speculation cannot but tend that way, and surely he
may be forgiven if he supposes that he reads and
scans, as by reflection in a faithful mirror, the mysteries of his own being in the manifested operations
of his who sympathises with him so wondrously.
This, at least, is clear : — if ever we hope to arrive at
a just notion of what has been called the nervous
energy, it must be through means like these.   Only
thus can we at once demonstrate its existence, and
analyse its nature.   The action on ourselves of
those powers which constitute our vitality is too
near us, too intimately blended with our being, to
form a subject of personal contemplation.   Our un-

consciousness of their agency is a law of our being, and
has hitherto been the great argument against admitting the very existence of a nervous medium, which,
however plausibly presented to reason, can in no way
be rendered palpable to sense. Here, however, the
mesmeriser, seeing that he is unconscious how he
actuates the person whom he sways, while that he
*does* actuate him through a medium is manifest by
visible signs, perceives that his ignorance, how he
feels or performs his own voluntary movements, is no
proof whatever that he may not unconsciously use,
both in sensation and in motion, a subtle, elastic, and
ethereal medium.

While physiologists have been engaged in surmis-
ing or disputing the existence of a nervous agency,
metaphysicians, on their parts, have transferred the
question into a more spiritual region, speculating on
the strange marriage between mind and matter, and
proposing, in order to effect a union between such
opposites, an intermediary partaking of the qualities
of both. That a link does really seem wanting between the mind and the grosser matter with which
we are visibly surrounded is to be presumed from
the subject having engaged men's attention from the
earliest ages; from the period when the ancient philosophy devised intermediary images of things, to
that in which Cudworth imagined a plastic mediator.
This palpable hiatus between the visible and the invisible — the subtle and the gross — mesmerism seems
calculated to supply. Let me not for one moment be
misunderstood. The notion of a half-intelligent me-

dium is as absurd a jumble as the idea of a half-thought would be. Mind cannot be cut up into fractions; and to divide it, or mingle it with matter, in any way, is just the same as if we were to talk of pounds of intellect or square inches of rationality. When I speak, then, of the possibility of a medium subsisting between the mind and the grosser manifestations of matter, I must by no means be conceived to intimate that this medium partakes in any degree of the qualities of spirit. In this respect I perfectly agree with Sterne. " The ancient philosophy materialised spirit, and the modern, in order to be even with it, has spiritualised matter. What extremes are men liable to run into, who depart one line from common sense." Besides, this *petitio principii* is altogether useless, and explains nothing. Thought and sensation are in the mind alone. The latter sensation, at least, seems brought there from afar, but we cannot shorten the road. When we have traced it to the brain, it has still a journey to perform, as long and as difficult as ever, — the transition, namely, from motion to thought, from a sign to an idea, from matter to spirit; and, however we may deem that we facilitate the passage by a medium, half-mind, half-matter, we in reality have left the opposing principles just where they were. We have only assumed the very point which we were desirous of proving. I would, then, plainly state that I re cognise only two grand divisions in nature, namely, of cogitative, and incogitative — that which thinks, and that which thinks not. Yet that, in a different sense, there may exist a medium between sensible objects and the mind, I trust to demonstrate is not impossible.

Striving to arrive at the true source whence arises so obstinate an idea in us of a medium between mind and matter, I imagine it, like most of our notions, to proceed from the ground of our own personal experience. We see and know that we never move matter external to ourselves except through the medium of our own bodies. Our minds we feel to be inadequate *per se* to stir one atom of extraneous matter. In the same way we feel and know that all information respecting external things comes to the mind through the medium of the body; and, apart from this medium, we feel that our minds are inadequate to take cognisance of external objects. Hence arises in the mind a general proposition that mind affects matter, or is affected by matter only through a medium; and we carry this piece of reasoning even into the precincts of our own corporeal frame — conceiving that this also, being matter, can only be moved by our minds through a medium, and can act on our thinking principle only through a medium. Now this reasoning, though imperfect, is not altogether null and baseless. That mind can only act on matter through a medium, and *vice versâ*, is, indeed, a false proposition; for allowing a medium, however ethereal, between the mind and brain, we cannot but allow that medium itself to be matter, unless indeed we fall into the old absurdity of a half-intelligent medium. Still there are media which manifestly exist, and are made evident in their effects; which, though they cannot be in any way confounded with mind, inasmuch as they are incogitative, do really differ from our common notions of matter in many particulars; — I speak of media, to which we may apply that which Dr. Maculloch has beautifully said concerning light. " Imponderable, intangible, incapable of being arrested and accumulated, ever under the most rapid motions, though we cannot discover a projectile force; coming we know not how, and vanishing we know not where, it is in all but the power of thought a spirit." Now to matter in general we certainly attach the ideas of weight and palpable solidity; and though, when we say of the ethereal medium that it is co-extensive with space, and composed of impenetrable, indivisible atoms, we really classify it with matter, still there is a lurking feeling in the mind which creates a distinction between matter of so ethereal a texture and that which we can see, weigh, and handle. Why matter of this nature should seem to us more adapted to act upon mind, and *vice versâ*, than any other sort of matter, we can perhaps give no very philosophical reason; yet the fact is that there is some such notion in the mind of most persons.

But, again, this notion is really in harmony with some striking facts. The grosser media, which are palpable to our senses, act upon and modify the finer media, which are in truth always the most forceful and pervading; but are either not palpable to our senses, or are perceived only by such a sense as sight, which is in truth a complicated intellectual operation. The atmosphere modifies the electric principle, and its insensible action on our bodies; the atmosphere modifies light, and adapts it to the visual capacities of man. Light itself has been recently conjectured to depend on electrical actions and reactions. Should this prove to be so, and should subsequent experiments demonstrate that man communicates with his own body through electric agency, the following proposition, now but a theory, would gradually evolve itself. Grosser media act on the finer, the finest on the mind of man; and again the mind of man reacts from finer to grosser media. I confess that to a system like this the phenomena of mesmerism incline me to lean; while, as it appears to me, the analogies of the visible universe tend also to the same conclusion. In the words of the poet —

> " Look nature through ; 't is just gradation all.
> By what minute degrees her scale ascends !
> Each middle nature join'd at each extreme,
> To that above is join'd, to that beneath :
> Parts into parts reciprocally shot
> Abhor divorce."

Again. Wherever we turn our contemplation, there is a manifest progression from finer to grosser media. An ethereal and infinitely subtle fluid, pervading the realms of space, is the first known step from Deity to matter. Distributed through this, and probably condensed from the prime element, are discerned masses of a luminous substance, which have received the name of nebulæ from astronomers.

Amidst these luminous masses, which constitute the
first traces of matter of which we have ocular evidence, are beheld certain nuclei, forming the germs of
future suns or worlds.  So, at least, we may safely
conjecture ; for all geology goes to prove that " all
this fair variety of things," which our earth displays,
was once in a fluid, or rather gaseous, state.  Nay,
there needs but a sufficient application of the universal solvent — heat, to melt down all again into
fluidity and vapour.  Thus, again, with science, we
arrive at a primal, universal element, such as is
now supposed to be the substratum even of the gases :
and this element (I speak it with reverence) is the
first demonstrated action of the Almighty mind.

If I err in thus expressing myself, I err with no
less an authority than Milton, who speaks of light as
the first effluence of divine power : —

 " Hail, holy Light, offspring of Heaven, first-born,
 Or of the Eternal co-eternal beam,
 May I express thee unblam'd? since God is light,
 And never but in unapproached light
 Dwelt from eternity, dwelt then in thee,
 Bright effluence of bright essence increate.
 Or hear'st thou rather, pure ethereal stream,
 Whose fountain who shall tell ?   Before the Sun,
 Before the Heavens thou wert, and at the voice
 Of God, as with a mantle, didst invest
 The rising world of waters dark and deep,
 Won from the void and formless infinite."

Now that, in mesmerism, a subtle medium is set
in motion by the mind has been proved, but whether

this be accomplished through the intermediate agency
of the brain, or whether the brain itself is moved by
the medium — whether, in brief, the brain or the
medium be most proximate to the mind, it is impossible for us to decide with absolute certainty, it being
evident that a man must be palpably proved to have
acted without a brain before the question can be
finally settled. There is, however, a sort of reluctance in the mind (instinctive, I grant, and not philosophical,) to conceive that spirit can act at once
and directly on nerve and fibre, and, above all, on
a torpid viscus like the brain, which may be cut and
pierced without the subject of such an operation
being conscious of any particular feeling. Rather
we incline to imagine that an impalpable ethereal
medium, such as the mesmeric, must be the next
thing to mind itself. A medium differing in so far
from ordinary matter, as it is set in motion by the
human mind immediately. In such a sense only
could we admit the existence of an intermediary
between mind and matter.

So much may be said without much thought or
careful induction : but, when we come to examine into
the principles of common muscular motion, we are
compelled to use stronger language. The physiologist here cannot fairly have recourse to his old
stalking-horse of *vital action,* on which he hobbles
away so often, when pressed for an explanation of
the mysteries of our frame. He himself must grant
that it is *will* which originates and governs the
duration of muscular contraction. He must also

z 4

grant that contraction cannot take place without the immediate *influence* of the brain and nerves — though what that *influence* is he very wisely forbears to tell us. Muscular motion, then, is performed by the will or mind (for will is but an attribute of the mind), through the medium of the nerves. But are the nerves of themselves a medium fitting for this work? Let us examine them, and we shall find nothing whatever to account for their marvellous capacities. They are, to appearance, the weakest of all weak instruments — mere soft white filaments, enveloped in a fine cellular membrane — and yet the strength and the power of all the body is derived from them! This simple fact reduces us to a dilemma. Either we must believe the nervous substance to be the channel of some ulterior and efficient energy, to us invisible, or we must admit, according to Pascal's generally received definition of a miracle — namely, " an effect exceeding the natural force of the means employed " — that a constant miracle is taking place in our mortal frames. Following the good old advice of the poet, who said " Nec Deus intersit," &c. we rather shrink from attributing to direct divine agency every minute corporeal motion which our fancy may incline us to make. It is not the mind itself which directly moves the muscle; if it did, we should have to seek no farther for an adequate agency, since we have never measured the power of mind to move matter. But, as we know, the mind moves the muscle *indirectly,* and through the medium of the brain and nerves. In this case there is a known sequence of events; and a sequence which, as it stands at present, is an absurdity. The weak soft fibre stirs the muscle, as we are told, so immediately, so mightily, that " its fibres shorten and become hard, without any preparatory oscillation or hesitation, and they acquire *all at once* such an elasticity, that they are capable of vibrating or producing sounds." * There is nothing in mesmerism more wonderful than this, more difficult of digestion to credulity itself, when we take the matter as it stands, without bringing into view an ulterior agency, — an efficient force. Why shrink we from so doing? When we look upon the heavens, and the magnificent system of sidereal worlds, harmoniously moving round a common centre, we acknowledge all this harmony to be the result of a force which we call gravitation. But what is the wonder of the universe compared to the mystery of the mind linked to matter and actuating it? It is a mystery; but we need not on that account render it an absurdity by inefficient causation — by denying the mind an adequate intermediate agency. Such denials betray physiologists into great inconsistencies; as, for instance, Magendie one moment censures every attempt to explain muscular contraction, and the very next explains it himself by the vaguest of vague terms — "cerebral influence," affirming, from the strength of maniacs, that " the muscular power may be carried to a *wonderful* degree by the action of the brain alone."

* Magendie.

z 5

What is this brain, what are the nerves, that they should possess such power of action? Is it inherent in them? If so, why does it not endure when man has ceased to think and live. Examine a nerve apart from the intelligent being it once served, pore over it with the microscope; then take the brain and weigh it well, or cut it up into sections: where is the innate power of action, where is all that may truly be called power — namely, the enduring and essential capacity of originating motion? If, then, it be an absurdity to attribute an inherent power of motion to flesh and fibre, it is also an absurdity to attribute to weak instruments like the nerves an inherent power of stirring, in an instant, the strong volume of a muscle. And if the power be not inherent, where is it? Granting all that we can grant as regards the efficiency of mind to stir its own organisation, we must beware how we charge inconsistency upon the Creator by supposing Him to violate those laws of action which he points out to us by manifest signs; and never has our experience shown that by a weaker agent we can move a stronger — in a case, too, where the disproportion is manifest, and where there are no connecting links to prepare the way for the action of the much weaker upon the much stronger. Again we ask — where are we to look for the force that effects such wonders? Now we fear not to explain muscular action in the dead by a force that we call galvanic. Why will we not explain it in the living by some efficient force at least — for here we have equally a sensible phenomenon that calls on us for an adequate explanation. I know not what philosophic doubt may make of this question, but certainly common sense decides, from the ordinary phenomena of muscular motion, that the mind actually dispenses and metes out a material force, of which the brain is, perhaps, the elaborator, and the nerves decidedly the appointed conductors. This renders the question, Whether the mind acts on the body, mediately or immediately? much more easy of decision; and, when we weigh probabilities, inclines the balance in favour of those who adhere to the doctrine of a medium.

Finally. When we agree to divide matter into ponderable and imponderable, we, in fact, create a distinction to the full as great and important as any which can be devised by supposing a third substance which is neither mind nor matter. And, this distinction once admitted, there is no inconsistency whatever in saying that by matter imponderable we communicate with matter ponderable.

Be it remembered, I assert no more than that mesmerism *bears* on this disputed point; — that it decides the question is not affirmed. But, when we reflect that the whole of mesmerism seems to consist in discarding ordinary intermediaries, such as the external mechanism of the senses and the media adapted to this, and in arriving at grounds and principles; above all, when we see, by the experiments of Dr. Elliotson, that mesmerism can be a succedaneum at least for muscular force, enabling the weak to raise heavy weights; and when again we combine this with

z 6

all that has been just advanced respecting an intermediary in muscular action,— surely, all this considered, we cannot be far from the belief that the
mesmeric medium is really the long-sought link that
was to render our system more complete and more
harmonious.

Should we come to this conclusion, we remove at
once the objection so often raised against the very existence of mesmerism, that its agent cannot — in our
usual state at least — be made the object of our senses.
It is a medium to enable us to perceive objects, and
therefore that it should not be itself a direct object of
sense seems natural.     Light itself is the true object of
the eye, and is, in certain cases, its sole object; and
whether we communicate even with this directly may
be made a question.  Should we establish a medium between the mind and all matter perceivable by us, still
the principle must here also be adhered to.  It would
only be in analogy with what we have ascertained in
other cases, should we decide that we only communicate with light through the medium of another and an
imperceptible agency.   Of course the same reasoning
will apply to air, and to the action of all media on
our senses.    Such, at least, as we have seen was the
view which Newton took of the economy of sensation — not giving all his reasons why he did so ; but
who can doubt but that such a mind as his kept back
within itself more reasoning on most subjects than
common thinkers have promulgated and found adequate to secure a very tolerable reputation ?

But, in truth, the objection to the reality of mesmerism, because it cannot be brought before our senses, is not worth much. We see it in its effects, and of these we can judge by more senses than one. The mesmerised, perhaps, being in a different state to ourselves, may see or feel *indications* at least of this subtle agent, when they speak of perceiving a kind of fine vapour, or electric light. If so, the essential difference between the mesmeric and the normal state would consist in the power of being sensible to the very medium which furnishes the means of sense.

But, in truth, what is the invisible? The microscope will prove our vision to be capable of a development, the limits of which will not easily be found. Revealed by this, "world within world enclosed" will burst upon our sight, where we had supposed a void. It would be, then, ridiculous to make the existence of a medium between mind and body dependent on the cognisance of our senses as at present constituted. That which is not discerned by us may be clearly seen by more spiritual optics. I should be sorry if my existence depended on the perfection or imperfection of some persons' organs of sight and general sensibility.

Indeed, if we lay to heart the deceptiveness and mutability of all the external species of matter, at the same time considering that we have no reason to deem it capable of change in its ultimate and imperceptible particles; if also we reflect that whatever is not palpable in itself yet is indicated by its effects, forces us on pure reason by withdrawing at once the aid and the illusion of our external senses; we shall

perhaps come to the conclusion that the invisible is
the only true, exclaiming with the old Latinist,
" Invisibilia non decipiunt."

These remarks will be received with indulgence or
not, by no means according to their merit, but according to the turn and character of each several
reader's mind ; and this observation involves a general truth not sufficiently attended to.   We all have
our own modes of reasoning, and, as our minds are
analytic or synthetic, we love those who never reach
or those who jump to a conclusion.   Again, some
delight in all theories, and some love no body's
theories but their own.   Now, whether I belong to
those who hastily snatch up an hypothesis, because it
so nicely accounts for things, or to those who draw
deductions, because their reason is convinced, I leave
to others to decide ; that is, provided they be of a
party who decide on any thing.   For my own part
I confess my weakness.   I like to come to a decision;
and herein I am at a manifest disadvantage, for in
all ages the doubters have lorded it greatly over the
deciders.   It is so grand, and so philosophic to doubt,
even of one's own senses.

I leave the world to apply this to mesmerism generally.   Whether the speculations into which this singular subject leads us almost without our wills be or
be not generally acceptable, it must be owned that,
in suggesting to us a mode of sensation more direct
and worthy of the mind itself than the common, they
do us no ill service ; and surely we cannot grossly
err in giving to thought so swift-winged a messenger

of volition, so subtle a vehicle of sensibility, as the mesmeric medium.

The idea of nature's constitution, which the apprehension of such a medium excites, are at once comprehensive and ennobling; and the philosopher, who desires to have large views of man and of creation in general, cannot take his stand on a better vantage-ground than the one which the inquirer into mesmerism is compelled to occupy. Thus expanded, his mind can scarcely fail to be cured of that dogmatism which cries out "impossible" to all that is not within the narrow round of its own experience. The first time of witnessing a successful experiment in mesmerism must be an era in the life of every thinking being; while the student in this science, who at length arrives at the knowledge of an ethereal medium, connected with all animated beings and with thought in particular, beholds a flood of light illuminating mysteries which have probably long weighed on his heart and brain. He beholds all life and intelligence at once connected and individualised — reciprocally connected in all its parts universally with God ; and he has a glimpse at least of the waves of the great ocean agitated by thought eternal, and tending to thought again in the limited portions of intelligence which the Almighty has gifted with individual consciousness.

Should his mind be less excursive, should he study mesmerism with practical severity; still he cannot but view it with interest as a link between sciences at present disunited or standing remotely

connected. Physics and metaphysics are now at dag-
gers-drawing; but here they may embrace and be
reconciled. It so evidently touches on all and blends
all. Without it the connection of the physical sciences is incomplete, for this alone concentrates them
personally in man. I cannot for a moment doubt
that a future generation will hail mesmerism as the
very note which was wanting to render all nature
harmonious, as the key which fits every ward of
knowledge and unlocks all the treasures of science.
The ornithorhyncus was deemed at first a fictitious
creature, but was found at last to be the very link
that was wanting to the animal creation.

Let me hope that if, in endeavouring to advance a
consummation so desirable, I have fallen into scientific errors or partial inadvertences, I may be forgiven on the score of motive at least; though I am
quite willing to be charged with the fault, provided
mesmerism be not blamed. In setting forth a new
science there is always a probability that errors will
be committed, especially in setting forth a science
like this, which to solve entirely would be to solve
the riddle of man.

There is one imputation, however, from which I
shall with difficulty escape, a charge of which I fear
not even my own conscience can acquit me. In the
first part of this work I uttered disrespectful words
regarding those who, treating on mesmerism, took
occasion to frame systems of the universe; and now,
behold, I am myself found guilty of soaring upwards
from my subject into the regions of space, and of

drawing from a few despised facts the highest conclusions. So easy is it to blame; so difficult to avoid
blame; so difficult, also, when treating of man, not
to touch upon lofty topics and to be led even unawares to speculate on the sublimities of creation.

It remains for me now to exhibit mesmerism in
such points of view as regard all men equally, whether
considered as material or immaterial beings.

First, as a remedial agent, mesmerism, when better
known, cannot but be universally important and interesting. That it should not already be so is only a
proof how inveterate prejudice can be, then most
when most irrational. When we see persons eagerly
running after every new or old new-revived remedy,
consenting to be starved, as is now the fashion in
Germany (where men are mad for what is called
the hunger-cure); or to have their insides drenched,
Sangrado-like, with water; or to sit smilingly plunged
up to the neck in mud. When we see such instances
of more than martyr fortitude displayed too on the
occasion of every trifling, or perhaps imagined, malady, we cannot but wonder that mesmerism, so
fraught with hope and healing to the human race,
should be neglected; more especially, too, as its
therapeutics are neither disagreeable nor disgusting.
" Blest be the man who first invented sleep," says
Sancho; and in this a man has only to sleep and to
be cured.

Not being medical myself, I cannot bring forward
a number of personally observed cases in which mesmerism has been useful. I have, however, sometimes

mesmerically treated slight indispositions in members
of my own family with success, and I have had occasion to be struck generally with the power of mesmeric agency to equalise the circulation and to
relieve pain. I have known it to be particularly
successful in cases of severe headach—by no means
one of the least ills " which flesh is heir to." On
one occasion, a friend of mine, the Baron de Karlo-
witz, whom I used to mesmerise sometimes for mere
purposes of experiment, came to me with a severe
cold and hoarseness which prevented him from
speaking above a whisper. We neither of us had any
idea of curing this affection; yet, after he had been
mesmerised for about an hour, during the greater
part of which time he slept, he rose from his chair
perfectly relieved from his catarrhal symptoms. He
also found, after I had mesmerised him a few times,
that he had got rid of a chronic rheumatism in one
arm, from which he had been suffering for many
weeks.

From medical friends I have received many particular accounts of the beneficial agency of mesmerism ;
but these the limits of my work do not allow me to
detail. I can only state generally that, as might
have been expected from the evident action of this
influence on the nerves, mesmerism is especially
useful in those derangements of the nervous system
which are generally beyond the reach of medicine.
Epilepsy, palsy, nervous depression, madness even,
have been successfully combated and subdued by
this, which may most emphatically be called the

natural remedy. If only in its power to calm, when opiates are hurtful and improper, this tranquillising influence is of incalculable value, enabling nature, during the periods of repose, to balance and recruit her disordered forces. Dr. Foissac, of Paris, has assured me that he has seen men restored to calmness and finally to health by mesmeric action, when suffering in the last stage of spasmodic cholera. Reasoning from analogy, we should conjecture that this is probably the only power available to counteract hydrophobia. At any rate, the instinct whereby persons in the mesmeric state prescribe remedies for themselves or others might here be turned to most valuable account. The insensibility to bodily suffering, which has been proved to characterise the mesmeric state, may also, as it has already in some few instances, prove of the greatest service in enabling patients to undergo the amputation of limbs or other painful operations. In reply to a question I addressed by letter to Dr. Foissac, I received the following communication : —

" Vous pouvez, en toute sureté, magnetiser des femmes enceintes ; vous leur ferez beaucoup de bien. Quelques somnambules ont accouché sans la plus petite souffrance ; dans tout cas le magnetisme rend les douleurs trés supportables."

The last and most important point of view in which we have now to consider mesmerism is in its reference to the future prospects of man, considered as an improveable being, capable of immortality. My object here is, as it has heretofore been, at once to explain and to utilise mesmerism. Already I have cast light on its apparent mysteries: first, by the laws of our normal consciousness; secondly, by the true nature of sensation; thirdly, by the native force and faculties of the mind; fourthly, by demonstrating that, in the mesmeric state, the mind is not only detached from its present restricted organs, but gifted with a medium of sensation more direct and more worthy of itself; and now, finally, I purpose to render evident whatever in my subject may yet appear to be unexplained, by considering man as a being capable of a development unsuited to his present era of existence. The subject is most important. The other topics on which we have been engaged are but speculative points, and are comparatively devoid of interest.

This comes home at once to our business and bosoms. Wretched, indeed, must be the view of man which confines him to this bank and shoal of time, which does not regard him, and all his glorious endowments, as intended for a series of existences.

> "——— For, we live by hope,
> And by desire; we see by the glad light,
> And breathe the sweet air of futurity." *

Here, only, when we doubt, we are vastly too humble; refusing to recognise our own dignity and the privileges accorded to humanity. That man is improvable and capable of a development, the limits

* Wordsworth.

of which have not yet been ascertained, is his high
prerogative which distinguishes him from the brute
creation. Even common life will teach us this : but
mesmerism, in an especial manner, proves the capacities of man to be susceptible of progression; for it
not only exhibits him in a state of moral and intellectual improvement, but in a condition of increased
sensitive power. Let this propitiate the religious
mind at least. Mesmerism is no miracle, but a development of faculties inherent in man. It is
fraught with instruction the most holy, so that we
may say of it that which Galen affirmed of anatomy,—
To examine it is to sing a hymn to the Creator of all
things. Here, indeed, " qui studet orat." It calls
attention to our being, frame, and structure ; and, by
proving that we have an immaterial principle which is
*not* dependent on the organs of sense, it is calculated
to reclaim to a purer faith the materialist, as it
actually did reclaim a young French physician,
(named Georget) of great promise, who, dying, left
on written record, desiring it by will to be made
public, that mesmerism had convinced him that man
was not a mere compound of organs, but a spiritual
as well as corporeal being.

The light which this phasis of our nature gathers
from and reciprocally sheds on revelation is remarkable, and to reason's eye most precious. By exhibiting man as at once delivered from the thraldom of
his exterior senses and enabled to attain to greater
heights of moral excellence, it leads one to see that
there is something more in the scriptural expression,

" Sins of the flesh " than a mere form of speech.  In mesmerism the mind recurs to its native character and fundamental endowments, seeming to cast aside the accidental differences induced by education, circumstances, and neglect of moral discipline.  We thus learn how great a part of the evil that clings to us is our own work, from our omitting to discipline and subdue the grosser principle and to combat those temptations which are actually engendered by our present position and manner of organisation.   I have been told that some writer has called mesmerism " a miserable compound of faith and fear."    I can assure the author of this sentence that *one* at least of these elements must be wanting to true mesmerism; and I would ask him why he should confound faith with base things, or identify that holy power with credulity?   I would send him to Thalaba for instruction, where, when all that is noble is to be performed, when all that is evil is to be vanquished,

" The talisman is faith."

With faith, then, with Christian faith, I would associate mesmerism ;  and with that which so largely promises another existence it can easily be allied.

As regards our future state (a question that concerns every mortal being), there is the greatest reason to believe that mesmerism is a boon granted by God to confirm our faith and to cheer us on our way. All its phenomena combine to identify it with that which Coleridge has called " the fundamental life."

As, however, in this expression, there may seem to be some ambiguity and vagueness, especially to such persons as have not turned their thoughts to the subject we are treating; and as I wish to be particularly clear on a point so important, I shall endeavour to explain precisely what is the fundamental life, and wherein it differs from the organic. If, in attempting this, I adopt, more than I have hitherto permitted myself, an *à priori* and abstracted mode of reasoning, there will be at least this benefit, — we shall have, from pure reason, an additional ground of confidence for our belief in an hereafter; and surely, in respect of this, we cannot build up the edifice of human hope on too strong a foundation, or protect it by too many buttresses against the inherent weakness of its own imperfection.

Already I have hinted at a distinction between sensation transitory and permanent — occasional and fundamental, and have insisted on the necessity of taking so large a view of man as to distinguish the accidents of his present situation from his immutable character, when considered as a being destined for immortality. I now proceed more particularly to develope this idea.

No man hath seen God at any time. In His works alone we can behold Him : they are, in the truest sense, His expressed thoughts; — as the poet says, addressing the Creator, —

" Glorious! Because the shadow of Thy might,
A step. or link, for intercourse with Thee!"

If by the word of revelation we learn to discern God spiritually and to discover our moral relationship towards Him, we no less by the universal manuscript of nature learn to view Him, as it were, with the eyes of the body, thus enlarging our views of His power, wisdom, and design.    Without this source of instruction, corporeal as well as intellectual, our relation to the author of all things, we being composed of body and spirit, would be incomplete.

Now, thus to hold converse with the attributes of God, as expressed in the visible creation, is not a temporary object ; for we have no reason to suppose that matter and its properties, whatever changes they may undergo, will ever be annihilated, matter, indeed, being the only thing respecting which we have sensible evidence that it perishes not : and from this the great advocate of man's immortality, the poet Young, draws a chief argument in favour of his subject : —

> " The world of matter, with its various forms,
> All dies into new life.   Life, born from death,
> Rolls the vast mass and shall for ever roll.
> No single atom, once in being, lost,
> With change of counsel charges the Most High.
> What hence infers our reason ?   Can it be ?
> Maker, immortal, and shall Spirit die ? "

Moreover, on matter hang all our hopes of individual existence, for, as I have before shown, the individuality of all spirit but God must consist in being impacted and restricted, — in being less than pure spirit, which the Absolute alone can be, — in having a body through which all information respecting the original ideas of God being transmitted should become proper to itself. Reason, then, proclaims external matter, not less than that portion of it connected with ourselves, to have been created for a durable existence; and, this being the case, it is evident that to bring the mind acquainted with all the qualities of the universe is an end and object which remains immutable, and which, unless we can suppose this world to be the be-all and the end-all, and God to be limited in creation and invention, can only be fulfilled here in a partial and imperfect manner. Supposing man to enter on individual existence for the first time in this our planet, it is here he takes his first lessons in natural divinity; and it would be indeed to charge the Almighty with "change of counsel" and vain expense of mighty means, were we to suppose that what man learns here is to be unlearned hereafter. Wise earthly parents deal not so with their children. Now by what means, at any time, shall we continue our process of instruction? Surely, as we began it; by means of a body, the great purpose of body being to connect mind individually with matter and its properties. A body, then, that shall endure seems essential alike to the continuation of our individual existence, and to our continued commerce with the qualities of the visible creation. Revelation is perfectly in accord with this idea. The Bible philosophically says, at man's creation, — " And man became a living soul," that is, an individual soul, a soul embodied, and capable of communicating with external matter. When

A A

we say, then, that man is to live hereafter, we also affirm that he is to continue embodied; and in this again revelation sanctions the deductions of our reason, one great object of Christ's mission on earth being to preach the resurrection of man with a body. Such a body then as would endure, and cleave to the mind through all external changes, is, in fact, the fundamental life; while the organic life is but that visible and grosser envelope which forms our tangible body in this world, and which is furnished with organs which have especial reference to this particular state of existence.

More briefly: the fundamental life is the *body* (for what is life, in our case at least, but mind incorporate?) that we are to retain throughout eternity. The organic life is the body, or means of communicating with matter, which we now palpably possess, and which may be imagined to be a temporary development of the other, just as leaves, flowers, and fruits are the temporary development of a tree. And in the same manner that these pass and drop away, yet leave the principle of reproduction behind, so may our present organs be detached from us by death, and yet the ground of our existence be spared to us continuously.

When St. Paul affirms that there is a spiritual body as well as a natural body, he not doubtfully declares the same thing. He does, in fact, draw a distinction between the body we now have for purposes connected with the economy of this world and the more lasting body which is to connect us with

the universe in general.   He was too good a logician
to have used the words "spiritual body" in a vague
and mystical sense.   Body is body, and the only
differences between one body and another which can
possibly obtain are gradations from grosser to finer,
or the contrary.

That we should rather evolve from our present
corporeal elements the body which is to be ours beyond the grave, than begin existence *de novo ;* that,
in other words, we should really possess a fundamental life, or (to speak more intelligibly) a fundamental
body, incapable of passing away with the grosser
covering that envelopes it; that, at death, we should
retain something physically from our actual condition,
seems pointed out to us by all the analogies of nature.

Everywhere we behold that one state includes the
embryo of the next, not metaphysically, but materi-
ally; and entering on a new scene of existence is not
so much a change as a continuation of what went
before.   A sudden leap from one condition to another is not compatible with nature's evident tendencies.   The very rudiments of organs, intended in a
higher stage of animal life to be useful, are found,
uselessly as it were, appearing in the lower classes of
animated creatures, or — stronger still — lying in embryo in the same creature in one state, only to be
developed in another.   It is an old allusion, but ever
beautiful, —

> "——— The wings that form
> The butterfly, lie folded in the worm."

We should then, *à priori*, expect to find the principle that individualises man, and is the true medium of his instruction, attached to him from the beginning, and that the germs of future capacities, physical not less than intellectual, should be discoverable in his constitution. Here again we are reminded of St. Paul's beautiful comparison between man's transitory body and a seed that loses in the ground only its exterior husk or covering, while the very circumstance that produces in it a partial loss and decay is the occasion of development to that germ which from the first included within itself all that it would be hereafter. Thus it appears manifest that the apostle supposes that, even from analogies before our eyes, we may draw conclusions respecting our future condition; — nay, he seems to find this truth so clear that he thinks it argues much want of intellect in a person not to perceive it, as it is evident by his saying, " Thou fool, that which thou sowest is not quickened except it die." In another place he strongly intimates that the doctrine of a permanent body was not unknown to his converts. " For we know," he says, " that, if our earthly house of this tabernacle (meaning evidently our present body) were dissolved, we have a building of God, a house not made with hands, eternal in the heavens." And with this he desires, at death, to be " clothed upon," or, in other words, embodied : not wishing to be " unclothed," or to be disembodied, which would be equivalent to annihilation, as if he were tired of existence; but only longing to depart and be with Christ — in other words to continue existence in a more elevated state of being, putting off the external weeds but not the unchanging vesture of the spirit.

From considerations like these we may surmise that the mind is capable of being organised in two ways — visibly and invisibly. It may have a subtle envelope, adapted generally to its own active and subtle essence, and, in addition to it, it may have a coarser covering, as our mortal body, adapted to its present state of existence and furnished with particular organs for particular purposes. The dissolution of this coarser covering is by us called death; that is, we seem unto men to die : but with our inner body we never part, and, consequently, by that we still retain our hold upon individual existence. As Leibnitz has remarked : " There is no such thing as death, if that word be understood with rigorous and metaphysical accuracy. The soul never quits completely the body with which it is united, nor does it pass from one body into another with which it had no connection before; a metamorphosis takes place ; but there is no metempsychosis." *

Let us now examine the validity of our *à priori* reasoning by a reference to the facts of mesmerim. Man is shown by these to be capable of increased sensitive power. *Cui bono* — to what end, if hereafter this increase of faculty become not permanent ? Would it be consistent with the goodness of Providence to tantalise us by imperfect glimpses of that

*Encyclopedia Britannica — Preliminary Dissertations.

which we shall never be permitted to realise ? Would
wings be folded in the worm if they were not one
day to enable it to fly ?    We cannot think so poorly
of creative wisdom or of thrifty nature.   Throughout
her realms there is no mockery of unmeaning displays of power ; and, if so, then is mesmerism a
pledge irrefragable of a future state of existence,
calculated for the exhibition of those energies which
are but a promise here.    Relative to this subject, I
have particular pleasure in remarking the coincidence
of thought between myself writing on mesmerism
and an esteemed author, who, in framing a " physical
theory of another life," has unconsciously borne witness by reason to the truth of mesmeric development.
Such unbought testimony is precious.   He is speaking of what an enlarged perception might effect.
" Doubtless," he says, " the mind might bring its
percipient faculty into contact with the properties of
matter more at large and under fewer limitations.
The medullary substance we may easily suppose to
be laid open to sensation otherwise than it actually
is, and also to be endued with a more refined or
exquisite sensibility.

" The mind, perhaps, in its next stage of life, and
when its active and higher principles have become
mature, may be well able to sustain and advantage-
ously to use a much more ample correspondence with
the material world than would now be good or possible.    Perception is at present a circumscribed faculty ; and we confidently anticipate an era, when it
shall throw off its confinements, and converse at

large with the material universe, and find itself fa-
miliarly at home in the height and breadth of the
heavens.   We may assume that it only needs to be
free from the husk of animal organisation to know
on all sides perfectly that which now it knows at
points only, and in an abated degree.   And besides
knowing effects, it would also know causes.   The
*inner* form of matter, as it has been termed, may, as
well as the external species, be discernible.   Instead
of looking only at the dial-plate of nature, we should
be admitted to inspect the wheel-work and the
springs."
    The above passage offers a delightful contrast to
those narrow-minded views of man which limit both
himself and his Almighty author to the *ipse dixit* of
some peevish sceptic.   The whole work, from which
it is extracted, is worthy of attention, and is in perfect accordance with mesmeric phenomena ; nor can I
refrain from believing that its clever author would rejoice to see realised in this life anticipations which
he only framed concerning the next.   From his remarks, which we have just cited above, we gather
that he looks forward to the future enlargement of
human perception in three several particulars —
namely, in an increased refinement or sensibility ; in
an extension of the sphere in which that sensibility
shall display itself; and in a power of discerning objects and their qualities through impediments which,
in our present state, form insurmountable barriers to
the action of our senses.   Now, in order to behold
these anticipated prerogatives really conferred on

human perception, we have only to behold the higher phenomena of mesmerism. Not even the commonest observer (supposing him to possess common honesty) can witness them without being struck with the conviction that a new and more pervading kind of perception has been opened to the mesmerised person; but he who should have reasoned previously on the subject, as the author from whom I have quoted, would at once perceive that every thing which was passing before his eyes was in beautiful harmony with science, philosophy, and truth. He would behold that fineness of perception — that enlargement of the mind's sphere — that power of acting through obstacles, which his own reason had told him must belong to man when the mechanical restrictions of the senses should be laid aside.

Let me, however, hasten to qualify whatever may appear exaggerated in the above statements, lest I prepare disappointment for the sanguine, or arouse the apprehensions of the timid. In pure compassion, I would appease the fears of those who deem mesmerism, if true, a perilous truth, that should be huddled up from knowledge, and, like the Bible in a Catholic land, be restricted only to the use of the initiated. Such persons (and there are such) actually dread lest the new science should make man too wise, and too powerful, and look on it with fear and trembling as a pernicious stepping-forward from our sphere, — as the very Babel-tower of this era. Let these apprehensions subside. In being permitted to view the mesmeric state, there is nothing to make man proud, but all to keep him humble, while he sees that the least step out of our present condition is an improvement upon our faculties, and that phenomena so despised, so vilified by common ridicule, are beyond the compass of our powers to achieve. Then, too, mesmerism has its restrictions, which keep it low to earth, even while it hints of heaven. Seldom does it occur in a degree to astonish. Many are the conditions it requires for its accomplishment. Carefully hedged about is it by the barriers of opposing will, by defective sensibility, even by a spirit of scepticism. Its highest capacities are exercised with difficulty, its loftiest wonders are few and fleeting, and exhaust evidently the person by whom they are displayed. Still, it is a rise upon our actual existence. Its true office is not to make us confident, yet still to encourage us to hope by the transient and imperfect display of faculties, the permanent continuance of which would be incompatible with our present state of being.

A state of such manifest development, once seen, would naturally be deemed, as I before have said, a kind of earnest of that which is to come — a corroboration of hope, a first-fruit of prerogatives, which, though temporary here, should be durable hereafter. And it would further be conceived that, as the transiently-developed faculties were to accompany man into a future state of existence, the means through which they were developed would also cleave to humanity and be assured to its possession. For, as we have seen, we do not merely behold in mesmerism

A A 5

a shadowy advantage, but a tangible benefit, a material good, working by material means; an actual physical improvement in sensation, brought about by the action of a more pervading and subtle medium than any of which we have ordinary cognisance. Hereby, then, we may be enabled to exchange the abstracted notions of theologians respecting a future state (which, like all other abstracted notions, touch the mind but feebly,) for something which may satisfy the natural longings of man after the real and the permanent, — a conclusion to which even the ideal tends.

We perceive the possibility of parting with the externals of sensation, yet retaining its inner ground; we reason from the partial and temporary abolition of the mechanical portions of our being, revealed in mesmerism, to the total and permanent abolition of those same portions of our being which is to be undergone in death. As one of my sleepwakers said to me, before I had myself conceived such a notion, "The mesmeric state is a far truer image of death than sleep is." All points in mesmerism to an invisible means of sensation, as real as it is invisible — a means of which the nerves themselves seem to be merely the subsidiaries and conducting channels; — all points in mesmerism to a finer organ of the mind than those which are palpable and external — an organ to which our visible body is but as a veil, or (as the author of the physical theory of another life calls it) a mere "husk of animal organisation." These are not, be it remembered, mere speculations.

They are all grounded on facts. Facts force on our
minds a fundamental medium in sensation; facts
identify it with mesmerism; facts lead us to infer
that it is the immediate organ of mind — the real
element in which we have our being; and facts lead
us to hope that it will serve us permanently in
another state of existence. If our feelings lead that
way, our reason not less accompanies and sanctions
their instinctive appeal. The demonstrated connection of mesmerism with the vital principle (or
action) in man; the striking manner in which sleepwakers have been shown to cling to their mesmeriser,
as if in governing the mesmeric medium he held the
talisman of their being, — as if, in touching this, he
roused all the instincts of self-preservation; the important, nay essential, offices which the medium has
been shown to fulfil in sensation, all proclaim forcibly
that the agency we have been considering may with
truth be called the fundamental life. It abounds
with instances wherein its own permanent and universal action is contrasted with the transitory and
partial operations of the organic life.

With facts like these before us, where, but to the
mesmeric medium, shall we look for indications of
that inner body — that germ of a better existence,
which analogy shows must be so intimate a part of
us, and which is nevertheless so much a stranger to
ourselves? Where, but in the medium of the fundamental sensation, shall we look for the fundamental
life; for are not, in truth, life and sensation identified in their purpose, at least, of bringing us into

sensible connection with the world of matter? This being admitted, there is no difficulty in conceiving that this action, as relates to the mesmeric medium, shall be continuous. As now it so manifestly appears to be an intermediate between us and our present body, so hereafter it may become the means of linking us to a new organisation; or it may itself remain as our spiritual body, appropriated to us in a certain portion, when this visible and fleshly tabernacle is dissolved to its primal elements. Where shall we find so obedient a servant to our commands, — where force to our strong desire after individual existence, — where an agency so swift, powerful, and penetrating, so near to our essence, so kindred to our thoughts? In proportion as we value whatever tends to bridge our way across the gulf of death, whatever tends to carry on a train of old familiar thought into the unknown void, let us esteem, cherish, and reverence this cheering manifestation of our being, which so beautifully exhibits a pre-existent harmony between our human hopes and their accomplishment. That the mesmeric medium should link science to science is comparatively but a trifling benefit. That it should connect this world with a future is its last and greatest service.

# SUPPLEMENT

## TESTIMONIES.

# TESTIMONIES.

TESTIMONY BY A. VANDEVYVER, OF ANTWERP, SURGEON.

JE, soussigné, Docteur en Médecine et en Chirurgie, certifie avoir eu le plaisir, vers le commencement de 1838, d'être invité, avec quelques uns de mes amis, à deux séances dans lesquelles M. Townshend s'est appliqué à nous démontrer les effets étonnants du magnétisme animal, auxquels jusques-là, faute de preuves personnelles, je n'avais ajouté qu'une croyance bien faible, et qui parvinrent à me donner une conviction inébranlable sur l'authenticité de tout ce que l'on rapportait de presque miraculeux à l'égard de cet agent physique.

Parmis plusieurs expériences également concluantes, je citerai entre autres celle faite sur la personne de Mademoiselle Mesdagh, que M. Townshend, après quelques passes de courte durée, parvint à plonger dans un véritable somnambulisme, et dans laquelle je pus me convaincre de la grande influence que produit le magnétisme sur les principales fonctions de notre

---

TRANSLATION.

I, the undersigned Doctor of Medicine and Surgery, certify to having had the pleasure of being invited, about the beginning of 1838, with some friends, to two sittings, in which Mr. Townshend purposed to show us the astonishing effects of mesmerism, — a matter that I had as yet given little credit to in the absence of personal observation, and which firmly convinced me of the authenticity of all that is reported, miraculous as it is, respecting this physical agent.

Amongst many equally conclusive experiments, I will adduce that made on Mademoiselle Mesdagh, whom Mr Townshend, after a few passes of short duration, threw into genuine sleepwaking, during which I had an opportunity of convincing myself of the great influence which mesmerism produces on the principal functions of our

économie, et notamment sur la circulation du sang. Cette demoiselle, avant le commencement de la séance, fatiguée d'avoir été la veille à un bal où elle avait dansé la plus grande partie de la nuit, se croyait peu propre, disait-elle, à ressentir, ce soir là, les effets du magnétisme. Aussi je lui trouvais la face colorée et toute sa personne agitée d'un véritable état de fièvre : le pouls rendait 120 pulsations par minute. Cependant l'influence magnétique ne se démentit point pour cela. M. Townshend parvint bientôt à la rendre somnambule, t à peine cet état avait-il été rendu complet, qu'un calme général s'établit chez elle, — que les battemens de l'artère de 120 étaient tombés jusqu'à 72.

Un autre phénomène qui excita ma surprise consistait dans les impressions reçues par le magnétiseur, et qui furent, au même instant, apperçues par la magnétisée. M. Townshend nous ayant un moment entretenus de ce phénomène, se mit en rapport avec la demoiselle en la tenant par la main, et il me prit l'idée de lui pincer différens points du tronc et des membres, en mettant le plus grand soin à ce que la magnétisée, qui d'ailleurs avait les yeux bandés, ne pût par aucun moyen découvrir quelles étaient les parties du corps de M. Towns-

---

economy, and particularly on the circulation of the blood. This young lady, from having been the previous evening at a ball, where she had danced the greater part of the night, was fatigued before the sitting, and said she did not think herself likely to experience the effects of mesmerism that evening ; accordingly I found her face flushed and her whole frame in really feverish excitement. Her pulse was 120 in a minute. Nevertheless the mesmeric influence did not fail to manifest itself. Mr. Townshend soon succeeded in producing sleepwaking, and scarcely was this state become complete, when a general calm came over her, so that her pulse, which had been 120, fell to 72.

Another phenomenon which excited my surprise was, that the impressions which the mesmeriser received were at the same instant perceived by the mesmerisee. Mr. Townshend, having conversed with us a short time respecting this phenomenon, placed himself in relation with the young lady by taking hold of her hand ; and the idea struck me, of pinching different parts of his body and limbs, taking, however, the greatest care that the mesmerisee, whose eyes, moreover, were bandaged, should have no means of discovering on what parts of Mr.

hend sur lesquelles je voulais faire mes expériences, et dissiper
ainsi le reste de mon incrédulité médicale ; mais, à mon grand
étonnement, je vis que chaque fois que je lui froissai tel ou tel
point, du bras par exemple, la magnétisée, tout en exprimant
par les traits de sa figure ainsi que par la parole, la douleur
qu'elle éprouvait, porta automatiquement la main qui lui
était restée libre à la partie de son autre bras, exactement cor-
respondante à celle que je venais de pincer sur le bras du même
côté du magnétiseur.

Une foule d'autres faits, non moins curieux, mais trop longues
à détailler, et dont M. Townshend eut la complaisance de me
rendre témoin, m'avaient, à la fin de la séance, procuré une
telle somme de preuves irréfragables sur les effets prodigieux
du magnétisme animal, que je me rétirai pleinement convaincu de tout ce que, ce soir là, mes yeux m'avaient si mani-
festement fait voir.

A. VANDEVYVER, Dr.

Anvers, le 26 Février, 1839.

---

Townshend's body I was making my experiments, and of thus dispel-
ling the remainder of my medical incredulity ; but, to my great astonishment, I saw that each time I bruised any particular part, of his
arm for example, the mesmerisee expressed by her features, as well as
by her speech, the pain which she experienced, and involuntarily
carried the hand which was at liberty to the part of her arm exactly
corresponding to that which I had pinched on the mesmeriser's arm
of the same side.

A host of other facts, not less curious, but too long to detail, and
which Mr. Townshend had the kindness to afford me an opportunity
of witnessing, furnished me, towards the end of the sitting, with such
an amount of undeniable proofs of the extraordinary effects of mesmerism, that I retired fully convinced of all that I had so clearly seen
that evening.

A. VANDEVYVER, Dr.

Antwerp, 26th February, 1839.

### Testimony to Mesmerism, by M. Van Owenhuysen, Antwerp.

Les effets du magnétisme m'avaient toujours semblé plus que problématiques et incertains ; je les avais souvent taxés de chimériques, et ma prévention allait si loin, que j'étais presque convaincu que les expériences magnétiques étaient le résultat d'une entente parfaite de deux personnes complètement éveillées. Aujourd'hui, mieux conseillé, et abdiquant les préventions d'un ignorant qui peut et ne veut s'instruire, j'ai eu le bonheur de reconnaître mes erreurs, et de me convaincre de nouveau que le jugement chez l'homme ne doit jamais dévancer les preuves.

Monsieur Hare Townshend ayant eu la bonté de m'inviter à une séance, dans laquelle il devait renouveller quelques expériences sur le jeune E. A——, j'acceptai avec empressement, et je veux consigner en sortant de chez lui, et sous l'impression de ce que je viens de voir, le résultat de ces expériences aussi intéressantes que réelles : en voici donc un fidèle récit.

E. A——, décidé à se soumettre à l'influence de Monsieur

---

TRANSLATION.

The effects of mesmerism had always appeared to me more than problematical and uncertain. I had often treated them as chimerical, and my prejudice went so far that I was almost convinced that mesmeric experiments were the result of a complete understanding between two persons perfectly awake ; but now, being better advised, and giving up the prepossessions of an ignorant person, who can, but will not, be instructed, I have had the happiness of seeing my errors, and convincing myself anew that man's opinion ought never to precede proofs.

Mr. Hare Townshend having had the kindness to give me an invitation to a sitting at which he was to repeat some experiments on the young E. A——, I eagerly accepted it, and, on leaving him, and under the impression of what I have just seen, am anxious to record the result of the experiments, which were as interesting as real :—here, then, is a faithful account of them.

E. A——, having consented to submit himself to the influence of

Townshend, s'assit a une heure moins huit minutes, et il se releva somnambule à une heure moins deux minutes.

Je retracerai maintenant sans commentaires les diverses expériences qui ont eu lieu, en déclarant que Monsieur Townshend m'avait autorisé d'indiquer et de faire toutes les questions et toutes les épreuves que mon imagination et ma curiosité pouvaient me suggérer ; j'ai largement usé de cette facilité, car je voulais me convaincre.

1º. Le magnétiseur prit un verre, le remplit de vin, et demanda ce qu'il contenait ; le magnétisé répondit, sans hésitation, " Du vin: " il en fut de même pour un biscuit que mangea le premier.

2º. Interrogé par où il voyait, E.'A—— répondit que c'était par le front ; il reconnaissait tous les assistants, et répondait à toutes leurs questions, après qu'il avait été mis en rapport avec eux ; ce rapport ne s'opéra que par la volonté du magnétiseur.

3º. Il lui fut présenté un petit dessin de Kremers représentant un chien : il en indiqua le sujet aussitôt qu'il l'eut en main. — Ensuite un portrait de femme, qu'il déclara ressembler à Mons. T—— : c'était celui de sa sœur ! — Interrogé s'il aurait pu dire si cette dame existait encore, — réponse, " Non." — En effet, elle était morte.

---

Mr. Townshend, sat down at eight minutes before one o'clock, and at two minutes before one got up in a state of sleepwaking.

I shall now describe, without comment, the different experiments which were made, remarking that Mr. Townshend authorised me to suggest and put any questions, and make as many trials as my imagination or curiosity could prompt. I made great use of this permission, for I wished to satisfy myself.

1st. The mesmeriser took a glass, filled it with wine, and asked what it contained ; the mesmerisee answered, without hesitation, " Wine : " the same thing occurred with a biscuit which the former ate.

2d. When asked with what part he saw, E. A—— answered that it was with the forehead he recognised all the bystanders, and answered all their questions, after having been placed in relation with them. This relation took place only through the will of the mesmeriser.

3d. A small drawing of a dog by Kremers was shown him : he declared the subject of it as soon as he had it in his hand. Then the portrait of a lady, which he said resembled Mr. T—— : it was that of Mr. Townshend's sister ! Being asked if he could say whether this lady was still living,— answer, " No." — In fact, she was dead.

4º. Je lui offris ma montre, avec invitation de m'indiquer l'heure : ma montre est à guichet. Il déclara ne rien y comprendre. Une autre montre lui fut passeé, et l'heure fut aussitôt indiquée avec justesse.

5º. On lui présenta les fables de La Fontaine, édition illustrée par Grandville ; il eut de la peine à reconnaître les animaux personnifiées des gravures, et cela devait porter confusion dans son imagination, mais il lut, sans hésiter un seul instant, et le titre du livre et les numéros indicatifs des fables ; il ne voulut cependant pas se donner la peine de chercher à lire les fables mêmes ; l'impression en était trop petite, disait-'il.

6º. J'écrivis son nom : il le lut, et rejetta le papier en disant, " Ce n'est que mon nom." Un autre nom, inscrit par un autre assistant, lui fut présenté ; il le lut aussi, mais en déclarant qu'il n'y reconnaissait pas mon écriture. J'inscrivis un nouveau mot, il reconnut ma main. Toutes les précédentes expériences ont été faites sans couvrir les yeux et le front du magnétisé ; ici Monsieur T—— les lui couvrit complètement d'un mouchoir, et les étonnantes épreuves suivantes eurent lieu.

7º. Conduit devant une glace, E. A—— déclara se voir ; re-

---

4th. I presented my watch to him, and desired him to tell the hour. My watch is double-cased, with a small central glass. He declared he could not understand it at all. Another watch was passed to him, and immediately he told the hour correctly.

5th. The illustrated edition of the fables of La Fontaine, by Grandville, was given him ; he could scarcely recognise the engravings of the animals, and this might have confused his thoughts ; but he read, without hesitating for an instant, both the title of the book, and the numbers referring to the fables ; yet he would not give himself the trouble of trying to read the fables themselves : the impression, he said, was too small.

6th. I wrote his name ; he read it, and threw the paper away, saying, " That is only my name." Another name, written by another bystander, was presented to him ; he read it also, declaring that he did not recognise my writing in it. I wrote a new word, and he recognised my hand. All the preceding experiments were made without covering the eyes and forehead of the mesmerisee. Mr. T—— now covered them completely with a handkerchief, and the following astonishing facts took place : —

7th. Being led before a glass, E. A—— declared he saw himself :

connut et désigna, en les nommant successivement, sans aucune hésitation, les assistans qui vinrent l'entourer, en indiquant les changements de leurs positions, et suivant du front tous leurs mouvemens.

8°. Je pris ma bourse ; je la serrai dans la main de manière à ce qu'il ne put rien reconnaître de ce que je tenais ; il déclara cependant, après un peu d'hésitation, que c'était ma bourse, et qu'elle renfermait de l'argent. Je renouvellai cette expérience avec un morceau de pain ; il indiqua bien vaguement ce que je tenais, mais il ne parvint pas à le préciser.

9°. Le magnétiseur lui donna du papier de musique. E. A—— composa, avec un aplomb admirable, et sans hésiter, deux différens motifs de musique pour la flute ; il les signa, et les exécuta sur son instrument, avec exactitude et précision. Il offrit l'un de ces motifs à Monsieur T——, et l'autre à moi. Je garde celui-ci comme un souvenir réel, irrévocable, de cette intéressant séance.

Pendant qu'il composoit, E. A—— répetait de la tête, du pied et de la bouche, tous les mouvemens d'une personne qui compose et qui compte la mesure.

Le magnétisé a été réveillé à 3 h. 5 m. de relevée ; il ne con-

---

he recognised, pointed out, and named the bystanders successively, without any hesitation, mentioning the changes of their position, and following all their movements with his forehead.

8th. I took my purse, and squeezed it up in my hand in such a manner that he could not tell what I was holding ; he declared, nevertheless, after a little hesitation, that it was a purse, and that it contained money. I repeated this experiment with a morsel of bread ; he described very vaguely what I held, and could not arrive at a knowledge of the precise thing.

9th. The mesmeriser gave him some music paper. E. A—— composed with admirable accuracy, and without hesitation, two different airs for the flute ; he signed them, and executed them on his own instrument, with exactness and precision. He offered one of these airs to Mr. T——, and the other to me. I keep it as a true and irrevocable remembrance of that interesting sitting.

While he was composing, E. A—— made, with his head, foot, and mouth, all the movements of a person who is composing and counting the time.

The mesmerisee was awakened at five minutes past three. On

servait aucun souvenir apparent de ce qui s'était passé, ou de ce qu'il avait fait pendant son somnambulisme.

JEAN VAN OWENHUYSEN.

Anvers, Mars, 1838.

---

waking he had, apparently, no recollection of what had happened to him, or of what he had done during his sleepwaking.

JEAN VAN OWENHUYSEN.

Antwerp, March, 1838.

## Testimony. (Mesmeric Vision.) Dr. Foissac.

Vous êtes plus heureux que moi ; je n'ai rien écrit depuis un siècle, tandis que vous, vous publiez un ouvrage sur le magnétisme, qui je n'en doute pas sera de nature à exciter à la fois l'intérêt des savants et des gens du monde. Si tous vos documens sont semblables à ceux qui concernent le jeune E. A——, ils feront certainement sensation. De tous les somnambules que j'ai vus, c'est l'un de ceux qui m'a plus attaché, et je dois dire étonné. C'est de grand cœur que je vais rappeler quelques uns des faits dont j'ai été témoin.

La vue sans le secour des yeux est l'un des phénomènes que j'ai le plus rarement rencontrés dans la pratique du magnétisme, aussi vous témoignai-je le vif désir de vérifier par moi-même les observations que vous aviez faites sur le jeune E. A——.

Vous eûtes la complaisance de le faire venir à Paris le printems dernier, et dans le peu de jours qu'il y resta, je pus m'assurer que vous n'aviez rien exagéré dans le récit que vous m'aviez fait des étonnantes facultés de cet intéressant somnambule. Je me rappelle parfaitement que dans les deux ou

---

### Translation.

You are more fortunate than myself. I have written nothing for an age, whereas you are publishing a work on mesmerism, which I doubt not will interest both learned men and the public at large. If all your documents are like those which relate to the young E. A——, they will certainly create a sensation. Of all the sleepwakers that I have seen, this is one of those who have the most interested, indeed I ought to say astonished, me. It is with great pleasure that I proceed to report some of the facts which I witnessed.

Vision, without the assistance of the eyes, is one of the phenomena that I have the most rarely met with in the practice of mesmerism ; you therefore found me exceedingly desirous of verifying the observations that you had made on young E. A——.

You had the kindness to bring him to Paris last spring, and, during the few days that he remained, I was able to assure myself that you had exaggerated nothing in the report that you had made of the astonishing faculty of that interesting sleepwaker. I remember perfectly

trois séances où je le vis pendant le sommeil magnétique, nous lui appliquâmes un bandeau sur les yeux, nous bouchames tous les interstices avec des linges fins et du coton, de manière à empêcher tout rayon lumineux de parvenir jusqu'à l'organe visuel, et cependant, malgré ces précautions, E. A—— distinguait les cartes qui lui étaient présentées, indiquait, sans hésitation, dans un livre ouvert au hasard, la lettre majuscule qui commençait un chapitre, le nombre des lignes d'une page, et même à la fin il pouvait lire quelques phrases.

Il n'avait pas toujours besoin de présenter le livre vis-à-vis des yeux ; il lui arrivait souvent de placer l'objet soumis à son exploration à la partie supérieure du front, qui semblait alors devenir l'organe de la vision, et toutes les indications étaient justes et précises. Chacun des assistants voulait à son tour faire une expérience ; sans le prévenir, et sans contact aucun, on lui présentait une carte à quelques pouces du sommet de la tète ; E. A—— levait brusquement le bras pour saisir la carte, et jamais sa main ne portait à faux.

Une fois, quelqu'un promenait sans rien dire le journal *l'Entre-acte* derrière la tête du jeune somnambule, et celui-ci dit, " C'est *l'Entre-acte.*" Il indiqua aussi l'heure de ma

---

that, at the two or three sittings when I saw him during the mesmeric sleep, we applied a bandage over his eyes, we stopped up all the interstices with fine linen and cotton, in such a manner as to prevent a single ray of light from arriving at his visual organs ; and yet, notwithstanding these precautions, E. A—— distinguished the cards which were presented to him, and pointed out, without hesitation, in a book opened at random, the capital letter which commenced the chapter, the number of lines in a page, and even towards the end he could read some passages.

It was not always necessary to place the book opposite his eyes ; he frequently placed the object submitted to his inspection at the upper part of his forehead, which seemed then to become the organ of vision, and all his remarks were just and acute. Each of the bystanders in his turn made an experiment. Without any hint, and without any contact, each placed a card at the distance of a few inches from the top of his head ; E. A—— abruptly raised his arm to seize the card, and his hand never missed its aim.

Once, somebody, without saying anything, presented the journal *Entre-acte* behind the head of the young sleepwaker, and he said, " That is *l'Entre-acte.*" He also told the hour by my watch, and

montre, et il renouvela cette épreuve avec le même succès, lorsque j'en eus à dessein dérangé les aiguilles. Ainsi je ne doute pas E. A—— voyait à la fois par toutes les parties de la tête ; il se levait de son siège, courait avec vîtesse, évitant très-bien les obstacles, et atteignant le but qu'il se proposait.

En finissant cette courte relation, je ne veux pas omettre l'un des faits de vision qui me frappa le plus vivement chez ce somnambule. Sur la table à côté de laquelle il était assis, il y avait une multitude de verres diversement coloriés ; E—— en saisit un, le porte à la partie supérieure du front en faisant le geste d'une personne qui regarde avec un lorgnon, et dit aussitôt, " Je vois tout bleu." C'était en effet un verre de cette couleur dont il faisait usage.

FOISSAC.

Paris, le 18 Mars, 1839.

---

repeated this proof with the same success when I had intentionally altered the hands. I have consequently no doubt that E. A—— saw at once with all parts of his head : he rose from his seat, ran quickly, avoiding obstacles very well, and reached the place he intended.

In concluding this short account, I cannot omit one of the facts of vision which struck me most forcibly in this sleepwaker. On the table, by the side of which he was sitting, there were a number of different coloured glasses; E—— took one, raised it to the upper part of his forehead, imitating the action of a person looking with an eyeglass, and immediately said, " I see all blue." It was in fact a glass of that colour that he had taken up.

FOISSAC.

Paris, 18th March, 1839.

B B

TESTIMONY BY VISCOUNT N——, RELATIVE TO MESMERIC
VISION (BEFORE E. A—— WAS FAR ADVANCED IN PER-
CEPTION).

ON the 26th July, 1838, I saw E. A—— to all intents and
purposes asleep, his eyes quite shut, and his countenance quite
in keeping with this state.   To make sure of his not seeing,
supposing he was shamming (I say this for security, not that
I doubted his being asleep), I bandaged his eyes with a large
silk handkerchief; to make more sure, the space between his
nose and the temples was filled with cotton ; and then, and
not till then, cards were produced, and he played a game of
" beggar my neighbour" with Mr. Townshend.   Notwithstanding all Mr. —— and I did by pinching E——, and talking
and laughing, he never made a mistake, and always detected
Mr. Townshend's faults without the least hesitation, but exactly as if his eyes were as unincumbered as those of the rest
of the party.   The next day I saw E. A—— put into the
same state of sleepwaking.   I then went into an armoire,
about five feet high, and just room enough for two persons to
stand in.   The door was shut, that I might be satisfied there
was no light admitted.   When I came out, Mr. Townshend
took E—— in, and I gave them different cards, some of which
E—— read ; also some letters on separate bits of paper; then,
for fear he might get time to see while the door was opened
to give in the cards, I wrote on a piece of paper, dog, cat, or
some such word, and drew a monster figure, and folded up the
paper like a note.   He could not read the words, but mentioned most of the letters, and their position on the paper, and
laughed at the figure, but could not make it out ; this was
enough to prove he saw in the dark.   I afterwards took a card
off the table, without seeing it myself, or letting any one else
see it.   I presented *the back of it to the back of his head*; he
said " he could not read it so :" it was a French card, all over
little stamped flowers, or other marks.   I then turned *its face
to the back of his head,* and he told me what it was.   I particularly remarked the change of his countenance from a tired

heavy look to his usual cheerfulness, on waking. E. A——
has told me cards placed in or *shut up* in a book. I have also
seen him write music, nib his pen, put his flute together, and
play as well as usual in a state of sleepwaking.

Viscount ————

Rome, 11th February, 1839.

Testimony (Mesmeric Vision, etc.) by the Baron de
Carlowiz. Berne, 1838.

Conformément à la plus stricte vérité, je, soussigné, atteste
le fait suivant : —

Mr. Hare Townshend avait chez lui un jeune homme,
d'environ 16 ans, qu'il avait magnétisé depuis quelque tems, et
qui se trouvait, à ce que j'avais ouï dire, en état de clair-voyance.
Quoique j'eusse lu plusieurs ouvrages sur ce sujet, et bien que
Mr. Townshend qui, ennemi de toute sorte de charlatanisme,
se distingue par une érudition profonde autant que par la vérité
de son caractère, m'inspirât de la confiance, mon incrédulité
à l'égard des merveilles du magnétisme animal était néan-
moins encore assez grande.   C'est pourquoi  je résolus d'ap-
porter une attention minutieuse à tout ce que je verrais pendant
que ce jeune homme serait en sommeil magnétique.   Mr.
Townshend l'y mit en ma présence.   L'opération même ne
durait que dix minutes, et le magnétiseur usait de peu de manipulations, qui pourtant faisaient une impression très-visible
sur le magnétisé.   Celui-ci eut enfin les yeux fermés, et répondit aux questions qu'on lui faisait par l'entremise de son

---

TRANSLATION.

In conformity with the strictest truth, I, the undersigned, attest
the following facts : —

Mr. Hare Townshend had at his house a young man, about sixteen
years of age, whom he had mesmerised for some time, and who had
become, as I was told, clair-voyant.   Although I had read several
works on this subject, and Mr. Townshend, who is opposed to every
kind of charlatanism, and distinguished by profound erudition, as well
as by truth of character, inspired me with confidence, my incredulity in
respect to the wonders of mesmerism was still very great.   I therefore
resolved to give close attention to every thing I might see during the
time this young man remained in the mesmeric sleep.   Mr. Townshend
sent him to sleep before me.   The operation continued only ten minutes, and the mesmeriser made but few manipulations, which, nevertheless, produced a very visible impression on the mesmerised.   His
eyes at length became closed, and he answered questions which we

magnétiseur. Il distinguait tous les objects comme s'il avait
les yeux ouverts, ce qui m'étonnait fort peu, parceque je n'étais
pas encore délivré de mes anciens soupçons.

Par cette raison j'étais bien curieux d'observer des preuves de
clair-voyance après que ce jeune homme fût mis dans l'impossi-
bilité parfaite et incontestée d'ouvrir les paupières. Pour cet
effet on alla lui bander les yeux. Ayant d'abord bien examine
le mouchoir qui devoit servir de bandage, je contribuai mon
possible pour le bien appliquer et serrer, et afin qu'aucun vide
ne restât entre le bandage et les paupierès, ce qui lui aurait
permis de lever les yeux, on y mit du coton. Mais je fus bien
surpris lorsqu'il voyait, malgré ces appareils, aussi-bien qu'au-
paravant. Je choisis parmi un grand nombre de livres qui se
trouvaient dans le salon, et désignai les passages qu'il devait
lire, ce qu'il exécutait avec vîtesse, et sans se tromper. Il
voyait aussi une pétrifaction que j'avais saisie clandestinement,
et que je tenais dans la main serrée. On lui montrait derrière
la tête, et à plusieurs réprises, l'étui qui renfermait sa flûte, et
il s'appercevait toujours si la flûte y était, ou si on l'en avait
ôtée.

J'ai assisté deux fois à l'acte où Mr. Townshend magné-

---

put to him through the medium of the mesmeriser. He distinguished
every object as if he had his eyes open, which surprised me very
little, for I was not yet freed from my old suspicions.

For this reason I was very curious to observe the proofs of *clair-
voyance*, after the young man was so placed that it was perfectly and
incontestably impossible for him to open his eyelids. For this purpose we proceeded to bandage the eyes. Having first carefully examined the handkerchief which was to serve as a bandage, I did all
I could to apply it effectually and fix it; and, to prevent any space
from being left between the bandage and the eyelids, which would allow
him to open his eyes, it was stopped with cotton. But I was much
surprised when he saw, notwithstanding these bandages, as well as
before. I chose a book from among a great number that were in the
room, and pointed out passages for him to read. This he did with
rapidity, and without mistake. He also saw a petrifaction which I
had taken secretly, and held in my closed hand. We repeatedly
showed him behind his head the case which contained his flute, and
he always perceived whether the flute was there, or had been taken
away.

I was twice present when Mr. Townshend mesmerised this young

tisait ce jeune homme, et j'ai fait à peu près les mêmes re-
marques.  La seconde fois, Mr. Townshend lui banda les yeux
dans la manière indiquée avant de le magnétiser.  Du reste,
il m'a paru que le magnétisé faisait tous ces expérimens avec
une certaine répugnance, et qu'il aurait plutôt préféré de rester
dans une situation tranquille et passive, attaché à son magnétiseur.

En foi de quoi, j'ai muni le présent certificat de ma signature.

GUSTAVE ADOLPHE, BARON DE CARLOWIZ.

Berne, le 13 Févr. 1839.

---

man, and I made nearly the same observations.  On the second occasion, Mr. Townshend bandaged his eyes in the manner described
before mesmerising him.

It appeared to me that the mesmerisee submitted to all these experiments with an evident repugnance, and that he would have preferred
remaining in a tranquil and passive situation close to his mesmeriser.

In testimony to which, I have signed the present certificate with
my name.

GUSTAVE ADOLPHE, BARON DE CARLOWIZ.

Berne, 13th February, 1839.

## Testimony to Mesmerism. By a Friend.

### *Credo* versus *Non-credo*.

" I think myself justified in suspending my assent to the facts related of mesmerism," said I one day to my friend C——, " till some opportunity of observing them *in propriâ personâ* shall offer itself."

" You shall have that opportunity this very morning," he replied ; " and it will be the best answer I can give to your arguments." So saying, C—— proceeded to an adjoining room, taking with him a youth about fifteen years of age, whom he had been in the habit of mesmerising for some weeks past ; he told me to remain where I was till he should call me. In about twenty minutes C—— called. On entering, I found them sitting *vis-à-vis ;* C—— held a book in his hand, and a pack of cards was placed by his side ; the boy (whose name was E. A——) had his eyes tightly bandaged with a pocket-handkerchief, which nearly enveloped his face. I was struck with the change which so short an interval had wrought in them both ; the mesmeriser looking rather wild and exhausted, whilst the patient before him, without seeming exactly asleep, yet had an air of such perfect resignation, abandonment, and even prostration, as made me think that he was without consciousness, and that some great, and to me unknown, influence was operating upon him. C—— then took up a card at hazard, begging me to observe accurately that he held it with its back always turned towards E. A——, and placing it behind a large octavo, he turned it again with its face towards E. A——, keeping the open volume exactly between it and the bandage. I positively affirm that there was no possibility of the boy's seeing the card during its whole progress, or of his being assisted either by the direction of light, or any mirror opposite. Now came the critical moment, and I had no doubt whatever of convincing my friend that he had been living under a delusion. C—— asked his waking-sleeper, or somnambule (as they term it), to tell him the number and suit of the card — to me the most unreasonable question possible, considering A——'s state of incapacity. The latter, without seeming to relish this request, bent a little forward, so as to bring his forehead nearly opposite the back of the book ; but remained silent, till, on C——'s gently chiding him for his slowness, he named the card quite correctly, to my no small astonishment and utter confusion, having kept strict watch the whole time to see that the experiment was properly conducted.

Without betraying my thoughts, I then asked my friend to repeat the experiment, and selected a card which he himself did not know. Upon this the patient showed signs of great uneasiness, and complained of fatigue. I took care that the same precautions should be taken. The sleepwaker, this time, showed a little obstinacy, declaring he could not tell *that* card, as he was sure it would be followed by another ; and in this he guessed rightly. The mesmeriser, however, becoming a little furious, urged his patient to tell the card instantly, as a third person was present for the express purpose of witnessing the experiment and making a report of it to a learned society. Upon this the patient turned full towards me, and looked as if he could stare at me through his bandage ; then he said that he could not see very clearly, but thought the card was the *six* of spades. In this, however, he was mistaken, it being the eight of spades ; but the approximation was too great to admit of my expressing any disappointment at this failure. I therefore only begged for a third trial, beginning now to feel a most lively interest, in spite of my previous incredulity, in what was taking place before me. C—— consented ; so pulling out a card from the middle of the pack, I carried it unseen, with its face downwards, to the book, and carefully turned it ; and after holding it there a few seconds, C—— asked E. A—— to tell him the card accurately. "The ten of clubs," cried the latter, with a tone of confidence and triumph that might have confounded a greater sceptic than myself ; — that was indeed the card that C—— held in his hand ! !

What could I say to this ? To doubt any longer would be doubting my own senses, — would, in short, be madness, for I could in no way justify my doubts ; yet to believe, on the other hand, would be another kind of madness, unless saying that I witnessed a person name a card which was intercepted by a solid opaque substance deserves to be called by any other name.

" The deuce is in it if you are not convinced *now*," said
my friend, turning about, and displaying a visage truly Me-
phistophelian.

" I *am* convinced that the deuce is in it, that's all," cried
I ; and forthwith tried to make my escape, but I found the
door locked.

" Will you have the goodness to wait a few minutes
longer ?" said C——, still insulting me by a look of the most
cruel satisfaction.

" Will you have the goodness to let me out of this illusive
atmosphere ? I will have none of it,"—I retorted.

" Come this way, then," said he, opening a door, through
which I hurried, bent only on my escape. He locked it upon me,
and left me, not as I had hoped to find myself, in the garden,
but in a dark, narrow, dismal closet, darker than the darkest
corner of a dungeon.

" What do you see in there ? " ironically asked my tor-
mentor.

" ' Blue spirits and white ;
Black spirits and grey,' —

all the spirits of darkness," I exclaimed.

" You see no spirits of LIGHT, then, it seems ? "

" No ! " groaned I : " let me out."

Satisfied with my answer, C—— opened the door, and
wished to make me believe that his sleepwaker could SEE in
that place.

Having consented to witness this new experiment, C——
took the boy A—— into the closet. I handed the former
some cards, unknown to him, which he held in his hand, concealed from E. A——, till I had locked and leant against the
door. This being duly accomplished, I desired that one of
the cards should be presented to the patient. It was done,
and in about one minute he named the card, *correctly,* and without the slightest hesitation. My astonishment, notwithstanding,
was as great as ever. " What ! " thought I ; " was I to be thus
ousted out of my conviction, that mesmerism was a thorough
humbug — a system of refined deception, (this was the straw
to which I vainly clung,) of the most puerile buffoonery, of
the lowest, most shameful, vulgar, highway charlatanism ? "

B B 5

And yet how could I avoid the humiliation of being at length convinced, after so long, obstinate, and, as I thought, rational a resistance. As a last hope, I determined to re-enter the closet, and investigate more closely. I did so, and the door was locked upon me. At first there seemed to flit before my eyes the indistinct forms of objects, which had attracted my notice in the room; these impressions became gradually fainter, till at length I was enveloped in perfect darkness; all endeavour to see the cards I had taken with me was fruitless. I poked into every corner, strained to catch the slightest ray of light, but it was almost *genuine* darkness — impenetrable obscurity to *ordinary* vision. I was glad to be liberated, and to let the mesmeriser and his sleepwaker take my place. I now waited in actual *fear*, lest the cards should be again discovered by E. A———, and stood listening in cruel suspense. After a few minutes the sleepwaker cried out that he held in his hand the knave of spades. On opening the door, I found that it was so. This experiment was repeated *three* times successfully; and I was at length obliged to congratulate my friend on the dexterity shown by his *clair-voyant* pupil; but I felt ashamed of using the term in a sense so different to that I had always attached to it, more especially when I remembered my scoffing incredulity of that very morning, which nothing but the most genuine and convincing proofs could have caused to give way.

D. C.

Berne, 1838.

TESTIMONY RELATIVE TO MESMERIC VISION. BY DR. WILD,
OF BERNE, WHEN E. A—— HAD GREATLY ADVANCED
IN CLAIR-VOYANCE.

J'ATTESTE par la présente d'avoir assisté, le 27 Sept. 1838,
à des expériences magnétiques, faites, par Monsieur Hare
Townshend, sur la personne d'E. A——, agé de 15 ans, né à
Boom, en Belgique, musicien distingué.

Quoique convaincu de l'existence de l'influence morale d'une
personne sur une autre, que l'on nomme ordinairement magnétisme animal, tant par des propres expériences sur mes malades, que par la complaisante amitié de feu M. Deleuze, à
Paris, je voulus pourtant procéder dans cette occasion comme
incrédule. Pour m'assurer personnellement de la réalité des
faits, je fis les expériences suivantes : —

1°. Je m'assis sur une chaise, tournant le dos à la fénêtre,
et me fis mettre sur la tête deux serviettes de toile épaisse,
qui me pendaient autour du corps jusqu'aux hanches ; ces
serviettes m'empechèrent de voir ma main, que je passai plusieurs fois devant mes yeux ; je n'en distinguai rien du tout.

2°. Je me fis fermer les yeux par les mains de M. Towns-

TRANSLATION.

I hereby certify to having been present, on the 27th of September,
1838, at some mesmeric experiments made by Mr. Hare Townshend
on the person of E. A——, aged fifteen years, born at Boom, in Belgium, a distinguished musician.

Although convinced of the moral influence of one person over another, which is commonly called animal magnetism, as much by my
own experiments upon my patients, as by the kindness of the late M.
Deleuze, at Paris, I wished, nevertheless, to proceed on this occasion
as if incredulous. To assure myself personally of the reality of the
facts, I made the following experiments : —

1. I sat down in a chair, with my back to the window, and put
upon my head two thick napkins, which hung around my body as
low as the hips ; these napkins prevented me from seeing my hands,
which I passed several times before my eyes. I could not distinguish
them at all.

2. I had my eyes closed by Mr. Townshend's hands, either by his

B B 6

hend, soit par les doigts serrés, soit par la pomme de ses mains, et mes yeux ne distinguèrent rien du tout.　M. Townshend plaça, avec le même résultat sur mes yeux, deux petits gobelets en porcelaine (qui servent pour baigner les yeux) remplis de coton.

3°. Je me fis enfermer dans une armoire vide, et, malgré mes recherches, je ne pus découvrir aucune ouverture, qui aurait laissé entrer un seul rayon de lumière.

Nous fîmes alors entrer le jeune A——.　M. Townshend le mit en sommeil magnétique, et lui adressa quelques questions pour me convaincre de son isolement magnétique.　Je lui dis inopinément à très-haute voix quelques mots, mais le calme parfait de sa mine me prouva assez, qu'il n'en entendit rien du tout.　Quelques minutes après, je me mis en rapport avec lui, et depuis ce moment il me répondit toujours.　Je l'assis, et lui posai les mêmes deux serviettes sur la tête, de la même manière qu'à moi-même un quart d'heure avant.　Je lui donnai dans la main un carte à jeu, que je tirai moi-même d'un jeu, placé sur une table deux pas de distance de nous;— après l'avoir tenue quelques moments dans la hauteur du front, de sorte que la plaine de la carte formait avec la partie supé-

---

fingers, brought into close contact with each other, or by the palms of his hands, and my eyes could discern nothing.　Mr. Townshend placed two small porcelain eye-goblets, such as are used for bathing the eyes, filled with cotton, on my eyes with the same effect.

3. I allowed myself to be shut up in an empty wardrobe, and, notwithstanding every effort, I could not discover any opening which would allow a single ray of light to enter.

Young A—— was then brought in.　Mr. Townshend threw him into mesmeric sleep, and put some questions to him for the purpose of convincing me of his mesmeric insulation.　I suddenly addressed a few words to him in a very loud voice; but the perfect quietude of his countenance fully convinced me that he heard nothing.　Some minutes afterwards I placed myself in relation with him, and from that moment he invariably answered me.　I set him down, and placed the same two napkins on his head just as they had been placed on myself a quarter of an hour before.　I gave him a playing card which I had myself taken from a pack lying on a table at two paces distant from us.　After holding it a few moments to the top of his forehead, so that the flat surface of the card formed with the upper part of his

rieure de son front un angle d'à peu près quarante-cinq dégrés, il me dit exactement les couleurs et figures imprimées dessous ; je changeai trois fois la carte avec une autre, avec le même résultat satisfaisant.

Alors je pris un livre, ouvrai une page, et lui indiquai avec le doigt une ligne qu'il devait lire, ce qu'il fit avec la même facilité ; je lui présentai un volume d'un autre ouvrage, avec le même résultat. Après lui avoir ôté les serviettes, M. Townshend lui ferma les yeux avec les mains, et plus tant avec les gobelets à bains d'yeux, remplis de coton, exactement de la même manière comme il l'avait fait à moi ; et le jeune A—— lisait dans les livres, et distingua les cartes à jeu que je lui présentai, et que je tirai toujours au hazard d'un jeu de cartes complète. Je l'enfermai enfin dans la même armoire ; il ne s'y trouvait qu'une robe de chambre ; il s'y assit dans un coin, me pria de la bien envelopper avec la robe de chambre, disant que plus il y avait obscurité complète autour de lui, mieux il pourrait voir, puisque les rayons de lumière partirent de son cerveau. Je lui donnai trois autres cartes à jeu, je fermai l'armoire à clef, et bouchai le trou de la clef avec la pomme de ma main ; dans peu de temps il me nomma les cartes. Alors M. T. me pria d'écrire quelque

---

forehead an angle of nearly forty-five degrees, he told me exactly the colours and figures printed on its underside ; I changed the card for another three times, and always with the same satisfactory result. I then took a book, opened a page, and pointed out a line which I wished him to read. This he did with the same facility. I then gave him a volume of another work with the same result. After having removed the napkins, Mr. Townshend closed his eyes with his hands, and afterwards with the goblets for bathing the eyes, filled with cotton, exactly in the same manner as he had done to me ; and young A—— read the books, and distinguished each playing card which was presented to him, and which I took always at random from a complete pack. I then shut him up in the same wardrobe, in which there was nothing but a morning-gown. He sat down in a corner, and begged me to cover him up well with the morning-gown, saying that, the more complete the darkness around him, the better he could see, because the rays of light came off from his brain. I gave him three other playing cards, and locked him up in the closet, stopping the key-hole with the palm of my hand ; in a short time he told me the cards. Mr. Townshend then begged me to write something on paper.

chose sur un papier. Je m'éloignai dans un autre coin de
la chambre, et j'écrivis sur un morceau de papier épais —
" Vous êtes complaisant aujourd'hui." Je pliai le papier deux
fois, le lui donnai, et fermai vîte de la même manière l'armoire
et le trou de la clef. Quoique le premier mot lui était bien
nouveau, il le lut exactement, ainsi que le reste du billet.
Sorti de l'armoire, je le priai de lire avec l'occiput. (Jusqu'a-
lors, il avoit toujours tenu les cartes et livres de la manière
décrite ci-dessus devant le front.) Je melai le jeu de cartes,
l'étendai sur une table éloignée de quelques pas de lui, en ayant
soin que le côté imprimé soit tourné vers la table. E. A——
s'approcha alors de la table, en marchant à réculons, prit une
carte, la plaça sur son occiput, et la nomma, en tournant toujours le dos à la table d'où il l'avoit prise. Je trouve superflu
de dire, que M. Townshend magnétisait le jeune A—— de
temps en temps, entre les différentes expériences, pour entre-
tenir la clair-voyance.

Si jamais expériences magnétiques ont été faites avec exactitude et méfiance, je crois que l'on peut mettre celles-ci en
première ligne.

CHAS. WILD, Docteur Médecin.

Berne, 10 Février, 1839.

---

I went away into another corner of the room, and wrote on a slip of
thick paper — " You are obliging to-day." I folded the paper twice,
gave it him, and quickly closed the closet and the key-hole in the
same manner. Although the first word was quite new to him, he
read it correctly, as well as the rest of the note.

On coming out of the closet, I begged him to read with the back
of his head (for hitherto he had always held the cards and books before
his forehead in the manner above described). I mixed the pack of
cards, and spread them on a table a few paces distant from him, taking
care that the printed side should be turned towards the table. E. A——
then approached the table, walking backwards, took a card, placed it
on the back of his head, and named it, keeping his back constantly
towards the table whence he had taken it. It is unnecessary to say,
that Mr. Townshend mesmerised young A—— from time to time,
between the different experiments, to keep up the lucidity.

If ever mesmeric experiments were made with accuracy and distrust, I believe that these may be placed in the first rank.

CHARLES WILD, Doctor of Medicine.

Berne, 10th February, 1839.

ACCOUNT OF SENSATIONS UNDER MESMERISM. BY PROFESSOR AGASSIS, OF NEUFCHATEL, SWITZERLAND.

*Notes relatives au Magnétisme Animal, 22 Février, 1839, au matin.*

Désireux de savoir à quoi m'en tenir sur le magnétisme animal, je recherchai depuis long-temps l'occasion de faire des expériences à ce sujet sur moi-même, afin d'éviter les doutes qui pourraient exister sur la nature des sensations que l'on entend rapporter par des personnes magnétisées. M. Desor, dans une course qu'il fit à Berne, engagea hier M. Townshend, qui l'avait magnétisé antérieurement, à l'accompagner à Neuchâtel, et à tenter de me magnétiser. Ces messieurs arrivèrent hier par le courrier du soir, et me firent prévenir. A huit heures j'étais auprès d'eux. Nous soupâmes jusqu'à neuve heures et demie, puis vers dix heures M. Townshend commença à opérer sur moi. Assis vis-à-vis l'un de l'autre, il ne fit d'abord que me prendre les mains en me regardant fixement. J'étais fermement résolu à arriver à la connaissance de la vérité, quelle qu'elle fût ; aussi le moment où je vis qu'il

---

TRANSLATION.

*Notes relating to Mesmerism, the morning of 22d February, 1839.*

Desirous of knowing what to think of mesmerism, I for a long time sought for an opportunity of making some experiments in regard to it upon myself, so as to avoid the doubts which might arise on the nature of the sensations which we have heard described by mesmerised persons. M. Desor, yesterday, in a visit which he made to Berne, invited Mr. Townshend, who had previously mesmerised him, to accompany him to Neuchâtel and try to mesmerise me. These gentlemen arrived here with the evening courier, and informed me of their arrival. At eight o'clock I went to them. We continued at supper till half past nine o'clock, and about ten Mr. Townsend commenced operating on me. While we sat opposite to one another, he, in the first place, only took hold of my hands and looked at me fixedly. I was firmly resolved to arrive at a knowledge of the truth, whatever it might be ; and, therefore, the moment I saw him endea-

tentait d'exercer une action sur moi, eut-il quelque chose de
solennel ; je m'addressai en silence à l'Auteur de toutes choses,
pour lui demander de me donner la force de résister à l'entraîne-
ment, d'être consciencieux vis-à-vis de moi-même comme
vis-à-vis des faits ; puis je regardais M. Townshend, attentif
à ce qui se passait. J'étais dans une position très-commode ;
l'heure peu avancée, à laquelle j'ai l'habitude de travailler,
était loin de me disposer au sommeil ; j'étais assez maître de
moi pour n'éprouver aucune émotion, et pour réprimer tout
élan d'imagination si j'eusse été moins calme. Aussi fus-je
assez longtemps sans éprouver aucun effet de la présence vis-
à-vis de moi de M. Townshend. Cependant, après un quart
d'heure au moins, je sentis comme un courant dans tous mes
membres, et dès ce moment ma paupière s'appésantit. Je vis
alors M. Townshend tendre ses deux mains devant mes yeux,
comme s'il eut voulu y enfoncer ses doigts, puis opérer divers
mouvemens circulaires autour de mes yeux, qui m'appésan-
tissaient toujours plus les paupières. J'avais par devers moi
le sentiment qu'il cherchait à me faire fermer les yeux, et
cependant ce n'était point comme si l'on eut menacé mes
yeux à l'état de veille, et que je les eusse fermés pour éviter
un attouchement ; c'était un poids insurmontable des paupières

vouring to exert an action upon me, I silently addressed the Author
of all things, beseeching him to give me power to resist the influence,
and to be conscientious in regard to myself as well as in regard to the
facts. I then fixed my eyes upon Mr. Townshend, attentive to whatever passed. I was in very suitable circumstances ; the hour being
early, and one at which I was in the habit of studying, was far from
disposing me to sleep. I was sufficiently master of myself to experience no emotion, and to repress all flights of imagination, even if I
had been less calm ; accordingly it was a long time before I felt any
effect from the presence of Mr. Townshend opposite me. However,
after at least a quarter of an hour, I felt a sensation of a current
through all my limbs, and from that moment my eyelids grew heavy. I
then saw Mr. Townshend extend his hands before my eyes, as if he were
about to plunge his fingers into them ; and then make different circular movements around my eyes, which caused my eyelids to become
still heavier. I had the idea that he was endeavouring to make me
close my eyes ; and yet it was not as if some one had threatened my
eyes, and, in the waking state, I had closed them to prevent him ; it
was an irresistible heaviness of the lids which compelled me to shut

qui me les faisait clore, et peu à peu je sentis que je n'avais
plus la force de les tenir ouverts ; mais je n'en conservais pas
moins la conscience de ce qui se passait autour de moi ; ainsi
j'entendais M. Desor parler à M. Townshend, je comprenais
ce qu'ils se disaient, j'entendais ce qu'ils me demandaient
comme étant éveillé, mais je n'avais pas la force d'y répondre.
J'essayai plusieurs fois inutilement de le faire, et lorsque j'y
parvenais, je sentais que je sortais de l'état de torpeur où je
m'étais trouvé, et qui m'était plutôt agréable que pénible.

Dans cet état, j'entendis le guet crier dix heures, puis sonner
dix heures et quart ; mais plus tard je me trouvais dans un
sommeil plus profond, quoique je n'aie jamais perdu entière-
ment connaissance. J'avais le sentiment que M. Townshend
cherchait à m'endormir complètement ; mes mouvemens me
paraissaient lui être soumis, car je voulus plusieurs fois changer
la position de mes bras, sans avoir assez de force pour le faire
ou pour le vouloir réellement, tandis que je sentais ma tête se
porter à droite ou à gauche sur mon épaule, et en arrière ou
en avant, sans que je le voulusse, et même malgré la résistance
que je cherchais à opposer, et cela à plusieurs reprises.

J'éprouvais même un sentiment de grand bien-être à céder

---

them ; and, by degrees, I found that I had no longer the power of
keeping them open, but did not the less retain my consciousness of
what was going on around me ; so that I heard M. Desor speak to
Mr. Townsend, understood what they said, and heard what questions
they asked me, just as if I had been awake, but I had not the power
of answering. I endeavoured in vain several times to do so, and,
when I succeeded, I perceived that I was passing out of the state of
torpor in which I had been, and which was rather agreeable than
painful.

In this state I heard the watchman cry ten o'clock ; then I heard
it strike a quarter past ; but, afterwards, I fell into a deeper sleep,
although I never entirely lost my consciousness. It appeared to me,
that Mr. Townsend was endeavouring to put me into a sound sleep ;
my movements seemed under his control, for I wished several times
to change the position of my arms, but had not sufficient power to do
it, or even really to will it ; while I felt my head carried to the right
or left shoulder, and backwards or forwards, without wishing it,
and, indeed, in spite of the resistance which I endeavoured to
oppose : and this happened several times.

I experienced at the same time a feeling of great pleasure in giving

à l'attrait qui m'entrainait tantôt d'un coté, tantôt d'un autre;
puis une sorte de surprise à sentir ma tête tomber dans la
main de M. Townshend, qui me paraissait dès-lors la cause
de cette attraction. A sa question si j'étais bien, ce que
j'éprouvais, je sentis que je ne pouvais pas répondre, mais je
souris ; je sentais que mon visage s'épanouissait malgré ma résistance ; j'étais intérieurement confus d'éprouver du bien-être
d'une influence qui m'était occulte. Dès-lors j'aurais voulu
me réveiller, j'étais moins à l'aise, et cependant sur la question
de M. Townshend, si je voulais m'éveiller, je fis un mouvement
d'hesitation des épaules. Alors M. Townshend fit de nouveau
quelques frictions qui me rendormirent davantage ; cependant
j'étais toujours sachant de ce qui se passait autour de moi.
Puis il me demanda si je voulais devenir clair-voyant, tout en
continuant ses frictions de la figure aux bras, que je sentais.
J'éprouvais alors un sentiment indéfinissable de joie, et un instant je vis devant moi comme une gerbe de lumière éblouis-
sante qui disparut au moment même. Je fus alors attristé
intérieurement de ce que cet état se prolongeait ; il me semblait
qu'on en avait assez fait avec moi ; j'aurais voulu me réveiller,
mais je ne le pus. Quand cependant M. Townshend et
M. Desor se parlaient, je les entendais ; j'entendis aussi son-

---

way to the attraction which dragged me sometimes to one side, sometimes to the other, then a kind of surprise on feeling my head fall
into Mr. Townshend's hand, who appeared to me from that time to
be the cause of the attraction. To his inquiry if I were well, and
what I felt, I found I could not answer, but I smiled; I felt that my
features expanded in spite of my resistance ; I was inwardly confused
at experiencing pleasure from an influence which was mysterious to me.
From this moment I wished to wake, and was less at my ease ; and yet
on Mr. Townshend asking me whether I wished to be awakened, I
made a hesitating movement with my shoulders. Mr. Townshend
then repeated some frictions, which increased my sleep; yet I was
always conscious of what was passing around me. He then asked
me if I wished to become lucid, at the same time continuing, as I
felt, the frictions from the face to the arms. I then experienced an
indescribable sensation of delight, and for an instant saw before me
rays of dazzling light which instantly disappeared. I was then inwardly sorrowful at this state being prolonged; it appeared to me
that enough had been done with me ; I wished to awake, but could
not. Yet when Mr. Townshend and M. Desor spoke I heard them.

ner, et le guet crier, mais je ne sus quelle heure; puis M. Townshend me présenta sa montre en me demandant si je voyais
l'heure, si je le voyais lui-même, mais je ne distinguais rien ;
j'entendis encore sonner un quart à l'horloge, mais je ne pouvais
sortir de mon état d'assoupissement.   Enfin, M. Townshend
m'éveilla par quelques mouvements rapides, transverses du
milieu de la face en dehors, qui m'ouvrirent instantanément les
yeux ; et au moment même je fus debout en lui disant, " Je
vous remercie." Il était onze heures et quart. Il me dit alors, et
M. Desor me le répéta, que le seul fait qui leur eut donné la
certitude que j'étais dans un état de sommeil magnétique, avait
été la facilité avec laquelle je suivais de la tête tous les mouvemens de la main de M. Townshend, bien qu'il ne me touchât
pas, et le bien-être que je paraissais éprouver dans les momens où,
après plusieurs frictions réitérées, il déplaçait ainsi ma tête à
volonté dans tous les sens.

AGASSIS.

---

I also heard the clock, and the watchman cry, but I did not know what
hour he cried.    Mr. Townshend then presented his watch to me, and
asked if I could see the time, and if I saw him ; but I could distinguish nothing : I heard the clock strike the quarter, but could not
get out of my sleepy state.    Mr. Townshend then woke me with
some rapid transverse movements from the middle of the face outwards, which instantly caused my eyes to open, and at the same time I
got up, saying to him, " I thank you."    It was a quarter past eleven.
He then told me, and Mr. Desor repeated the same thing, that the
only fact which had satisfied them that I was in a state of mesmeric
sleep, was the facility with which my head followed all the movements of his hand, although he did not touch me, and the pleasure
which I appeared to feel at the moment when, after several repetitions
of friction, he thus moved my head at pleasure in all directions.

AGASSIS.

M. l'Avocat Valdrighi avait l'ouie tellement fin et exalté,
qu'il put entendre des mots prononcés à la distance de deux
chambres, dont les portes étaient fermées, quoique prononcés à
voix très-faible et basse.

L'exaltation de la vie qui s'observe chez quelques individus
malades, arrive au point qu'un d'eux voyait les objets les plus
minces et fins dans la plus grande obscurité.   Cela est re-
marquable dans les personnes nerveuses et très-délicates.

DR. FILIPPI.

Milan, Juillet, 1839.

M. Valdrighi, advocate, had his sense of hearing so exquisite and
exalted that he could hear words pronounced at the distance of two
rooms, the doors of which were shut, although pronounced in a weak
and low voice.

The exaltation of life which is observed in some patients attains
such a height, that one of them could see the most delicate and
minute objects in the greatest darkness.   This is noticed in nervous
and very delicate persons.

DR. FILIPPI.

Milan, July, 1839.

### Account of Sensations during Mesmerisation. By Signor Ranieri, of Naples.

AYANT été magnétisé par mon honorable ami, M. Hare Towns-hend, j'exposerai tout simplement les phénomènes que j'ai éprouvés avant, pendant, et après ma magnétisation. M. Towns-hend commença par me faire asseoir sur un sofa : il s'assit sur une chaise vis-à-vis de moi, et ayant pris mes mains dans les siennes, il les plaça sur mes genoux. Il me regardait fixé-ment, et de temps en temps il laissait mes mains, et plaçait les pointes de ses doigts en droite ligne de mes yeux, à un pouce, je crois, de mes prunelles ; puis, en décrivant une espèce d'ellipse, il redescendait ses mains sur les miennes. Après dix minutes qu'il eut porté comme cela alternative-ment ses mains de mes yeux sur mes genoux, j'éprouvai un besoin irrésistible de fermer mes ſpaupières. Je continuai cependant d'entendre sa voix, et celle de ma sœur (qui était dans la même chambre), lorsqu'ils me faisaient des questions. Je leur répondais même toujours pertinemment, mais tout mon système musculaire était dans un état d'affaiblissement singulier, et de désobéissance presque complète à ma volonté ; et par conséquent la prononciation des mots que je voulais répondre m'était devenue d'une difficulté extrême. Tandis

---

TRANSLATION.

Having been mesmerised by my honourable friend Mr. Hare Townshend, I will simply describe the phenomena which I experi-enced before, during, and after my mesmerisation. Mr. Townshend commenced by making me sit upon a sofa : he sat upon a chair oppo-site me, and, having taken my hands in his, placed them on my knees. He looked at me fixedly ; and from time to time let go my hands, and placed the points of his fingers in a straight line opposite my eyes, at an inch, I should think, from my pupils; then, describing a kind of ellipse, he brought his hands down again upon mine. After he had moved his hands thus alternately from my eyes to my knees for ten minutes, I felt an irresistible desire to close my eyelids. I continued nevertheless to hear his voice, and that of my sister, who was in the same room, whenever they put questions to me. I always answered him correctly, but the whole of my muscular system was in a state of peculiar weakness, and of almost perfect disobedience to my will ; and consequently the pronunciation of the words with which I wished to answer had become extremely difficult.

que j'éprouvais jusqu'à un certain point les effets du sommeil,
non seulement je n'étais pas étranger à tout ce qui se passait
autour de moi ; mais aussi j'y prenais plus de part qu'à
l'ordinaire. Toutes mes conceptions étaient plus rapides ;
j'éprouvais des secousses nerveuses qui ne me sont pas habi-
tuelles ; enfin, tout mon système nerveux était dans un état
d'exaltation, et semblait avoir acquis tout le surplus des forces
que le système musculaire avait perdu.

Voilà les principaux phénomènes qu'il me fut possible de
ressentir d'une manière incontestable. M. Townshend ne
manquait pas de me demander de temps à autre, se je pouvais
voir lui ou ma sœur sans ouvrir les paupierès ; mais c'est ce
qui me fut toujours impossible, et tout ce que je pus dire
d'avoir vu, ce fut une lueur entrecoupée par les images noires
et confuses des objets qu'on me présentait, lueur qui me
parut un peu plus claire que celle qu'on voit ordinairement
lorsqu'on ferme les paupierès vis-à-vis du soleil ou d'une
chandelle.

Enfin M.Townshend résolut de me démagnétiser. Il commença à fairè de ses mains des mouvements elliptiques in-
verses à ceux qu'il avait faits dans le commencement ; je pus
ouvrir les paupières sans aucune espèce d'effort, tout mon

---

Whilst I experienced to a certain point the effects of sleep, not only
was I not a stranger to all that was passing around me, but I even took
more than usual interest in it. All my conceptions were more rapid ;
I experienced nervous startings to which I am not accustomed ; in
short, my whole nervous system was in a state of exaltation, and appeared to have acquired all the superabundance of power which the
muscular system had lost.

The following are the principal phenomena which I was able to feel
distinctly. Mr. Townshend did not fail to ask me occasionally if I
could see him or my sister without opening the eyelids ; but this was
always impossible, and all that I could say I had seen was a glimmering
of light interrupted by the black and confused images of the objects presented to me, a light which appeared to me a little less clear than that
which we commonly see when we shut the eyelids opposite the sun
or a candle.

Mr. Townshend at last determined to demesmerise me. He began
to make elliptical movements with his hands, the reverse of those
which he had made at the commencement ; I could now open my
eyes without any kind of effort, my whole muscular system became

système musculaire devint très-obéissant à ma volonté ; je pus me lever debout, et je fus parfaitement éveillé, mais je demeurai presque une heure dans une espèce de stupéfaction très-semblable à celle qui me prend quelquefois le matin, si je me lève deux ou trois heures plus tard qu'à l'ordinaire.

ANTOINE RANIERI.

Naples, 15 Juin, 1839.

---

perfectly obedient to my will ; I was able to get up, and was perfectly awake, but I remained nearly an hour in a kind of stupefaction very similar to that which sometimes attacks me in the morning if I rise two or three hours later than usual.

ANTOINE RANIERI.

Naples, 15th June, 1839.

THE END.

## ERRATA.

Page 214. last line but one of note, and page 227. line 8. from top,
for " *Leherin,*" read " *Seherin.*"

406. line 11. from bottom, for " colour : only the rest," read
" colour only : the rest."

506. line 14. from bottom, for the " latter sensation," read the
" latter, — sensation."